CREATING A WINNING E-BUSINESS

H. Albert Napier

Philip J. Judd

Ollie N. Rivers

Stuart W. Wagner

COURSE
TECHNOLOGY
THOMSON LEARNING

Australia • Canada • Mexico • Singapore • Spain • United Kingdom • United States

COURSE
TECHNOLOGY
—★— TM
THOMSON LEARNING

Creating a Winning E-Business
by H. Albert Napier, Philip J. Judd, Ollie N. Rivers, and Stuart W. Wagner

Publisher
Kristen Duerr

Production Editor
Elena Montillo

Text Designer
Books by Design

Managing Editor
Jennifer Locke

Marketing Manager
Toby Shelton

Cover Designer
Betsy Young

Product Manager
Margarita Donovan

Editorial Assistant
Janet Aras

Development Editor
Deb Kaufmann

Manufacturing Coordinator
Alec Schall

Brief Contents

Contents

CHAPTER **4**

Creating an E-Business Plan   113

CHAPTER **5**

Getting Your E-Business off the Ground   153

CHAPTER **6**

Building Your E-Business   187

CHAPTER **7**

Designing an E-Business Web Site   233

Defining Security Issues   271

CHAPTER **9**

Understanding Back-End Systems   309

Preface

Creating a Winning E-Business provides general business students, graduate students, continuing education students, executive education seminar participants, and entrepreneurs with practical ideas on planning and creating an e-business. We assume that readers have no previous e-business knowledge or experience. This book is designed to help you learn about the key business elements of planning and starting an e-business from the ground up.

When we began teaching both an executive education seminar and an MBA-level course in planning and starting an e-business, we found many textbooks that emphasized the technological aspects of electronic commerce, but we did not find an existing textbook that focused on the planning and startup phase of an e-business. *Creating a Winning E-Business* is our attempt to fill that void.

Organization and Coverage

Creating a Winning E-Business takes a practical case-based and hands-on approach to planning and starting an e-business. Numerous real-world e-business examples are used in each chapter to illustrate important concepts. Additionally, a real-world startup e-business, foodlocker.com, is followed throughout the book from the idea stage through the marketing stage.

Important topics covered in the book include:

- The new e-business economy and current e-business models (Chapter 1)
- The e-business entrepreneurial process and how to exploit e-business advantages (Chapter 2)
- Traditional and electronic payment methods (Chapter 3)
- How to create an e-business plan (Chapter 4)
- E-business startup financing options (Chapter 5)
- "Back office" issues such as selecting office space, hiring employees, building the e-business brand, and selecting appropriate e-business technologies (Chapter 6)
- Web site design and usability (Chapter 7)
- E-business security and risk management (Chapter 8)

- Integrating front-end Web-based systems with back-end enterprise resource planning (ERP), customer resource management (CRM), and order fulfillment systems (Chapter 9)
- Testing and marketing a Web site and measuring Web site results (Chapter 10)

Features

Creating a Winning E-Business is unique in its field because it includes the following features:

- **Opening and Closing Case:** A real-world e-business case opens and closes each chapter and provides a unifying theme for the chapter. The case establishes background elements and introduces relevant issues at the beginning of the chapter. The case concludes at the end of the chapter with a discussion of whether those relevant issues were resolved and how they were resolved.
- **E-Case In Progress:** A startup e-business, foodlocker.com, is used throughout the text to illustrate the processes involved with planning, starting, and marketing a new e-business.
- **E-Cases:** Other real-world e-business examples are used throughout the text to illustrate key concepts. A table of these e-cases appears in the front of the book.
- **Numerous illustrations:** The text of the chapter is well supported with many conceptual figures and screenshots of e-business Web sites.
- **Tips:** Each chapter has multiple margin tips that contain useful additional information about individual topics.
- **Summary:** Each chapter concludes with a Summary that concisely recaps the most important concepts in the chapter.
- **Checklist:** A Checklist of the major concepts discussed in the chapter is provided following the Summary. Students can use this list to establish reference points from the chapter concepts to their own e-business—whether it is a real-world or classroom-based e-business.
- **Key Terms:** Following the Checklist is a list of key terms used in the chapter. These key terms are bolded in the chapter text and defined in the Glossary located at the end of the text.
- **Review Questions and Exercises:** Every chapter concludes with meaningful review materials that include both objective questions and hands-on exercises. The exercises involve experiences that result in a computer output or a typed paper. One of the exercises in Chapter 10, for example, asks students to use search tools and Web browser features to review the Meta tags included in the HTML coding of multiple e-business Web pages and then compile a list of the keywords included in those Meta tags. Exercises throughout the text ask students to research an issue using the Web, answer questions, and produce a summary of research results and answers to those questions.
- **Case Projects:** Every chapter contains three case projects, each project based on a given startup e-business scenario. Students are required to apply concepts discussed in the chapter to the scenario and then write a short paper describing the application of those concepts to the case project solution. These three projects can be completed individually or in groups.

- **Team Project:** A specially designed team project is included that allows a team of two to three students to work together on the project solution and then make a formal presentation of that solution to others. Team members must work together to name and describe the startup e-business and complete the project by implementing a key concept from the chapter. This format allows multiple teams to work on the same project and arrive at different solutions. An important aspect of the team project is the requirement to prepare a 5-10 slide presentation illustrating the project solution and then use the presentation materials to formally present the project solution to other students. This allows students to both understand the practical application of key concepts and to experience and practice important presentation skills.
- **Useful Links:** A list of Web site names and URLs for Web sites providing additional information on the chapter topics is at the end of the chapter.
- **Links to Web Sites Noted in This Chapter:** An additional list of all the e-business Web sites used to illustrate chapter concepts is also included at the end of the chapter.
- **For Additional Review:** Every chapter contains a comprehensive list of references to online magazine articles and reports, print magazine articles, newspaper articles, journal papers, and books that students can read to learn more about topics discussed in the chapter. The text's MyCourse Web site will be periodically updated to include references to new reference materials available after the book is published.
- **Glossary:** A glossary containing the key terms and their definitions appears at the end of the text.
- **Appendix–Microsoft FrontPage Tutorial:** A short appendix is included that introduces how to use a Microsoft FrontPage Wizard to create a Web site. This appendix can be used to help students create a simple e-business Web site to complement their case project or team project solutions, if desired.

Teaching Tools

When this book is used in an academic setting, instructors may obtain the following teaching tools from Course Technology:

- **Instructor's Manual:** The Instructor's Manual has been carefully prepared and tested to ensure its accuracy and dependability. The Instructor's Manual is available through the Course Technology Faculty Online Companion on the World Wide Web. (Call your customer service representative for the exact URL and to obtain your username and password.)
- **ExamView®** This textbook is accompanied by ExamView, a powerful testing software package that allows instructors to create and administer printed, computer (LAN-based), and Internet exams. ExamView includes hundreds of questions that correspond to the topics covered in this text, enabling students to generate detailed study guides that include page references for further review. The computer-based and Internet testing components allow students to take exams at their computers, and also save the instructor time by grading each exam automatically.

- **Classroom Presentations:** Microsoft PowerPoint presentations are available for each chapter of this book to assist instructors in classroom lectures. The Classroom Presentations are included on the Instructor's CD-ROM.
- **MyCourse:** MyCourse.com is an online syllabus builder and course-enhancement tool. Hosted by Course Technology, MyCourse.com adds value to courses by providing additional content that reinforces what students are learning. Most importantly, MyCourse.com is flexible, allowing instructors to choose how to organize the material—by date, by class session, or by using the default organization, which organizes content by chapter. MyCourse.com allows instructors to add their own materials, including hyperlinks, school logos, assignments, announcements, and other course content. Instructors using more than one textbook can build a course that includes all of their Course Technology texts in one easy-to-use site. Instructors can start building their own courses today at *www.course.com/instructor.*

Acknowledgments

Creating a quality text is a collaborative effort between author and publisher. We work as a team to provide the highest quality book possible. The authors want to acknowledge the work of the seasoned professionals at Course Technology. We thank Jennifer Locke, Managing Editor; Margarita Donovan, Product Manager; Elena Montillo, Production Editor; Toby Shelton, Marketing Manager; and Janet Aras, Editorial Assistant, for their tireless work and dedication to the project. We also thank Deb Kaufmann, our terrific Development Editor, for her insightful suggestions and unflagging support.

We want to thank the following reviewers for their very helpful comments and suggestions at various stages of the book's development: Diane Lockwood, Seattle University; Hermann Gruenwald, University of Oklahoma; and Thomas Case, Georgia Southern University.

H. Albert Napier
Philip J. Judd
Ollie N. Rivers
Stuart W. Wagner

Dedication

To Liz, my wonderful wife

H. Albert Napier

To Valerie, for her support, understanding, and assistance

Philip J. Judd

To Laura and Lucy, my two terrific daughters

Ollie N. Rivers

Thank you, Celeste, for your insight, wisdom, and inspiration

Stuart W. Wagner

About the Authors

H. Albert Napier is the Director of the Center on the Management of Information Technology and a Professor in the Jones Graduate School of Management at Rice University, where he teaches graduate and executive development courses related to information technology and e-business. Dr. Napier also makes numerous management development program presentations on e-business and related topics. Additionally, he is a principal of Napier & Judd, Inc., a company engaged in computer training and consulting. Dr. Napier is on the board of directors of an e-business and a media distribution company where he advises their e-business group. Additionally, he consults with clients from a variety of industries including construction, legal and accounting, financial, real estate, energy, agricultural, and manufacturing. Dr. Napier holds a Ph.D. in Business Administration, an M.B.A., and a B.A. in Mathematics and Economics, all from The University of Texas at Austin. He is the author of more than 20 articles related to management information systems and applications of computer-based decision processes in business and the co-author of over 60 textbooks.

Philip J. Judd is a principal of Napier & Judd, Inc. His consulting activities include the analysis and design of automated business systems, planning for large-scale computer operations, the selection and implementation of office automation systems, the development of corporate modeling systems, and the design and implementation of personal computer network systems including Web site integration. Mr. Judd was previously an instructor in the Management Department at the University of Houston and the Director of the Research and Instructional Computing Service at the university. He received his M.B.A. and B.B.A. degrees from the University of Houston.

Ollie N. Rivers is an associate at Napier & Judd, Inc. where she develops materials for software applications training courses and e-business courses and seminars. Additionally, she has more than 20 years business experience in financial and administrative management and holds an M.B.A. and a B.A. in Accounting and Management from Houston Baptist University.

Stuart W. Wagner is an e-commerce marketing manager with Compaq Computer Corporation and a lecturer on e-commerce at the Jones Graduate School of Management at Rice University. He is a founder and Vice President/Chief Technology Officer of foodlocker.com. Prior to founding foodlocker.com, he worked as the information technology manager for a subsidiary of a multi-billion dollar real estate operations company. Additionally, he has experience working with a number of startup businesses and has experience in commercial Web site design. He also conducts training classes on using on Web development software and executive education seminars on starting and operating an e-business. He has an M.B.A. from Rice University.

CHAPTER 1

Understanding the New Internet Economy

In this chapter, you will learn to:

Discuss electronic commerce basics

Describe the Internet and World Wide Web

Discuss the role of e-business in the new economy

List e-business advantages and disadvantages

Identify e-business models

Egghead Software was born in 1984, in Spokane, Washington. Throughout the remainder of the 1980s and into the early 1990s, Egghead enjoyed great success, rising to prominence as the first nationwide software-only chain. By 1992 Egghead had 250 stores, 2,500 employees, and $750 million in revenue. Along with Egghead's success, however, came heavy competition, as mass merchandisers such as Wal-Mart and CompUSA began selling software at discounted prices. As software prices fell, Egghead ran into trouble. In 1996 Egghead had a $250 million operating loss and began closing its stores.

In desperation, the Egghead board of directors turned to George Orban, a longtime Egghead investor with a track record for turning around failing businesses; he joined Egghead as chairman and CEO. Orban believed that Egghead's problems stemmed from failure to identify which customer segment to target and failure to carve out a clear position for itself in the market. Also, Orban thought that Egghead had followed an overly aggressive expansion plan by opening stores too quickly and in too many geographic markets. He decided that the only way to save Egghead was to simplify the business.

Orban began Egghead's turnaround with a massive cost-cutting program that included closing 77 stores, closing its Lancaster, PA, distribution center, selling real estate holdings, and laying off employees. Several of the remaining stores switched to a more profitable format that included sales of computer systems, accessories, and services. Egghead also began selling software, computer systems, and services online.

Would these changes be enough to save Egghead?

Electronic Commerce Basics

Commerce, the exchange of valuable goods or services, has been conducted for thousands of years. Today, we think of commerce as conducting business, in which buyers and sellers come together in a marketplace to exchange information, products, services, and payments. Traditionally, business has been conducted in physical buildings, often referred to today as **brick-and-mortar** marketplaces. When the marketplace is electronic, business transactions occur across a telecommunications network where buyers, sellers, and others involved in the business transaction—such as the employees that process transactions—rarely see or know each other and may be physically located anywhere in the world. This process of buying and selling of products and services across a telecommunications network is often called **electronic commerce**, and the electronic marketplace is sometimes called a **marketspace**.

The initial development of electronic commerce began in the 1960s and 1970s, when banks began transferring money to each other electronically, using electronic funds transfer (EFT), and when large companies began sharing transaction information electronically with their suppliers and customers via electronic data interchange (EDI). Using EDI, companies exchange information electronically with their suppliers and customers, called trading partners, including information traditionally submitted on paper forms such as invoices, purchase orders, quotes, and bills of lading. These transmissions generally occur over private telecommunications networks called value-added networks, or VANs. Because of the expense of setting up and maintaining these private networks and the costs associated with creating a standard interface between companies, implementing EDI has usually been beyond the scope of small and medium-sized companies. Now small and medium-sized companies (and many large companies) are beginning to use the Internet, which is a less expensive network alternative to VANs for the exchange of information, products, services, and payments.

The Internet and World Wide Web

Millions of people use the Internet to shop for products and services, listen to music, view artwork, conduct research, get stock quotes, keep up to date with current events, and send e-mail. More and more businesses are using the Internet to conduct their business activities.

What is the Internet?

To understand the Internet, you must understand networks. A **network** is simply a group of two or more computers linked by cable or telephone lines (Figure 1-1). The group of linked computers includes special computers called **servers** that provide user access to shared resources such as files, programs, and printers.

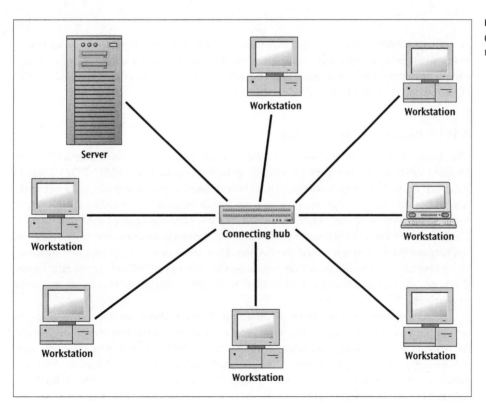

Figure 1-1
Computer network

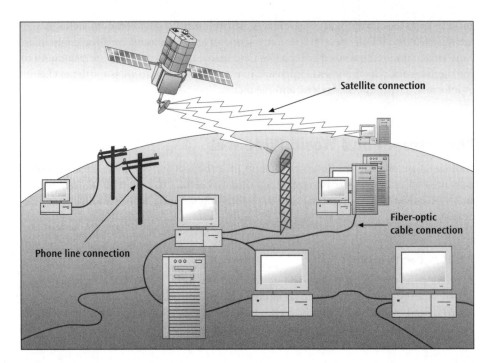

Figure 1-2
The Internet

The **Internet** (Figure 1-2) is a public worldwide network of networks, connecting many small private networks. Computers on the Internet use a common set of rules, called protocols, for communication. The primary Internet protocol is called TCP/IP (Transmission Control Protocol/Internet Protocol).

A brief history of the Internet

The Internet originated in the late 1960s, when the United States Department of Defense developed a network of military computers called the ARPANET (Advanced Research Projects Agency Network). Quickly realizing the usefulness of such a network, researchers at colleges and universities soon began using it to share data. In the 1980s the military portion of the early Internet became a separate network called the MILNET. Meanwhile the National Science Foundation (NSF) began overseeing the remaining non-military portions, which it called the NSFnet. Thousands of other government, academic, and business computer networks began connecting to the NSFnet. By the late 1980s, the term Internet had become widely used to describe this huge worldwide "network of networks."

Businesses need some physical communications medium, such as network cable or a modem, connected to their computers, to connect to the Internet. Increasingly, individuals and businesses are using wireless connections to connect to the Internet. Additionally, to access the Internet, many small and medium-sized businesses must make arrangements to connect to a computer on the Internet called a **host**. An **Internet service provider (ISP)** provides access to a host computer. Large businesses, colleges, universities, and government institutions may already have a computer network that is part of the Internet.

There is a wide variety of services that users can access on the Internet. Table 1-1 explains just some of these services.

Table 1-1
Internet
services

Name	Description
E-mail	Electronic messages sent by one computer and received by another
Newsgroups	Electronic "bulletin boards" or discussion groups where people with common interests (such as hobbies or professional associations) post messages that participants around the world can read and respond to
Mailing Lists	Similar to newsgroups except that participants exchange information via e-mail
Chat	Online conversations in which participants key messages and receive responses on their screen within a few seconds
FTP	Sending (uploading) or receiving (downloading) computer files via the File Transfer Protocol (FTP) communication rules
Telephony	Sending and receiving telephone calls by computer

The World Wide Web

In 1989, a software consultant named Tim Berners-Lee was at CERN (the European Laboratory for Particle Physics) in Switzerland, where he was working on ways to improve information sharing and document handling between the research scientists both at CERN and throughout the world. Using the concept of hypertext, whereby text on one page links to another page, he developed the first program that allowed pages containing hypertext to be stored on a computer in a way that allowed other computers access to the pages. Berners-Lee called his system of documents linked by hypertext the World Wide Web.

The **World Wide Web** (WWW), also called simply the **Web**, is, therefore, a subset of the Internet where computers called **Web servers** store documents that are linked together by hypertext links, called hyperlinks. A **hyperlink** can be text or a picture that is associated with the location (path and filename) of another document. The documents, called **Web pages**, can contain text, graphics, video, and audio as well as hyperlinks. A **Web site** is a collection of related Web pages. A **Web browser** is a software application used to access and view Web pages stored on a Web server. The two most popular Web browsers at this writing are Microsoft Internet Explorer and Netscape Navigator.

Internet and Web demographics

Determining how many individuals are online and which Internet services they are using can be difficult. Research and marketing groups publish various estimates on a regular basis. However, because of the dynamic growth of the Internet, these estimates are quickly outdated. Additionally, there are differences among the various estimates because of differences in how Internet access and Web content are defined, the survey and calculation methods used, and how the data are gathered. Furthermore, the rapid growth in Internet access and online content makes estimating current growth and predicting future growth difficult at best.

However, there is one thing all growth estimates and predictions have in common: They indicate that the remarkable growth of Internet access and online content from year to year shows no sign of abating. For example, Nua Internet Surveys provides an "educated guess" of the growth in Internet access by aggregating data from multiple surveys. According to Nua Internet Surveys, Internet access worldwide increased from 171 million people in March 1999 to 378 million people in September 2000, an increase of more than 100 percent. Figure 1-3 illustrates the rapid growth of the Internet and the Web, based on data compiled by Nua Internet Surveys in September 2000.

Figure 1-3
Number online (millions) September 2000

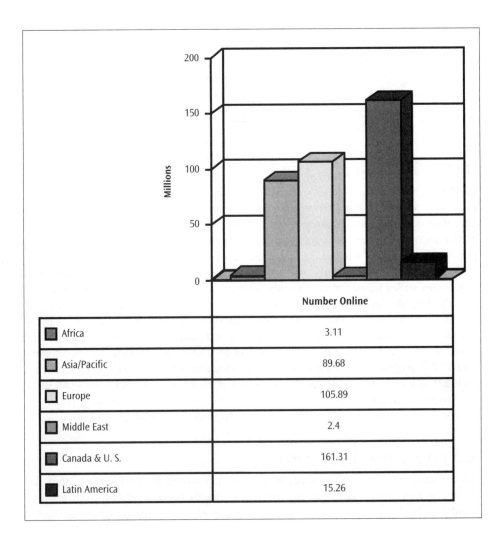

	Number Online
Africa	3.11
Asia/Pacific	89.68
Europe	105.89
Middle East	2.4
Canada & U. S.	161.31
Latin America	15.26

Along with increased Internet access, the amount of information available on the Web is growing very rapidly. A study by Inktomi and NEC Research Institute, Inc. indicated that in January 2000 the Web contained more than one billion unique pages, compared to 100 million in October 1997.

E-commerce versus e-business

With the commercialization of the Internet in the early 1990s, electronic commerce became **e-commerce**, the buzzword for buying and selling products and services on the Internet. Today, many people use the term e-commerce in a broader sense, encompassing not only buying and selling but also the delivery of information, providing customer service before and after the sale, collaborating with business partners, and enhancing productivity within organizations. Others prefer the term **e-business** to indicate the broader spectrum of business activities that can be conducted over the Internet. Most people today use the term e-commerce, in its broadest sense, interchangeably with e-business. In this book we use the term e-business to indicate the widest spectrum of business activity using Internet technologies.

E-Business and the New Economy

The widespread electronic linking of individuals and businesses has created a new economic environment in which time and space are much less limiting factors, information is more important and accessible, traditional intermediaries are being replaced, and the consumer holds increasing amounts of power. The Internet is both an effect and a cause of this new economy. The Internet is a product of the tremendous technological and economic changes driving the new economy, and increasingly, it is the medium of the new economy. Although, in the past, large companies were able to conduct their business electronically using EDI and private networks, the high costs associated with EDI prevented most businesses from using the technology. The Internet has leveled the playing field by making it easier and cheaper for companies of all sizes to transact business and exchange information electronically.

In this revolutionary economic environment, many of the limitations of space and time are disappearing. Businesses that once had geographically limited customer and competitor bases are finding that the whole world is now both customer and competitor. Companies that previously conducted business activities during traditional hours can now conduct those activities online 24 hours a day, 7 days a week, 365 days a year.

A popular independent Denver, Colorado, bookstore, The Tattered Cover, with two brick-and-mortar locations, has been doing business successfully for over 20 years. The Tattered Cover has traditionally competed with other local Denver bookstores. Now, the Tattered Cover also competes with online bookstores such as Amazon.com and Barnes&Noble.com. To meet this new competition, the Tattered Cover added an Internet bookstore. From its Internet bookstore, the Tattered Cover (Figure 1-4) sells books and

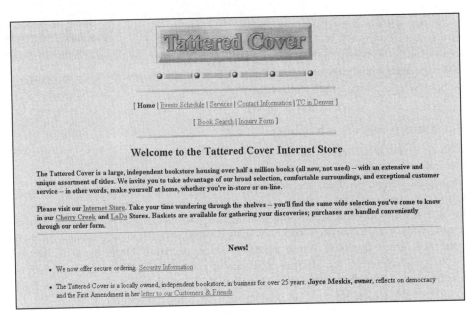

Figure 1-4
Tattered Cover

gifts and provides special customer services around the clock, including a schedule of upcoming special events and personalized search requests for hard-to-find books.

In the new economy, processing information is more powerful and cost-effective than moving physical products. Increasingly, the new economy is becoming less about the transfer of goods and more about the transfer of information. Information companies, such as Yahoo!, which have relatively few physical assets and employees, can enjoy disproportionately large market values in the new economy. A leading information company in the automobile industry is Edmunds.com. Edmunds, founded in 1966, is the publisher of a number of automobile and truck reviews and pricing guides. In 1994, Edmunds introduced its Internet site, Edmunds.com (Figure 1-5), which has become a valuable online data source for information about new and used automobiles, including pricing, dealer cost and holdbacks, reliability, buying advice, and product reviews.

Because information is easier to customize than hard goods, many companies are finding that the information portion of their products or services is becoming a larger part of the total value they offer customers. Office product suppliers, such as Staples, create customized product catalogs for their large online customers that list only those items and prices negotiated by contract. Online grocers automatically retain previous shopping lists, allowing buyers to quickly add frequently purchased items to their grocery cart. Webvan.com, which began selling groceries online in 1999 in the San Francisco Bay area, automatically creates a My Personal Market for each customer, listing all of the items the customer has purchased in the past. My Personal Market is created with the customer's first order and updated with each subsequent order. Customers can log in to Webvan.com (Figure 1-6) and go directly to My Personal Market, where they can view a list of their 50 most frequently purchased items sorted by

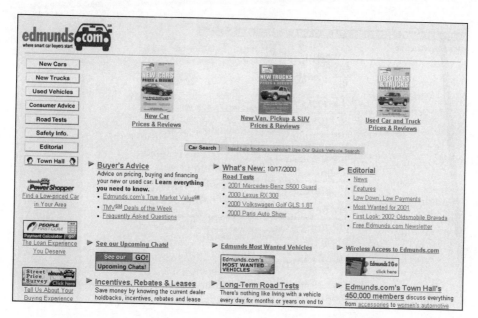

Figure 1-5
Edmunds.com

purchase frequency, items from their most recent order, or every product they ever purchased sorted by product category. Finally, customers can create, save, and view their own custom grocery lists.

Traditional business intermediaries, such as distributors and agents, are being threatened by the new Internet economy in which buyers are linked directly to sellers.

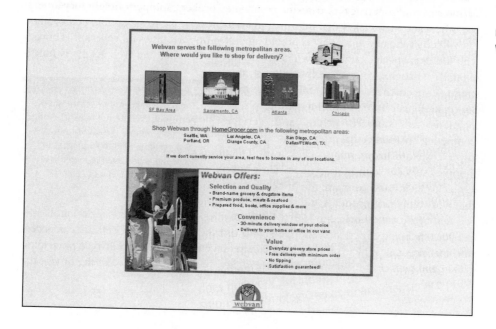

Figure 1-6
Webvan.com

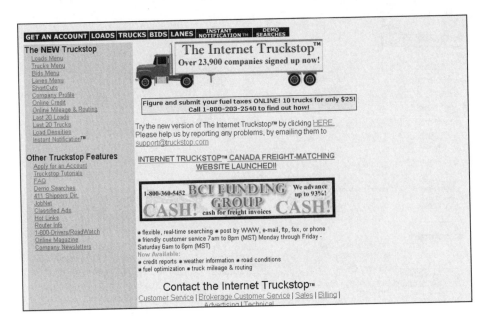

Figure 1-7
The Internet Truckstop

Travel Web sites such as Travelocity are replacing travel agents, and online trading services such as E*TRADE are taking business from traditional brokerage firms. In place of these traditional intermediaries, the new economy has encouraged the growth of a new kind of middleman, sometimes called an "infomediary," that organizes information on the basis of customer needs—for example, The Internet Truckstop, an online load, truck, route, freight matching, and bid posting service. The Internet Truckstop (Figure 1-7) gathers information from truckers, trucking companies, brokers, shippers, freight forwarders, and others and then makes that information available for a fee to subscribing customers.

Although the new economy is providing online opportunities for sellers, it is buyers who are dramatically gaining new economic power. Internet and Web access is funda-mentally changing buyers' expectations about speed, convenience, comparability, and price. Buyers no longer have to travel to various physical locations to compare prices and services. Competing businesses that offer unique services or lower costs are just a mouse click away. Online shopping services such as Shop4.com allow buyers to quickly locate and compare the prices and availability of competing products and services.

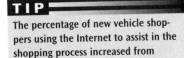

TIP

The percentage of new vehicle shoppers using the Internet to assist in the shopping process increased from 25 percent in 1998 to 40 percent in 1999 and is projected to increase to 65 percent by the end of 2000.

Autobytel.com, launched in March 1995, provides shoppers with a haggle- and hassle-free auto shopping experience. Shoppers who visit the autobytel.com Web site can access auto specifications, vehicle reviews, manufacturer incentives, and dealer invoice price infor-mation, and then elect to submit a purchase request to a local accredited dealer or visit the

AutobytelDirect virtual car lot to purchase a car and have it delivered. In 1999, autobytel.com (Figure 1-8) generated more than $13 billion in car sales through its Accredited Dealer Network and had more than seven million registered users. There were more than five million unique visitors to autobytel.com and its wholly owned subsidiary, CarSmart.com, during the first quarter, 2000.

The new economy is changing the rules of business: Time is collapsing, distance has vanished, information has greater value, traditional intermediaries are being replaced by "infomediaries," and buyers hold more power than ever before.

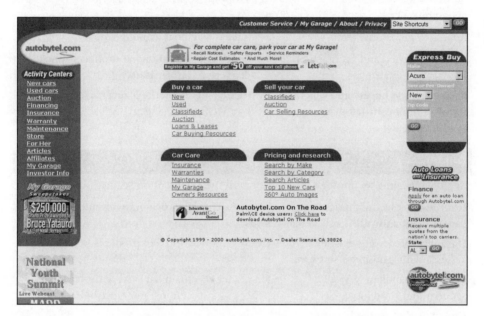

Figure 1-8
Autobytel.com

E-business advantages and disadvantages

Sellers are finding tremendous advantages in doing e-business. They can increase sales and operations from local to worldwide, improve internal efficiency and productivity, enhance customer service, and increase communication with both suppliers and customers. Buyers are enjoying greater access to markets. However, there are some disadvantages for both buyers and sellers in the new economy. Tables 1-2 and 1-3 illustrate some of the e-business advantages and disadvantages for buyers and sellers.

Table 1-2
E-business
advantages

Sellers	Buyers
Increased sales opportunities	Wider product availability
Decreased transactions costs	Customized and personalized information and buying options
Operate 24 hours a day, 7 days a week from one virtual marketspace	Shop 24 hours a day, 7 days a week
Reach narrow market segments that may be widely distributed geographically	Easy comparison shopping and one-stop shopping for business buyers
Access to global markets	Access to global markets
Increased speed and accuracy of information exchange	Quick delivery of digital products; quicker delivery of information
Bring multiple buyers and sellers together in one virtual marketplace	Participate in auctions, reverse auctions, knowledge exchanges

Table 1-3
E-business
disadvantages

Sellers	Buyers
Rapidly changing technology	Concern over transaction security and privacy
Insufficient telecommunications capacity or bandwidth	Lack of trust when dealing with unfamiliar sellers
Difficulty integrating existing systems with e-business software	Desire to touch and feel products before purchase
Problems maintaining system security and reliability	Resistance to unfamiliar buying processes, paperless transactions, and electronic money
Global market issues: language, political environment, currency conversions	
Conflicted legal environment	
Shortage of skilled technical employees	

E-business value chains

In his 1985 book, *Competitive Advantage*, Michael E. Porter of the Harvard Business School first introduced the concept of a value chain. A company's **value chain** consists of all the primary and support activities, called value activities, performed to create and distribute its goods and services. Primary activities include all the activities necessary to produce, sell, and support its products. Support activities include purchasing, human resources, technology, and other functions necessary to support the primary activities. Figure 1-9 illustrates a generic value chain.

Figure 1-9
Generic value
chain

Value chains are also used to represent the components, or value activities, of any transaction that starts with a product or service and ends with a customer. Because the growth of Internet access is redefining relationships among manufacturers, suppliers, distributors, and customers, many companies are rethinking their traditional value chains. Increasingly, a company's value chains can be seen as value networks of multiple relationships (Figure 1-10) needed to produce and sell their products and services.

The process of rethinking and redefining value chains is enabling companies to develop new ways of conducting business in the new economy.

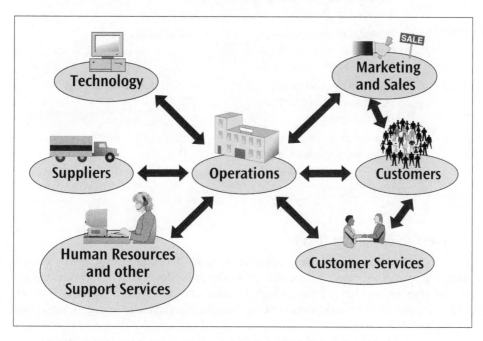

Figure 1-10
Value
network

E-Business Models

A company's business model is the way in which it conducts business in order to generate revenue. In the new economy, companies are creating new business models and reinventing old models. Although there are many different ways to categorize e-business models, they can be broadly categorized as business-to-consumer (**B2C**), business-to-business (**B2B**), business-to-government (**B2G**), consumer-to-consumer (**C2C**), and consumer-to-business (**C2B**). Within these broad categories, there are a number of variations in the way the models are implemented. Table 1-4 summarizes some of the current e-business models.

Model	Description	Examples
B2C	Business-to-consumer: sells products or services directly to consumers	Amazon.com, autobytel.com, The Sunglass City, eDiets.com, Pets.com
B2B	Business-to-business: sells products or services to other businesses or brings multiple buyers and sellers together in a central marketplace	Chemdex, MetalSite, VerticalNet, SHOP2gether, CATEX, HoustonStreet.com
B2G	Business-to-government: businesses selling to local, state, and federal agencies	eFederal, iGov.com
C2C	Consumer-to-consumer: consumers sell directly to other consumers	eBay, American Boat Listing, InfoRocket
C2B	Consumer-to-business: consumers name own price, which businesses accept or decline	priceline.com, ReverseAuction.com

Business-to-consumer (B2C)

Consumers are increasingly going online to shop for and purchase products, arrange financing, arrange shipment or take delivery of digital products such as software, and get service after the sale. B2C e-business includes retail sales, often called **e-retail** (or e-tail), and other online purchases such as airline tickets, entertainment venue tickets, hotel rooms, and shares of stock.

Many traditional brick and mortar retailers—from nationwide companies such as Barnes & Noble (Figure 1-11) and the Gap to regional or local stores such as The Sunglass City—are now e-tailers with a Web storefront. These combined brick and mortar/online businesses are also known as **brick-and-click** companies.

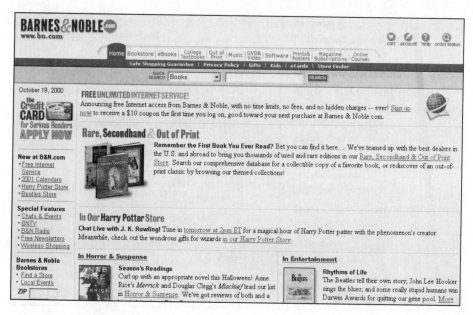

Figure 1-11
Barnes&Noble.com

Some B2C e-businesses provide high-value content to consumers for a subscription fee. Examples of e-businesses following this subscription model include the *Wall Street Journal* (financial news and articles), *Consumer Reports* (product reviews and evaluations), and eDiets.com (nutritional counseling) (Figure 1-12).

Figure 1-12
eDiets.com

Figure 1-13
Yahoo!
Shopping

B2C e-business models include **virtual malls**, which are Web sites that host many online merchants. Virtual malls typically charge online merchants setup, listing, or transaction fees and may include transaction handling services and marketing options. Examples of virtual malls include Excite Stores, Choice Mall, World Shopping Network, Women.com Network Web store, Amazon.com Zshops, and Yahoo! Shopping (Figure 1-13).

E-tailers that offer traditional or Web-specific products or services only over the Internet are sometimes called virtual merchants, and provide another variation on the B2C model. Examples of virtual merchants include Amazon.com (books, electronics, toys, and music), eToys.com (children's books and toys), and ashford.com (fine personal accessories) (Figure 1-14).

Some businesses supplement a successful traditional mail-order business with an online shopping site, or move completely to Web-based ordering. These businesses are sometimes called catalog merchants. Examples of catalog merchants include Avon.com (cosmetics and fragrances), CHEF'S (cookware and kitchen accessories), Omaha Steaks (premium steaks, meats, and other gourmet food), and Harry and David (gourmet food gifts) (Figure 1-15).

Business-to-business (B2B)

Although B2C is the most familiar form of e-business, transactions between and within businesses account for a large share of commercial activity. Business activities within a company are increasingly transacted online via the company intranet. An **intranet** uses Internet technology to allow employees to view and use internal Web sites that are not accessible to the outside world.

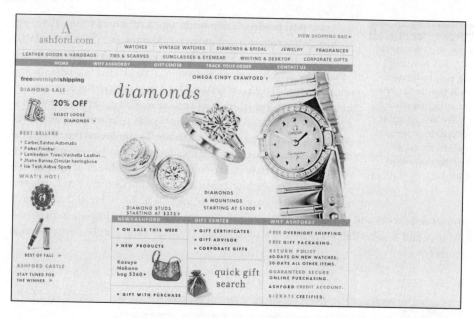

Figure 1-14
ashford.com

Business activities between companies can be transacted over an extranet. An **extranet** consists of two or more intranets connected via the Internet, where participating companies can view each other's data and complete business transactions such as purchasing.

Figure 1-15
Harry and David

Like B2C models, B2B models take a variety of forms. There are basic B2B Internet storefronts such as Staples and Office Depot that provide business customers with purchase, order fulfillment, and other value-added services. Another B2B model is a business trading community, also called a "vertical Web community," that acts as a central source of information for a vertical market. A **vertical market** is a specific industry in which similar products or services are developed and sold using similar methods. Examples of broad vertical markets include insurance, real estate, banking, heavy manufacturing, and transportation. The information available at a vertical Web community can include buyer's guides, supplier and product directories, industry news and articles, schedules for industry trade shows and events, and classified ads. MediSpeciality.com (healthcare), Hotel Resource (hospitality), and NetPossibilities (building trades, Figure 1-16), are examples of virtual vertical marketspaces.

Figure 1-16
NetPossibilities

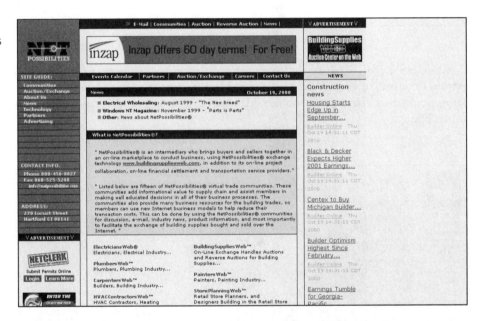

B2B exchanges are Web sites that bring multiple buyers and sellers together in a virtual centralized marketspace. In this marketspace buyers and sellers can buy from and sell to each other at dynamic prices determined by the exchange rules. Table 1-5 illustrates some common elements of B2B exchanges.

B2B exchanges can be further categorized in several ways: for example, aggregators, trading hubs, post and browse markets, auction markets, and fully automated exchanges.

Element	Benefit
Centralized marketspace	Neutral and nonaligned with either sellers or buyers
Standardized documentation	Users are prequalified and regulated
Price quotes, price history, and after-the-sale information provided	Pricing mechanism is self-regulating
Transactions between businesses are confidential	Clearing and settlement services provided

Table 1-5
Common elements of B2B exchanges

B2B aggregators provide a single marketspace for company purchasing by providing many like-formatted supplier product catalogs in one place. Examples of B2B aggregators include e-Chemicals (industrial chemicals), Chemdex (chemicals), MetalSite (steel and other metals), freightquote.com (shipping services), VIPAR (truck parts), and .commerx PlasticsNet (plastics) (Figure 1-17).

Trading hubs are B2B sites that provide a marketspace for multiple vertical markets. Horizontal trading hubs support buyers and sellers from many different industries. VerticalNet is an example of a horizontal trading hub. VerticalNet (Figure 1-18), a pioneer in providing virtual vertical marketspaces, maintains business trading communities for many different industries, including communications, energy, healthcare, food service, and manufacturing. Diagonal trading hubs support specific types of buyers or sellers, or specific types of products across multiple industries. SHOP2gether.com (Figure 1-19) is an example of a diagonal trading hub.

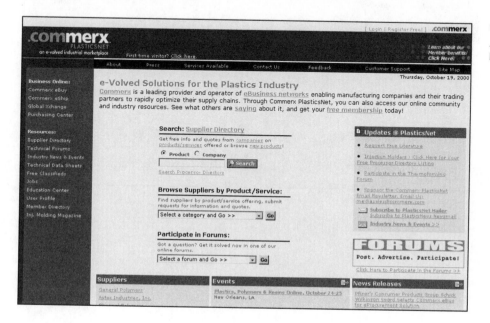

Figure 1-17
.commerx PlasticsNet

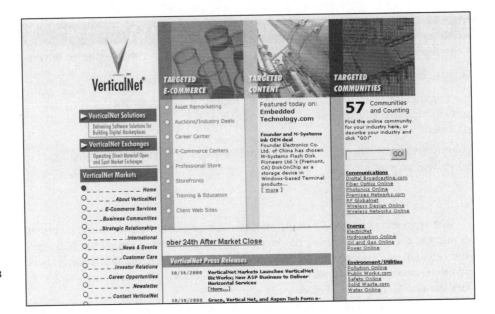

Figure 1-18
VerticalNet, Inc.

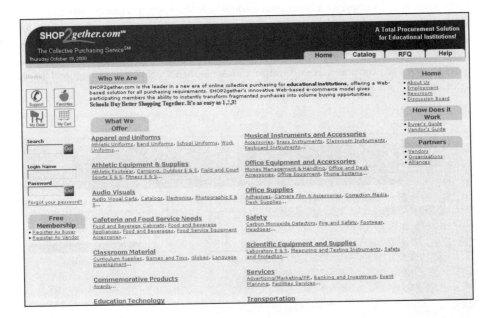

Figure 1-19
SHOP2gether.com

Post and browse markets such as CATEX (insurance, reinsurance, and risk management, Figure 1-20), CreditTrade (credit derivatives), and TechEx (life sciences intellectual property) enable participants to post buy or sell opportunities on an electronic bulletin board. Interested parties meet through the postings and negotiate transactions for themselves.

Figure 1-20
CATEX

B2B auction markets enable multiple buyers or sellers to enter competitive bids on a contract. Examples of B2B auction markets include e-STEEL (steel and other metals), HoustonStreet.com (energy) (Figure 1-21), Altra (energy), and Manheim Online (auto dealer auctions). Auction markets may include reverse auctions or "name your price" auctions. In a reverse auction, a product's selling price continues to decline until the product is purchased. "Name your price" auctions, which allow buyers to enter a bid for a product or service that a seller can then provide at the bid price, are also called reverse auctions. FreeMarkets (Figure 1-22) is an example of a B2B site conducting reverse auctions.

Figure 1-21
Houston
Street.com

At a fully automated B2B exchange, multiple buyers and sellers competitively bid on commodities or standardized products, and the buy and sell orders are matched automatically. PaperExchange.com (paper, Figure 1-23) is a fully automated B2B exchange.

Table 1-6 summarizes some different types of B2B e-businesses.

Figure 1-22
FreeMarkets

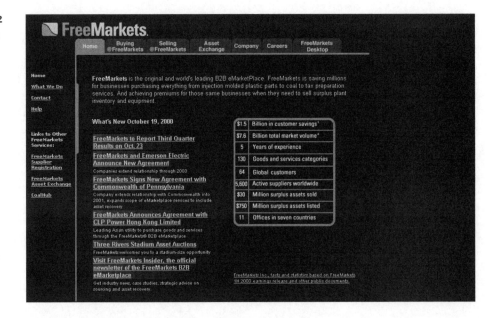

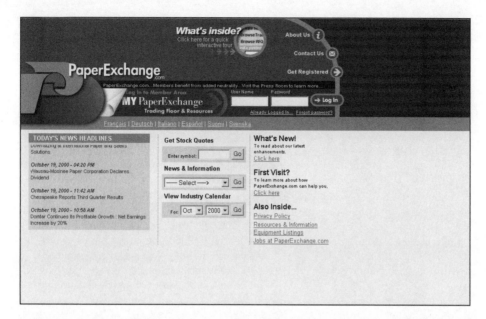

Figure 1-23
PaperExchange.
com

Table 1-6
B2B summary

Type	Description	Examples
B2B storefronts	Provide businesses with purchase, order fulfillment, and other value-added services	Staples, Office Depot
B2B vertical markets	Provide a trading community for a specific industry	MediSpeciality.com, NetPossibilities, HotelResource.com
B2B aggregators	Provide a single marketspace for business purchasing from multiple suppliers	Chemdex, MetalSite, VIPAR
B2B trading hubs	Provide a marketspace for multiple vertical markets	VerticalNet
B2B post and browse markets	Provide a marketspace where participants post buy and sell opportunities	CATEX, CreditTrade, TechEx
B2B auction markets	Provide a marketspace for buyers and sellers to enter competitive bids on contracts	e-STEEL, HoustonStreet.com, Altra, Manheim Online, FreeMarkets
B2B fully automated exchanges	Provide a marketspace for the automatic matching of standardized buy and sell contracts	PaperExchange.com

Some B2B exchanges combine business models in their operation. XSAg.com (Figure 1-24), a marketspace where participants buy and sell agricultural chemicals, equipment, and seeds, uses auctions, reverse auctions, and fixed price listings methods.

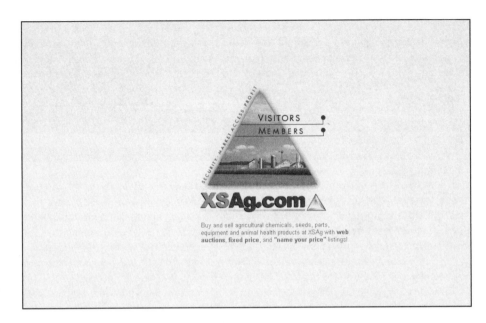

Figure 1-24
XSAg.com

Another business model similar to the B2B exchange model is the business-to-government, or B2G, procurement model. B2G e-businesses such as eFederal.com (Figure 1-25) and Gov.com are hoping to tap into the $18 billion per year market for government procurements that do not require a bid.

Figure 1-25
eFederal.com

Not only do businesses sell directly to consumers and other businesses online, but consumers are now interacting with each other to buy, sell, or trade products, personal services, or information.

Jake Winebaum and Sky Dayton, entrepreneurs formerly associated with Disney Online and Earthlink, set a record by paying $7.5 million for the Business.com Internet domain name. Their plan is to build a business Web site, or portal, that can be the first place a businessperson looks for information on the Web. Business.com (Figure 1-26), launched in late Spring 2000, provides news, statistics, company profiles, financial data, and product/service directories for 57 industries.

Figure 1-26
Business.com

Consumer-to-consumer (C2C)

With the C2C e-business model, consumers sell directly to other consumers via online classified ads and auctions, or by selling personal services or expertise online. Examples of consumers selling directly to consumers are American Boat Listing (boat listing service), eBay (auction), and TraderOnline.com (classified ads, Figure 1-27).

Figure 1-27
Trader
Online.com

There are also a number of new consumer-to-consumer expert information exchanges that are expected to generate $6 billion in revenue by 2005. Some of these exchanges, such as AskMe.com and abuzz, are free, and some allow their experts to negotiate fees with clients. InfoRocket.com, one of the first question and answer marketplaces, is driven by a person-to-person auction format. The InfoRocket.com bidding system allows a person who submits a question to review the profiles of the "experts" who offer to answer the question. When the person asking the question accepts an "expert" offer, InfoRocket.com (Figure 1-28) bills the person's credit card, delivers the answer, and takes a 20 percent commission.

The power shift from sellers to buyers in the new Internet economy has led to another business model, consumer-to-business.

TIP

In addition to these broad categories of e-business models, many nonbusiness organizations such as government agencies, not-for-profit institutions, and social or religious organizations are reducing expenses and improving customer service using e-business models. NPR Online, the National Public Radio Web site, is an example of a not-for-profit institution using an e-business model.

Consumer-to-business (C2B)

The C2B model, also called a "reverse auction" or "demand collection model," enables buyers to name their own price, often binding, for a specific good or service generating demand. The Web site collects the "demand bids" and then offers the bids to participating sellers. ReverseAuction.com (travel, autos, consumer electronics) and priceline.com (travel, telephone, mortgages) (Figure 1-29) are examples of C2B e-business models.

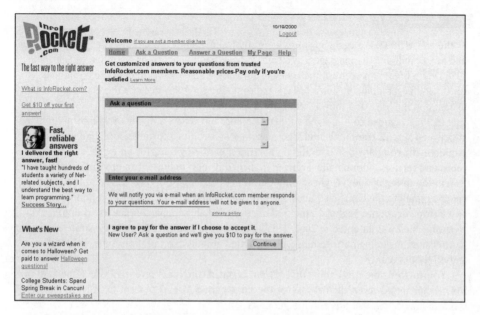

Figure 1-28
InfoRocket.com

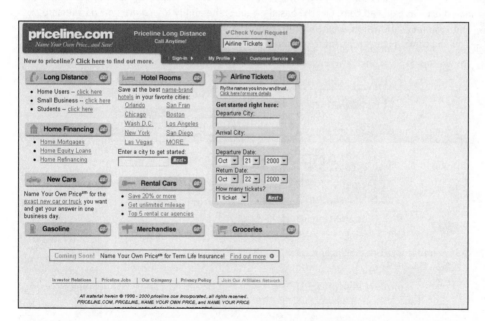

Figure 1-29
priceline.com

In this chapter, you learned about e-business basics and the changing role of e-business in the new economy. You also learned how the new economy is providing opportunities for a company to create a new e-business model or revamp its traditional business model. In the next chapter, we introduce a new virtual merchant located in Houston, TX, foodlocker.com. In subsequent chapters, we will follow foodlocker.com's progress from idea to conception.

By the end of 1997, the Egghead stores were still losing money, but online sales had grown five-fold to $12 million. The message was clear: Egghead had to change or go under. In February 1998 Orban fired 80 percent of the Egghead Software employees, closed all the remaining stores, and began doing business as Egghead.com. The company became the first major chain to abandon the brick and mortar retail world for the Web.

In 1999 Egghead.com merged with Onsale, Inc. and is now a highly rated Internet retailer of new and surplus computer products, consumer electronics, sporting goods, and vacation products. The company also operates an auction site offering bargains on excess and closeout goods and services. Additionally, Egghead launched its new online Office Products Superstore.

In the spring of 2000, Egghead announced new customer offerings designed to meet the needs of small to medium-sized businesses, including new software licensing options and volume pricing discounts. Egghead's method of automatically supplying order status updates via electronic mail was patented in 2000. Finally, Egghead announced several new strategic reseller partnerships, including partnerships with Compaq Computer Corporation, Sony, Telstreet.com, and AllBusiness.com.

For the three months ending on 3/31/00, Egghead.com Inc.'s revenues rose 34 percent to $147.8 million; however, the net loss for the same period rose 37 percent to $25.1 million as a result of lower gross margins and merger costs.

By changing its business model and rethinking its value chains, Egghead Software reincarnated itself as Egghead.com. Clearly, based on sales, the ability to acquire new partnerships, and customer satisfaction, Egghead.com is successful. However, the jury is still out on Egghead.com's (Figure 1-30) future financial success.

Figure 1-30
Egghead.com

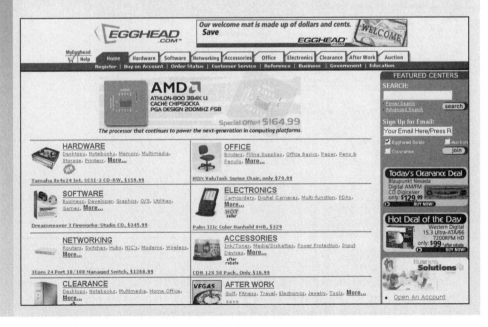

Summary

- In traditional commerce, buyers and sellers come together in a physical marketplace to exchange information, products, services, and payments.
- In electronic commerce, business transactions occur across a telecommunications network in a virtual marketplace called a marketspace.
- The earliest forms of electronic commerce were electronic data interchange (EDI) and electronic funds transfer (EFT).
- The Internet is a network of networks.
- The World Wide Web (Web) is a subset of the Internet where Web servers store hypertext documents called Web pages.
- The Internet is both a product of and a medium for the new economy.
- "E-commerce" and "e-business" both refer to the process of conducting a broad spectrum of business activities over the Internet.
- The growth of the Internet, which electronically links individuals and companies worldwide, is a major factor in the development of the new economy.
- Old business limitations of time, space, and the necessity of moving physical products are being overcome by the flexibility of doing business over the Internet, an intranet, or an extranet.
- Market power is shifting from seller to buyer in the new economy.
- A company's value chain is all the primary and support activities necessary to create and distribute its goods and services.
- Because of the growth of the Internet and its effect on redefining business relationships, many companies are rethinking their traditional value chains and considering new ways of doing business.
- A company's business model is the way in which it conducts business in order to generate revenue.
- In the new economy, companies are creating new e-business models and reinventing old ones.
- E-business models can be broadly categorized as business-to-consumer (B2C), business-to-business (B2B), business-to-government, consumer-to-consumer (C2C), and consumer-to-business (C2B).

CHECKLIST ✓   Thinking about Doing Business Online?

☐ Consider your current costs of providing information and services that your customers could get for themselves by accessing a Web site.

☐ Determine what additional online information or transaction services you can provide to your existing customer base.

☐ Think about how to use your current customer information to make it easier for customers to continue to do business with you online.

☐ Identify ways to assist your customers online, using the expertise of your employees or other customers.

☐ Consider the competitive disadvantages if your competitors offer online customer access to services and information before you do.

☐ Discover ways to improve your business by rethinking your value chains.

☐ Identify the abilities you need to take over the functions provided by others in your value chains.

☐ Consider new ways to generate revenue by enhancing sales or attracting new customers to your Web site.

☐ Develop ideas for ways to repackage current information to attract new customers or create new business opportunities on your Web site.

☐ Determine whether online competitors can significantly harm your business by providing some of the value you currently offer customers in the traditional way.

Key Terms

B2B	brick-and-mortar	electronic commerce
B2B exchange	C2B	e-retail, e-tail, e-tailer
B2C	C2C	extranet
B2G	e-business	host
brick-and-click	e-commerce	hyperlink

Internet
Internet service provider (ISP)
intranet
marketspace
network

server
value chain
vertical market
virtual malls
Web browser

Web pages
Web servers
Web site
World Wide Web (Web)

Review Questions

1. A traditional business environment in a physical building is called a:
   a. brick-and-click marketplace.
   b. brick-and-mortar marketplace.
   c. brick-and-mouse marketplace.
   d. brick-and-Web marketplace.

2. The Internet is a:
   a. network of networks.
   b. Web site.
   c. host.
   d. server.

3. Which of the following e-businesses follows the B2B exchange model?
   a. autobytel.com
   b. Amazon.com
   c. AskMe.com
   d. freightquote.com

4. An e-business that allows consumers to name their own price for products and services is following which e-business model?
   a. B2B
   b. B2G

   c. C2B
   d. C2C

5. Which of the following is not an advantage of doing business online?
   a. Increased sales opportunities
   b. Rapidly changing technology
   c. $24 \times 7 \times 365$ operations
   d. Access to global markets

6. The World Wide Web is a subset of the Internet. **True or False?**

7. An electronic marketplace is sometimes called a marketspace. **True or False?**

8. It is easy to keep track of how many people are using the Internet and how many Web sites exist on the World Wide Web. **True or False?**

9. In the new economy, time, space, moving physical products, and location are less limiting factors for businesses. **True or False?**

10. One disadvantage to online buyers is the lack of trust when dealing with unfamiliar sellers. **True or False?**

Exercises

1. Using Internet search tools or other relevant resources, research the origins and history of the Internet and the World Wide Web. Then create a one- or two-page paper describing at least five major events and how the events led to the growth of e-business in the new economy.

2. Using Internet search tools and other relevant resources, locate information about two men or women whose contributions have had a significant impact on the growth of the Internet and the World Wide Web as a business medium.

Write a one- or two-page paper describing each person and his or her contribution.

3. Define the subscription, virtual mall, and virtual merchant e-business models and provide two examples of each model.

4. Define the terms Internet, intranet, and extranet, and explain the role each plays in e-business.

5. Explain the differences between the traditional reverse auction e-business model and the "name your price" reverse auction e-business model, and include an example of each model.

◆ 1 ◆

While cleaning out your grandfather's storage shed, you find several old items, including carnival glass plates, china dolls, baseball cards, and soft drink signs. You think some of the items may be valuable as collectibles. First, you want to determine the value of the items, and then you want to find a place to sell them. Locate at least five C2C Web sites that might be helpful in determining the items' value and in selling the items. Write a brief summary of each site and how it could be useful to you.

◆ 2 ◆

You are the executive assistant to the president of a company that sells extreme sports equipment, such as mountain bikes, from two brick-and-mortar locations. The president is considering pursuing e-business and asks you to prepare a report on existing extreme-sports-related e-business sites. Locate five extreme sports business Web sites and write a brief summary of each site, including an explanation of its e-business model. Make a recommendation to the president on the type of e-business model he should consider.

◆ 3 ◆

You maintain a file of Internet statistical data for your supervisor, the online sales manager. She asks you to prepare a report containing current estimates of the number of people who are online in the United States and worldwide, online sales estimates, and other relevant data for the next sales meeting. Using the Nua Internet Surveys, e-Marketer, and other relevant Web sites, gather useful data estimates. Then write a brief report containing the data estimates and their sources for your supervisor.

TEAM PROJECT

You and two classmates are eager to start your own e-business. Meet with your classmates and use brainstorming and other applicable techniques and resources to decide on an e-business idea, including the e-business model your e-business will follow. Then create a 5–10 slide presentation, using Microsoft PowerPoint or another presentation tool, describing the e-business and its e-business model. Include in your presentation an analysis of advantages (and/or disadvantages) you expect to experience by doing business online. Present your e-business idea to a group of classmates selected by your instructor.

Useful Links

autoX.com – Collectors Auto Exchange
www.autox.com/

Business Directory – Global Biz Directory.Com
www.globalbizdir.com/

Center for Research in Electronic Commerce - University of Texas at Austin
cism.bus.utexas.edu/

E-Business Research Center – *CIO Magazine*
www.cio.com/forums/ec/

E-Commerce Times
www.ecommercetimes.com/

eCompany
www.ecompany.com/

Inktomi WebMap – Web Statistics
www.inktomi.com/webmap/

Internet History
www.isoc.org/internet-history/brief.html

Managing the Digital Enterprise – North Carolina State University
ecommerce.ncsu.edu/topics/index.html

NetAcademy – *Electronic Markets Journal*
www.electronicmarkets.org/

OracleExchange.com
https://www.oracleexchange.com/US/home.jsp

The Navigational Hub of E-commerce News and Information
www.allec.com/

The Standard – B2B and B2C Industry Updates
www.thestandard.com/index.html

Tim Berners-Lee Biography
www.w3.org/People/Berners-Lee/

VARBusiness Online – E-business Information
www.varbusiness.com/

Links to Web Sites Noted in This Chapter

abuzz
www.abuzz.com/

Altra
www.altranet.com/home.html

Amazon.com
www.amazon.com/

American Boat Listing
www.ablboats.com/

ashford.com
www.ashford.com/

AskMe.com
www.askme.com/

autobytel.com
www.autobytel.com/

avon.com
shop.avon.com/avonshop/campaigns/
    linkshare_index.html

Barnes&Noble.com
www.bn.com/

Business.com
www.business.com/

CATEX
www.catex.com/

CHEF'S
www.chefscatalog.com/

Chemdex
www.chemdex.com/

Choice Mall
mall.choicemall.com/

.commerx PlasticsNet
www.commerxplasticsnet.com/

Consumer Reports
www.consumerreports.org/

CreditTrade
www.credittrade.com/

E*TRADE
www.etrade.com/cgi-bin/gx.cgi/AppLogic+Home

eBay
www.ebay.com/

e-Chemicals
www.e-chemicals.com/

eDiets.com
www.ediets.com/

Edmunds.com
www.edmunds.com/

eFederal.com
www.efederal.com/

Egghead.com
www.egghead.com/

e-Marketer
www.emarketer.com/

e-STEEL
www.e-steel.com/home.shtml

eToys
www.etoys.com/html/welcome.shtml?etys=welcome

ExciteStores
www.excitestores.com/

foodlocker.com
www.foodlocker.com/

FreeMarkets
www.freemarkets.com/

freightquote.com
www.freightquote.com/

Gap.com
www.gap.com/asp/home.asp?wdid=0

Harry and David
www.harryanddavid.com/

Hotel Resource
hotelresource.com/

HoustonStreet.com
www.houstonstreet.com/

iGov.com
www.igov.com/

InfoRocket.com
www.inforocket.com/

Manheim Online
www.manheim.com/

MediSpeciality.com
www.medispecialty.com/

MetalSite
metalsite.net/

National Public Radio
www.npr.org/

NetPossibilities
netpossibilities.com/

Nua Internet Surveys – Directory of Internet Surveys
www.nua.ie/surveys/

Office Depot
www.officedepot.com/

Omaha Steaks
www.omahasteaks.com/

PaperExchange.com
www.paperexchange.com/

Pets.com
www.pets.com/

priceline.com
www.priceline.com/

ReverseAuction.com
www.reverseauction.com/site/Main.jsp

SHOP2gether
edu.shop2gether.com/

Shop4.com
www.shop4.com/

Staples
www.staples.com/

Tattered Cover
www.tatteredcover.com/

TechEx
www.techex.com/public/

The Internet Truckstop
www.truckstop.com/

The Sunglass City
www.thesunglasscity.com/

The Wall Street Journal
interactive.wsj.com/ushome.html

TraderOnline.com
www.traderonline.com/merch/

VerticalNet
www.vertical.net/

VIPAR
www.vipar.com/

Webvan
www000111.webvan.com/default.asp

Women.com Network Web Store
www.women.com/shopping/

World Shopping Network
www.wsnetwork.com/

XSAg.com
xsag.com/

Yahoo! Shopping
shopping.yahoo.com/

For Additional Review

Aldrich, Doug. 1998. "The New Value Chain." www.planetit.com/techcenters/docs/.

Amor, Daniel. 2000. *The E-Business (R)evolution: Living and Working in an Interconnected World.* Upper Saddle River, NJ: Prentice Hall PTR.

Berners-Lee, Tim. 1999. *Weaving the Web.* New York: HarperCollins.

Boston Consulting Group. 1999. "New BCG Study Re-Evaluates Size, Growth, and Importance of Business-to-Business Commerce." December 21. Available on the Web at: www.bcg.com/media_center/media_press_release_archive2.asp.

Brown, Eryn. 1999. "9 Ways to Win on the Web: Sure, Cisco, Dell, and Yahoo Get It. But So Do FedEx, Ford Motor, and Pitney Bowes," *Fortune,* 139(10), May 24, 112.

Business2.0. 2000. "The 10 Driving Principles of the New Economy," March. Available online at: www.business2.com/content/magazine/indepth/2000/03/01/11757.

Business Week. 2000. "The Be-All and End-All of B2B Sites?" (3684), June 5, 56.

Fields, Robin. 2000. "More E-Tailers Are Vying for Government's Business," *Los Angeles Times,* July 31, Home Edition, Business Section, C-1.

Financial Executive. 2000. "The E-business Transformation: Managing for Value in an Internet Economy," 16(3), May, 1.

Ghosh, Shikhar. 1998. "Making Business Sense of the Internet," *Harvard Business Review,* 76(2), March-April, 126–135.

Guglielmo, Connie. 1997. "He's Unscrambling Egghead," *PC Week,* 14(8), February 24, A1(2).

Haylock, C. F. and Muscarella, L. 1999. *NET Success: 24 Leaders in Web Commerce Show You How to Put the Internet to Work for Your Business.* Holbrook, MA: Adams Media.

Hutheesing, Nikhil. 1998. "Last Chance for a Software Vendor," *Forbes,* 161(12), June 15, 130(1).

Industry Week. 1999. "Getting Ready Your Company: How to Prepare Your Organization to Adopt a Value-chain Model," 248(16), September 6, 54.

J. D. Powers and Associates. 1999. "More Than Five Million New-Vehicle Shoppers Nationwide Use the Internet to Shop for New Vehicles," Press Release, August 23.

Jupiter Communications. 2000. "U.S. Internet B-to-B Trade Soars to $6 Trillion in 2005," June 26. Available online at:jup.com/company/pressrelease.jsp?doc=pr000626.

Kalakota, Ravi and Robinson, Marcia. 1999. *E-business: Roadmap for Success.* Reading, MA: Addison-Wesley Longman, Inc.

Kosiur, David. 1997. *Understanding Electronic Commerce.* Redmond, WA: Microsoft Press.

Office.com. 2000. "Establishing an E-commerce Business Model," May 31. Available online at: www.office.com/ob_tools/sales/establishing_an_ecommerce_business_model.html.

Porter, Michael E. 1985. *Competitive Advantage: Creating and Sustaining Superior Performance.* New York: Free Press.

PR Newswire. 2000. "E-Commerce Pioneer Predicts Winners of B2B e-Marketplace Consolidation: Report Outlining the Model Characteristics Now Available," May 31, 1327.

Rappa, Michael. 2000. "Business Models on the Web," June 26. Available online at: ecommerce.ncsu.edu/business_models.html.

Rayport, Jeffery F. 1999. "The Truth About Internet Business Models," Third Quarter. Available online at: www.strategy-business.com/briefs/99301/page1.html.

Richards, Bill. 2000. "Dear Supplier: This Is Going to Hurt You More Than It Does Me," *eCompany*, 1(1), June, 136.

Schneider, Gary P. and Perry, James T. 2000. *Electronic Commerce.* Cambridge, MA: Course Technology.

Schonfeld, Erick. 2000. "Corporations of the World Unite! You Have Nothing to Lose But Your Supply Chains," *eCompany*, 1(1), June, 123.

Sculley, Arthur B. and Woods, W. William A. 1999. *B2B Exchanges.* USA:ISI Publications.

Segaller, Stephen. 1999. *Nerds 2.0.1: A Brief History of the Internet.* New York. TV Books, L.L.C.

Taninecz, George. 2000. "Forging the Chain: Value-chain Initiatives," *Industry Week*, 249(10), May 15, 40.

The Economist (US). 2000. "New Economy, Old Problems," 355(8168), April 29, 23.

U. S. Department of Commerce. 2000. "Digital Economy 2000," June. Available online at: www.esa.doc.gov/.

Walker, Leslie. 2000. "Tapping Expertise on the Net," *The Washington Post Online*, May 4. Available online at: www.washingtonpost.com/wp-dyn/articles/A1055-2000May3.html.

CHAPTER 2

Defining Your E-Business Idea

In this chapter, you will learn to:

Identify entrepreneurial abilities

Describe the entrepreneurial process

Understand the factors affecting e-business success

Identify ways to exploit e-business advantages

Kelby D. Hagar, a 29-year-old fifth-generation Texan, grew up in Hereford, Texas, a small town in the Texas panhandle. Kelby graduated magna cum laude from Angelo State University in 1991, with a degree in prelaw and accounting. After becoming a CPA, he decided to study law and was accepted at Harvard Law School. Before leaving for Harvard, Hagar married his childhood sweetheart.

At Harvard, both Hagar and his wife were extremely busy. His wife was busy with her first teaching position, and Hagar was spending most of his time in class or studying. Late one night, Hagar opened the refrigerator looking for a snack and found nothing but a chunk of old cheese and a bottle of olives. Clearly, grocery shopping was in order, but who had time? Hagar, who used the Internet to order books, make travel arrangements, and send e-mail, wondered why he couldn't simply order his groceries online and have them delivered to his doorstep, just like the books he ordered.

After Hagar graduated from Harvard cum laude in 1995, the Hagars moved to Dallas, where he began working for a major law firm, specializing in mergers and acquisitions. However, the idea of enabling individuals with hectic schedules to save time by ordering groceries online and having them delivered just wouldn't go away.

The Entrepreneurial Process

Do you have an e-business idea? Are you excited about taking your e-business idea to market? One of the first things you should do is consider the entrepreneurial abilities needed to start any business, including an e-business.

The entrepreneur

An **entrepreneur** is someone who assumes the risks associated with starting and running his or her own business. Entrepreneurial abilities can include leadership abilities, a high-energy personality, plenty of self-confidence, a high level of organizational skills, and the ability to act quickly and decisively. Successful entrepreneurs are usually independent, goal-oriented, creative, and competitive.

While it is difficult to define leadership, most people would agree that a list of leadership traits would include intelligence, determination, integrity, listening skills, and the ability to relate to others in a positive way. As a leader, an entrepreneur must be able to articulate to others the mission and goals for his or her business idea. In the leadership role, an entrepreneur must also put together, energize, and direct the high-performance teams necessary to pursue the new business' mission and goals. To successfully put together, run, and grow any business, including an e-business, leadership skills are essential.

Starting any new business requires considerable energy as well as the ability to focus that energy on accomplishing objectives. Because starting and running a business is more time-intensive than working for someone else, it is likely that long hours are the rule. Adequate exercise, a good diet, and stress management are critical for an entrepreneur.

A successful entrepreneur must believe in his or her business idea and have the self-confidence to attain business goals. When an entrepreneur's business idea is closely related to an area of his or her interests or expertise, that self-confidence is enhanced. Because of time demands, an entrepreneur must also be able to successfully organize his or her business activities. This allows the entrepreneur to get things done on time, locate business-critical information quickly, and be on time for appointments.

All business, including e-business, is very competitive. Most successful entrepreneurs display a competitive nature early in life in their winning approach to school, hobbies, or sports. Finally, an entrepreneur must be ready to make short-term sacrifices, such as limited time spent with family and friends, in return for long-term success.

TIP

Some people consider starting their own business so that they can set their own working hours. As an entrepreneur, you may well be able to set your own work schedule, but be aware that it will likely be a long one! The typical entrepreneur works an average of 12 hours a day, 6 or 7 days a week.

TIP

Before quitting your current job and devoting yourself to an entrepreneurial adventure, you should consider your ongoing financial needs. Various sources suggest that you should set aside funds to cover from six months' to two years' living expenses.

E-CASE No Stigma Attached to Failure

A few years ago, Jason Zasky, a survivor of three failed music magazines, was walking with his cousin in Manhattan, when his cousin stopped and said: "I have a great idea for a magazine—failure." Zasky, who also thought the idea was great, began working "serious 18-hour days" with his partner, Kathleen Ervin, in a small building in Scarsdale, New York, where they launched the online magazine *Failure* in July 2000.

Targeted to adventurous, risk-taking individuals between 20 and 40 years old, for whom the word "failure" carries little stigma, *Failure Magazine* is about not being successful. According to Zasky, *Failure Magazine* is designed to be thought-provoking rather than provocative and will include articles and features on topics such as the world's greatest golfer never to succeed on the pro tour, an interview with the player who made one of the greatest baseball errors in history, and an interactive page that will allow viewers to decide which movie ideas sound like bombs and which ideas are appealing.

Zasky believes that because failure is a universal experience, people will be interested in failure-related stories. When asked if it were possible to create a successful e-business by publicizing the failures of others, Zasky responded that he and his partner believe the greatest failure would be in not trying to make *Failure Magazine* (Figure 2-1) a success. Even with a history of three failed entrepreneurial attempts, Zasky was not afraid to use his entrepreneurial abilities and hard work to try a new e-business idea!

(Continued on the next page)

Figure 2-1
Failure Magazine

After you consider the abilities needed by an entrepreneur, the next step is to understand the entrepreneurial process.

The entrepreneurial process

The **entrepreneurial process** can be divided into several steps or stages:

- Step 1: Deciding whether or not you are an entrepreneur. It is critical that you assess your entrepreneurial abilities and evaluate the time and effort required by you and the effect on your family, before going to the next step.
- Step 2: Deciding whether to buy or start a business. If you determine that you are an entrepreneur, the next step in the entrepreneurial process is to decide whether to start a new business or purchase an existing business.
- Step 3: If you start a new business, you must first define the business idea or concept, including the products or services to be sold. Then you must create a business plan that will include your assessment of the business environment, an identification of the business need, estimates of anticipated profitability, and a description of the legal form the business will take. After developing the business plan, you must then secure financing from various sources such as "sweat" equity, financing from friends and family, angel investors, and venture capitalists.

- Step 4: After purchasing or starting a new business, you then operate the business.
- Step 5: At some point you will harvest the business by letting the business become a "cash cow," by going public, by selling the business, or by liquidating the business.

Figure 2-2 illustrates the stages of the entrepreneurial process.

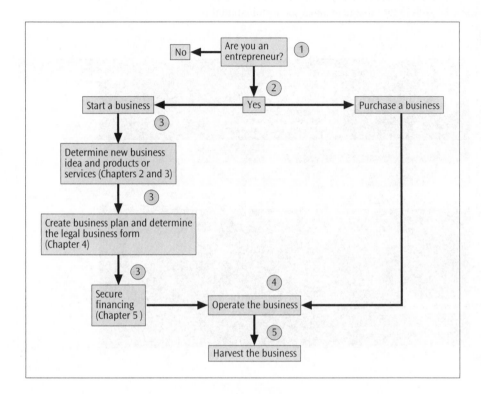

Figure 2-2
Entrepreneur-ial process

How an Entrepreneur Gets It Done

When David M. Tolmie, a 20-year marketing veteran, became the chief executive and president of WebPromote in December 1998, he quickly realized that he would need all his energies and entrepreneurial abilities. He needed to think beyond traditional solutions, take calculated risks, and make quick decisions in order to direct WebPromote's 30 employees, refocus WebPromote's services, acquire funding, put together a new management team, launch new software, expand to a full-service marketing company, and take the company public. In a little more than a year, Tolmie accomplished all that and more.

WebPromote was launched in 1995 as a Web marketing company that sent a weekly newsletter containing promotional messages to subscribers. When the newsletter started generating revenue, the founders began searching for funds to expand the business. Tolmie, who met with the founders on behalf of a venture capital firm, liked their idea and, because he wanted the challenge of running a business, agreed to join WebPromote. His goal was to quickly take the

(Continued on the next page)

company public. In slightly more than one year, Tolmie changed the company name to Yesmail, refocused the company on its permission e-mail business, doubled the company's member base to 11 million, and took Yesmail public. Yesmail reported $15.6 million in revenue for 1999, compared with $4.6 million for the previous year, and, in March 2000, Tolmie harvested the Yesmail (Figure 2-3) business by selling it to GMCI. Tolmie recognized a great idea and used his entrepreneurial skills to turn that idea into a successful e-business.

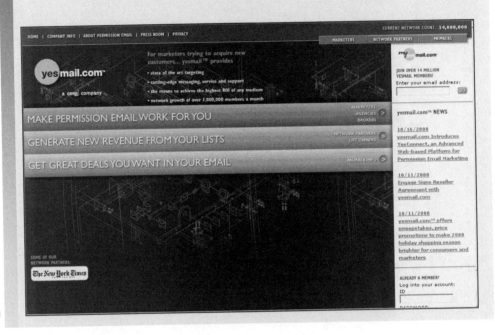

Figure 2-3
yesmail.com

After reviewing the entrepreneurial process and determining that you have the entrepreneurial abilities to take your e-business idea to market, you should next consider some of the major factors that could affect the long-term success of your e-business idea.

Factors Affecting E-Business Success

Ideas for new e-businesses span the spectrum from selling books and records directly to consumers, to auctioning excess oil and gas capacity, to creating software and hardware for the e-business infrastructure. As the new Internet economy continues to grow, change, and mature, many of today's e-businesses that are based on viable business models will also grow, change, and mature. Unfortunately, many e-businesses based on flawed business models won't be able to survive, including, perhaps, some of the e-business examples used in this text.

Although it is too soon to tell who will be the ultimate e-business winners and losers, it is important to think about the many varied e-business ideas—grocery order and delivery, offbeat online magazines, e-mail advertising, awards programs, garage sales, electronic prescription writing, regional foods sales, gift registries, excess energy sales, real estate sales, and so many more—that fuel the new Internet economy.

Many factors related to Web-based businesses can affect the success of a new e-business idea, including the network effect, scalability, the need for innovative marketing ideas, and ease of entry into electronic markets for you and your competitors.

The network effect

A primary factor in the growth of e-business in the new economy, where businesses and consumers are linked together via the Internet, is the network effect. The **network effect** means that the value of a network, such as the Internet, to each of its participants grows as the number of participants on the network grows. For example, a telephone network with a single telephone has little value to the user because there is no one else to call. However, when you add a second telephone, the value of the telephone network increases because there are now other people to call. As additional telephones are added, the telephone network's value continues to increase. For each Internet user, the value of being online grows as the number of people online grows. The Web-based network of businesses and consumers becomes more valuable as more and more businesses and consumers participate, increase contact, and conduct business transactions.

E-businesses that take advantage of network effects to get new customers include NetMind (automatic notification of changes to subscriber Web pages), NetZero (free Internet service), and Third Voice.

Third Voice, launched in May 1999, gained worldwide attention by offering a free personal Web assistant service that allows users to post Web "sticky notes" on any Web page. Users simply download the Third Voice software that installs as a plug-in to their Web browser. Users can then post comments or view comments by others, join discussion groups or communities related to specific Web pages, and import their e-mail address book into Third Voice, using it to generate a message that includes a link to the Web page being viewed. If message recipients don't have Third Voice, a quick click on a link takes them to the Third Voice Web site where they can download the software. By importing the user's address book, Third Voice immediately creates a network of potential users.

In April 2000, Third Voice released a software version that enables all words on any Web page to become "active words" related to additional Web content. Users can select frequently used, underlined "active words" to instantly get relevant content. Other less popular non-underlined words can be used to search Third Voice for relevant content. Users can post "sticky notes" related to the specific words. Although all Third Voice transactions occur between the users' browsers and Third Voice servers, and nothing happens on a viewed Web site itself, the Third Voice service is controversial with some Internet privacy groups. However, the more users download Third Voice (Figure 2-4) and participate in the network, the more valuable it becomes through user-generated postings and new Web page communities.

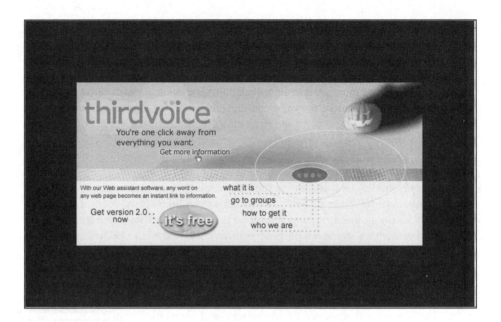

Figure 2-4
Third Voice

Take a close look at your e-business idea and decide whether or not the value of the product or service you plan to offer increases with greater use or distribution over the Internet.

Scalability

Scalability refers to the ability of your e-business idea to continue to function well, regardless of how large the business gets. The rapid growth of an e-business can cause problems if that e-business idea does not scale with unexpected growth. AllAdvantage, a self-styled "infomediary," launched its Web site in March 1999. AllAdvantage, which as of this writing pays members $.53 per hour to surf the Web, also pays members for referrals. Members who refer other members are paid an additional $.10 an hour while the new members surf the Web. Original members are also paid a bonus if the member they refer refers someone else.

TIP

Scalability is also used to refer to the ability of servers to handle increased traffic loads to a Web site without crashing.

AllAdvantage underestimated the power of the network effect on their e-business idea. In the first week, 100,000 members enrolled. Expecting to have 20,000 members within three months, AllAdvantage instead found itself with one million members. By late spring 2000, AllAdvantage had 6.7 million registered members and about 2.5 million active members, with membership continuing to grow at 15,000–16,000 members per month. The payments and referral fees to active members in the first quarter of 2000 alone were approximately $40 million.

In February 2000, AllAdvantage announced an upcoming IPO. But problems with AllAdvantage's e-business idea were becoming apparent. AllAdvantage's revenue model was based on selling demographic data to advertisers. However, the idea of targeting online advertising to specific individuals was met with some resistance by advertisers, and AllAdvantage had trouble selling them on targeted ads. In the first quarter of 2000, advertising revenue was only $9 million, and total advertising revenue since its launch was only $14.4 million. Altogether, AllAdvantage lost $102.7 million in its first year of operation.

In an attempt to reduce membership costs, AllAdvantage cut back the maximum number of paid browsing hours for U.S. members from 25 to 15, while raising the hourly rate from $.50 to $.53. The company also changed its payment schedule from monthly to every 45 days. But some analysts questioned the viability of giving away millions of dollars in the hope of earning money sometime in the future. In July 2000 AllAdvantage (Figure 2-5) withdrew its IPO, citing adverse market conditions.

AllAdvantage's e-business idea failed to anticipate the network effect, and the company found that its business model did not scale with unanticipated customer growth. When evaluating your e-business idea, consider its scalability.

 TIP

There are many other e-businesses similar to AllAdvantage, such as MyPoints, FreeRide.com, ClickRewards, Cybergold, beenz, Spedia, Jotter, and MoneyForMail. These pay-as-you-surf e-businesses are trying to find ways to differentiate themselves from AllAdvantage. For example, Spedia provides its members with a variety of ways to make money, including games, surfing, shopping, and sweepstakes. Jotter also provides its members with a digital wallet, form filler, customizable news ticker tape, and media player.

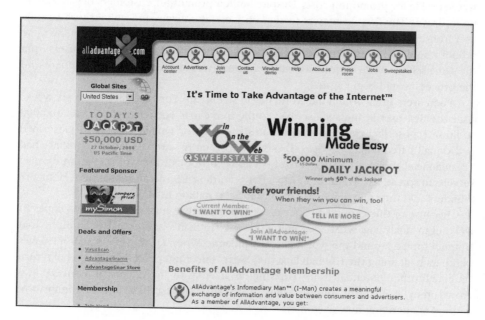

Figure 2-5
AllAdvantage.
com

Innovative Web marketing ideas

Just as the growth of e-business encourages the development of new business models, it also encourages the growth of new marketing ideas to promote the e-businesses based on those models. One of the most famous examples of an e-business whose business model successfully exploits the network effect by using an innovative marketing idea is Hotmail.

Hotmail, a wildly successful e-business that offers free e-mail services, almost didn't get launched (see Figure 2-6). As with many e-business startups, Hotmail's founders Sabeer Bhatia and Jack Smith had a problem interesting venture capitalists in their idea. When they met with Tim Draper, the founding partner of Draper Fisher Jurvetson (DFJ), Draper became interested in the idea of free e-mail but insisted that text be added to the bottom of each outgoing e-mail message, encouraging recipients to get their free e-mail at Hotmail. Draper wanted the text to be linked to the Hotmail site so that message recipients could click the link, view the Hotmail Web site, and immediately sign up for the service. At first Bhatia and Smith resisted the idea of adding the linked message, considering it too much like spam, which consists of commercial messages or advertisements blindly posted to a large number of unrelated and uninterested parties. At this time, the contents of e-mail were generally considered to be completely private. Finally, Bhatia and Smith agreed with Draper on the inclusion of the marketing message link.

Hotmail was quietly launched on July 4, 1996. Because of the holiday, the launch generated little press coverage. Also, because of limited seed capital, less than $50,000 was set aside for promotion costs. Despite limited promotion, customers began signing up in droves. The first customers to sign up were at colleges and universities. A single user from a school would sign up, and then, the next day, a hundred users from the same school would sign up, and by the end of the week there would be a thousand users from the same school. Next, information about Hotmail would spread to another school, and the process would begin again.

Soon users from around the world began signing up. For example, within six weeks after the first user in India subscribed to the free e-mail service, Hotmail had 100,000 additional subscribers in India. In less than six months, one million users had subscribed to Hotmail's free e-mail services. Less than 18 months after its launch, Hotmail had 12 million subscribers. The network effect allowed Hotmail, the world's first free e-mail service, to spread like wildfire. Shortly after recording 12 million subscribers, Hotmail's founders sold the company to Microsoft for $400 million in Microsoft stock.

Because of the astounding success of Hotmail, Steve Jurvetson and Tim Draper of DFJ began analyzing what was unique about Hotmail's success. They realized that Hotmail spread around the world like an epidemic. When current users sent messages to others containing the Hotmail link, they were "infecting" potential users, who in turn "infected" other potential users. Drawing an analogy to the effects of a sneeze in a crowd, they coined the phrase "viral marketing" to describe Hotmail's innovative method of attracting new customers.

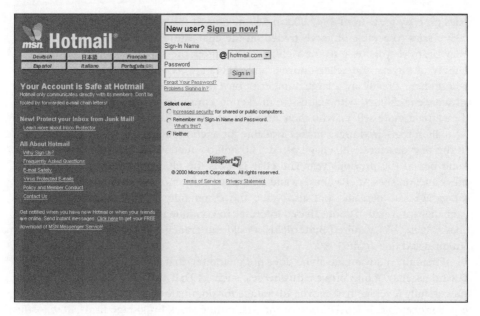

Figure 2-6
Hotmail

As part of the evaluation of your e-business idea, you should identify the innovative marketing techniques you can use to promote it.

Ease of entry into electronic markets

The Internet opened up market access for both buyers and sellers. Low-cost Internet technology encourages the growth of e-businesses and enables many e-businesses to easily enter a market and be competitive. A good example is the B2C auction market.

Two major reasons for the proliferation of online auction sites are the low barrier to entry in the auction market and the potential to earn fat commissions on sales without dealing with inventory. The cost of entry into the online auction arena is so low that soon it may be possible for every online newspaper, directory, portal, storefront, and perhaps personal Web page to become competitors in the auction market. eBay, one of the few e-businesses to be profitable from the beginning, remains the Web's top auction site despite this intensified competition. eBay provides auction services for more than 4,300 item categories, including automobiles, collectibles, sports memorabilia, toys, computers, stamps, jewelry, and antiques.

eBay has two business advantages that may be competitive barriers for other auction sites: **first-mover advantage**, or first to market, and name identification. eBay, one of the first auction Web sites, was launched on Labor Day 1995, as a central location to buy and sell unique items. Within two years, sales on eBay grew from $347,000 to $47 million. At the end of the first quarter 2000, eBay's sales (the value of goods traded on the eBay site) exceeded $1 billion. Because eBay beat everyone else to the Web auction market, the site developed a strong community of buyers, and sellers. Buyers and sellers are naturally attracted to a Web site where there already are a large number of buyers and sellers. eBay, with its first-mover advantage combined with aggressive marketing, continues to attract more buyers, who then attract more sellers, who then attract more buyers. Each day, eBay, which has over 12 million registered users, makes available for auction more than 450,000 new items. In the first quarter 2000, eBay hosted 53.6 million auctions, up 113 percent from the same period the previous year. The eBay label is the ultimate competitive barrier for other auction sites. At eBay, buyers and sellers not only enjoy the excitement of the hunt for garage sale bargains, but also value the community of buyers and sellers who exchange information about shared interests. To continue to add credibility to its name, eBay (Figure 2-7) acquired Butterfields, an old San Francisco auction house known for antiques and fine art, in 1999.

First-mover advantage alone does not guarantee that an e-business will be as successful as eBay. While other e-businesses, such as Dell Computer, have managed to successfully leverage first-mover advantage for dominant market position, many have not. For example, E*TRADE, one of the first online stock brokerage firms, now jostles for first place with Charles Schwab and Fidelity Investments, in a market that E*TRADE pioneered.

Figure 2-7
eBay

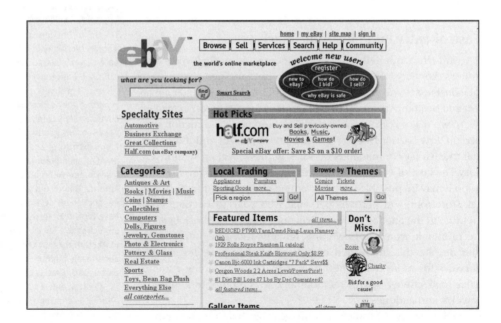

It Doesn't Happen by Itself

Margaret C. "Meg" Whitman's first job was as a snack bar cook and general manager at Valley Ranch in Valley, Wyoming, where she made brownies and cookies, and bought other items to sell at the snack bar. When asked by the *Detroit News* what she learned from that first job, Whitman replied, "You have to work hard to make things work right. It doesn't happen by itself." Whitman, with a BA in Economics from Princeton and a Harvard MBA, more than proved her point by tackling a succession of high-profile jobs with companies such as Proctor and Gamble and Disney.

As president and CEO of Florists Transworld Delivery (FTD) from 1995 to 1997, Whitman led the launch of FTD's Internet strategy. Whitman joined eBay as President and CEO in 1998. Whitman recognized that eBay was a great e-business idea because it was unique, and also because it had no brick-and-mortar equivalent. Her instincts told her eBay was going to be huge. Just how huge no one anticipated. Analysts estimate that eBay controls between 85 percent and 88 percent of an online auction market that may reach $41.6 billion by 2003.

You should identify whether your e-business idea has a first-mover or other competitive advantage, and, if not, consider the competitive barriers your e-business idea faces.

Adaptability to change

Rapid knowledge transfer and the need to make decisions quickly in order to act on new ideas, take advantage of new opportunities, and resolve new challenges characterize the e-business environment. For example, Amazon.com, the e-tailer that started as an online store selling books, quickly added music, toys and games, videos, and other items, as consumer interest in purchasing these items online grew. When the online garage sale craze hit, Amazon.com (Figure 2-8) quickly added its own auction site and automatically registered its online store customers to participate in the auctions.

Because the rapid pace of change affects the most fundamental rules about how your e-business idea functions, you must consider how well your e-business idea could respond and adapt to change.

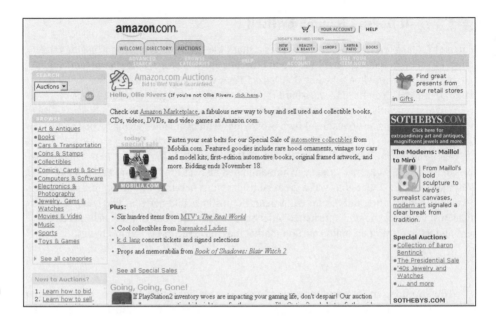

Figure 2-8

Amazon.com Auctions

Exploiting E-Business Advantages

There are certain prospective advantages inherent in doing business online. Many companies are extending their existing businesses or creating new e-businesses by utilizing these built-in advantages, including the potential to quickly expand their market, acquire greater visibility, maximize customer relationships, create new services, and reduce costs.

Expand the market

One of the major advantages of doing business online is the ability to reach a larger market of potential customers who are no longer bound by the constraints of physical location or time. Many companies are taking their existing business model and revamping it to create a complementary e-business. Examples of brick-and-mortar businesses revamping their existing businesses to create complementary e-businesses include Egghead.com, Costco, and Ticketmaster. The example of Ticketmaster, a giant company that is a world leader in the traditional market for live event ticket sales, demonstrates how a very successful business idea can be revamped as an e-business idea.

USA Network acquired Ticketmaster in 1998 and then combined the Ticketmaster Web site business with its Citysearch Web site to create Ticketmaster Online - Citysearch, Inc. The new company's Ticketmaster.com Web site quickly carved out its own live event ticket sale market. While Ticketmaster targets large stadium events, Ticketmaster.com expands Ticketmaster's traditional market by also providing access to small venues with fewer than 200 seats. Ticketmaster waits for customers to come to it to purchase tickets for a specific event. Ticketmaster.com again expands the market by actively soliciting repeat customers via e-mail. Ticketmaster.com e-mails customers the playlist from the most recent concert they attended and invites them to purchase concert memorabilia. Ticketmaster.com (Figure 2-9) also solicits sales by sending subscribing customers a newsletter containing a list of events they might be interested in attending.

If your e-business idea is based on a successful traditional business model, you should determine whether your idea takes advantage of the Internet to expand your market.

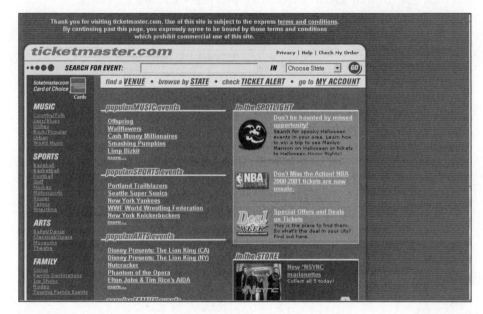

Figure 2-9
Ticketmaster.com

Acquire greater business visibility

The Internet enables companies to get their name and products or services in front of more potential customers than ever before. For many retailers and manufacturers, having a presence online is now essential. In the early days of e-business, many companies created Web sites that were little more than a company brochure. Today, companies want to use the power of the Internet to generate sales and increase their customer base. One longtime traditional company that plans to harness the power of the Internet to gain greater business visibility is Fingerhut.

Anyone who has ever ordered products from e-tailers such eToys or Pier1.com may be surprised to know that their shipment may have been shipped from a Fingerhut warehouse. Fingerhut, a 51-year-old company that was traditionally a home furnishings and clothing catalog retailer, is now one of the Web's top distributors. Fingerhut fulfills orders for e-tailers that would rather use Fingerhut's warehouse space than build their own.

In addition to order fulfillment for other e-tailers, Fingerhut also sells its own goods online. One of Fingerhut's objectives is to gain greater business visibility and become one of the top five e-tailers by 2003. In 1999, 10 percent of Fingerhut's sales were generated online. Fingerhut's goal is to generate 100 percent of its sales online by creating a Web superstore featuring both enormous selection and personalized service. To provide that personalized service, Fingerhut (Figure 2-10) has data on 31 million customers that it can use to create highly personalized Web pages and targeted e-mail.

Figure 2-10
Fingerhut

E-CASE I'll Buy Half

Half.com, Inc. provides a marketspace for selling used books, CDs, video games, movies, and other items at a fixed price no more than half the retail price. Half.com, Inc., which never touches the merchandise, takes a 15 percent commission from the seller and requires the buyer to pay for shipping. Half.com, Inc. has over 4 million listed products and 250,000 registered users.

When the marketing team at Half.com, Inc. wanted a sure-fire way to get national attention and create greater business visibility for their new startup, they hit upon the idea of renaming a small town "Half.com." In December 1999, a representative of Half.com, Inc. met with the mayor of Halfway, Oregon, a small town with a population of 345, to propose that the town change its name to Half.com. After talking with several of Halfway's residents, Half.com, Inc. offered to donate computers to the city's elementary school, provide a raffle prize for the County Fair, and donate funds for civic improvements. The town accepted and issued a proclamation changing its name to Half.com, Oregon. On January 19, 2000, Half.com, Inc. launched its Web site during the live broadcast of NBC's *Today Show* from Half.com, Oregon, and Conshohocken, Pennsylvania, Half.com, Inc.'s corporate headquarters.

Meg Whitman, eBay's CEO, heard about Half.com, Inc. by reading eBay customer comments raving about the site. Intrigued by what the eBay customers had to say about Half.com, Inc., Whitman browsed the Half.com, Inc. Web site, where she bought a book about fly-fishing. Recognizing that Half.com, Inc.'s (Figure 2-11) business model was complementary to eBay's, she then bought the whole company for more than $312 million.

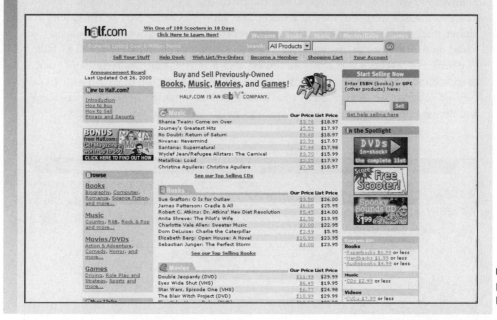

Figure 2-11
Half.com, Inc.

You should consider how your e-business idea could use the Internet to gain greater business visibility.

Maximize customer relationships and improve responsiveness

The Internet is the ultimate tool for communication. Companies that do business online can use the Internet to stay in touch with their customers' needs, build one-on-one relationships, provide information to enrich their customers' online experience, and build customer loyalty. With consumers increasingly in control, e-businesses can build customer loyalty by creating warm customer relationships using personalized e-mail, displaying welcome messages, keeping track of customers' interests, and delivering what the customers want. One company that has been winning the war for customer loyalty both offline and online is Southwest Airlines.

Southwest Airlines, headquartered in Dallas, Texas, is famous for its "no frills" and low-cost approach to air travel. Southwest understands that simplicity, low airfares, and customer service are what keep its customers coming back, and the company kept this in mind when developing its Web site. The Southwest.com Web site is simple for customers to use. With just a few clicks, a customer can review flight and fare options and book a low-cost flight. Customers who participate in Southwest's frequent flyer program, Rapid Rewards, can also monitor their accounts and get double trip credit when they book a flight at the Web site.

Understanding exactly what their online customers want and then using the power of the Internet to build on that understanding is paying off for Southwest Airlines. The August 1999 Nielsen/NetRatings' survey indicated that 13.8 percent of people who visited Southwest's Web site booked a ticket. This ratio was twice that of Travelocity and higher than those of other Web retailers. In February 2000, Southwest announced that, based on the increase in revenues from its Web site, it was on track to exceed $1 billion in e-business revenues for 2000. Southwest also reported that Deutsche Bank, which tracks online ticket bookings, placed Southwest Airlines at the top of its list, with 20 percent of Southwest's (Figure 2-12) total bookings being made online. This was almost double the number of online ticket bookings for the second airline on the list.

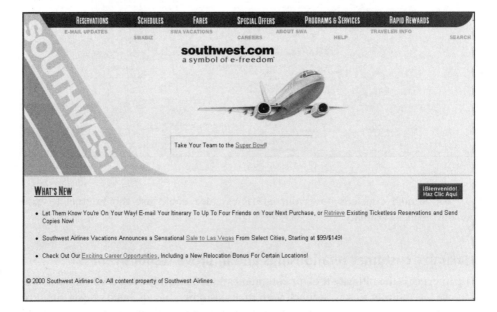

Figure 2-12
Southwest
Airlines

W. W. Grainger, a 72-year-old, $5-billion-a-year distributor of electrical merchandise such as power tools, electric motors, and light bulbs, seems an unlikely candidate for e-business. Grainger, which maintains a comfortable lead over its competitors in the sales of machine maintenance and repair supplies, at first seems to be exactly the kind of traditional intermediary many people thought e-business would destroy.

By November 1999, however, Grainger's Web-based sales were thriving. Although many of Grainger's customers are not yet ordering from its Web site, orders that are placed from the Web site are averaging twice as large as those submitted via the traditional methods of phone or fax. Additionally, customers ordering online are spending 20 percent more annually than they did ordering by traditional methods.

A major reason for Grainger's success online is improved responsiveness to their customers' need to quickly locate product and pricing information. Traditional customers must browse through a huge, red, 4,000-page catalog that lists 70,000 products. Online customers can quickly search through 220,000 products, get immediate information on the availability of a product, and get up-to-date prices customized for any contractual discounts. Grainger (Figure 2-13) anticipates a time in the near future when more than 50 percent of its orders will be placed from its Web site. As with many other e-businesses, Grainger's experience with e-business continues to evolve. As of this writing, a major revamping of its Web site is under way.

Figure 2-13
W. W. Grainger

Create new services

The Web is both an environment that allows companies to expand and refine their traditional businesses and an incubator for hundreds of new e-business ideas. Many of these new e-business ideas follow a B2B or B2C infomediary business model, providing a single, central location for users to get advice, tips, and recommendations for online activities. One example of an infomediary whose e-business idea is based on the proliferation of other successful Web-based businesses is TrailBreaker.

During the 1998 holiday season, Noah Eckhouse, TrailBreaker's CEO and founder, became increasingly frustrated as he tried to shop for gifts. He discovered that the size of the Web led to too many confusing choices for the average shopper, making it difficult to decide where to shop for quality and price. With his frustrating online shopping experience in mind, Eckhouse founded TrailBreaker in 1999 as one of the first Web sites to use a staff of experts to provide content for a shopping advisory service.

TrailBreaker has about 40 consulting experts with 10–20 years' experience in their respective fields. These experts provide information to consumers on buying products or services in industries such as travel, auto, home mortgage, gourmet food, furniture, personal training, and sporting goods. TrailBreaker also sends a free weekly newsletter to subscribers, providing information on the best deals on the Web.

TrailBreaker makes money in two ways: by receiving a small percentage on sales from companies who pay referral fees, and by selling customized content to partnering Web sites. To maintain an unbiased approach, TrailBreaker (Figure 2-14) does not accept advertising on its site, and its consulting experts do not know which companies pay referral fees. TrailBreaker has taken a new idea and created a new service that takes advantage of the increase in consumer online purchasing.

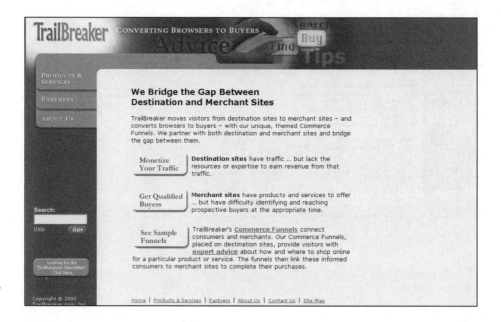

Figure 2-14
TrailBreaker

The growth of Internet access and Web-based businesses creates an environment in which new business ideas, such as TrailBreaker, have the potential to flourish. When evaluating your business idea, consider whether or not it provides a new, unique, or revamped product or service suited to the Internet environment.

Reduce costs

One of the most important factors inherent in doing business online is the potential to reduce the costs associated with gathering and maintaining information on customers, processing transactions, and providing customer service. In a traditional business, it can cost $1 or more to gather information on a single customer. With an online business it can cost much less, perhaps as little as 1¢ per customer. Allowing customers to get quotes and place and track orders from a Web site can reduce order handling and sales support costs. Other cost savings can be achieved by providing customer support online.

Technology companies such as Dell Computer, Sun Microsystems (computer workstations and software), and Cisco Systems (networking products) save many millions of dollars each year by allowing customers to access support services from their Web sites. By making their Web sites the starting point for customer support, Cisco Systems, Dell, and Sun Microsystems (Figure 2-15) greatly reduce the number of calls to their technical assistance centers—and that translates into major cost savings.

When evaluating your e-business idea, consider how it exploits any inherent online advantages by expanding your market, gaining greater business visibility, improving customer relationships, providing new services, and reducing the costs associated with maintaining customer relationships and processing transactions.

Figure 2-15
Sun
Microsystems

E-CASE IN PROGRESS foodlocker.com

Michael Bates, a graduate of Rice University's Jones Graduate School program, traveled frequently between Texas and New Jersey and found that he couldn't find his favorite Texas food items in New Jersey grocery stores. He rapidly tired of purchasing the items in Texas and carrying them to New Jersey. Bates also missed the favorite foods he had learned to love during his childhood in Memphis.

Bates decided that what was needed was a Web store where buyers around the world could purchase the regional food specialties they loved, such as Stubbs BBQ Sauce from Austin, Texas, coffee and beignet mix from Café Du Monde in New Orleans, Louisiana, and rugelach and strudel pastries from Brooklyn, New York.

Bates discussed the regional food e-tailer idea over breakfast with fellow Rice University Jones Graduate School graduate Stuart Wagner and Professor H. A. Napier, and the foodlocker.com (Figure 2-16) e-business idea was born. In subsequent chapters we will follow the progress of foodlocker.com as it grows from idea to a fully realized e-business.

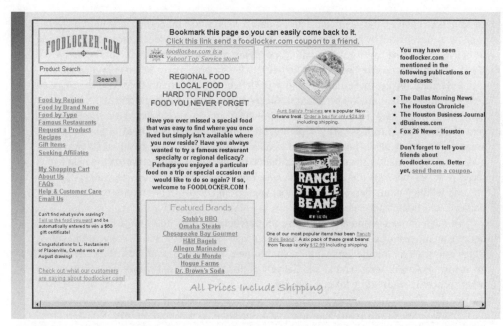

Figure 2-16
foodlocker.com

. . . KELBY HAGAR'S VISION

While working for the Dallas law firm, Kelby Hagar quickly realized that he enjoyed the deal-making process much more than creating the documentation that followed, so he decided to explore the excitement and potential of creating an e-business. In his free time he began to discuss his online grocery service idea with business experts and advisors. The more he discussed his idea, the more sense an online grocery business made.

Hagar did his homework. He studied various industries and came to the conclusion that, without the fixed costs associated with a brick-and-mortar storefront, a "virtual store" could operate more efficiently. He also learned that most people spend 100 hours a year at the grocery store. Hagar believed that many people didn't want to spend their free time at the grocery store and were ready for the convenience of online shopping and home delivery. His marketing studies also showed that 85 percent of grocery purchases are repeat purchases for basic items such as laundry powder, ice cream, and disposable diapers. Hagar decided that consumers, knowing from experience what they were getting with these kinds of products, would be comfortable purchasing them from a Web page. In early spring, 1999, Hagar left the law firm to co-found a grocery delivery e-business. After four years of planning, research, and putting together a group to provide financial support, operating expertise, transportation and delivery logistical experience, and grocery experience, GroceryWorks.com opened for business in the Dallas area in November 1999 and expanded to the Houston area in May 2000.

Customers place their orders online at the GroceryWorks.com Web site (see Figure 2-17). The orders are then fulfilled from a distribution center and delivered by specially equipped trucks. Each distribution center stocks 15,000–20,000 dry goods items, gets produce, meat, and fish from local restaurant suppliers, and obtains prepared food from Eatzi's, a gourmet takeout store.

(Continued on the next page)

Customers can place orders 24 hours a day, 7 days a week, and can specify a one-hour delivery window. Orders placed by noon can be delivered that evening, and orders placed after lunch can be delivered the next day. As of this writing, prices are comparable to those of local grocery stores, and there is no order minimum, no delivery fee for orders over $60, and no tipping.

In January 2000, Gary Fernandes, a former Electronic Data Systems Corp. vice chairman and venture capitalist, became the president of GroceryWorks.com for a 15 percent stake in the company. Kelby Hagar became its chairman. In April 2000, grocery giant Safeway purchased 50 percent of GroceryWorks.com as part of a plan to offer online grocery services in 16 metropolitan areas by the end of 2001. The purchase of a stake in GroceryWorks.com positions Safeway to be a leader in what analysts say may be a $50 billion online grocery market within the next five to seven years. By July 2000 the GroceryWorks.com Web site was receiving approximately 25,000 hits and fulfilling more than 5,000 orders per week from two distribution centers in the Dallas/Ft. Worth area and one in Houston.

Through hard work and an unflagging entrepreneurial spirit, Kelby Hagar's vision first became an interesting e-business idea and then a viable e-business. Now GroceryWorks.com, which calls itself an "online home fulfillment business," plans to expand into other geographical areas such as Austin, Texas, and to add delivery of flowers, books, magazines, dry cleaning, and other services that save time for busy people.

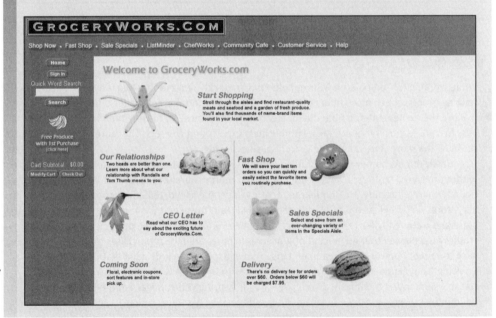

Figure 2-17
Grocery
Works.com

Summary

- An entrepreneur is someone who assumes the risks associated with starting and running his or her own business.
- The entrepreneurial process includes (1) deciding whether you are an entrepreneur, (2) deciding whether to buy or create a business, (3) planning the business, (4) operating the business, and (5) harvesting the business.
- The network effect means that the value of a network to its participants grows as the number of participants grows.
- Scalability refers to the ability of a business idea to continue to function well regardless of how large the business becomes.
- The growth of the Internet encourages the growth of innovative marketing ideas.
- The growth of the Internet has lowered the cost of entry into markets for many e-businesses.
- Two competitive barriers are failure to secure first-mover advantage and lack of name awareness and loyalty.
- The inherent advantages of doing business online include the potential to quickly expand the market, acquire greater business visibility, maximize customer relationships, create new services, and reduce costs.

CHECKLIST   You and Your E-Business Idea

❑ Do you have the energy, self-confidence, organizational skills, and ability to focus on objectives needed to start an e-business?

❑ Are you a risk taker: can you make risky decisions and cope with the consequences?

❑ Do you have at least six months of personal financial resources, should you have to leave your current employment? If not, how many hours of your own time can you devote to starting your e-business?

❑ Does your e-business idea offer a unique or revamped product or service suitable for the Internet environment?

❑ Does your e-business idea take advantage of positive network effects by increasing the value of the product or service you plan to offer through faster or wider distribution over the Internet?

❑ Is your e-business idea scalable?

❑ Can your e-business idea adapt quickly as changes to the dynamic Internet environment occur?

❑ Does your e-business idea have a first-mover advantage or other competitive advantage in your market?

❑ Have you identified ways your e-business idea can exploit the inherent advantages of doing business online?

Key Terms

entrepreneur
entrepreneurial process

first-mover advantage
network effect

scalability

Review Questions

1. Which of the following may be a disadvantage to someone planning to start his or her own e-business?

    a. High level of energy
    b. Strong belief in the e-business idea
    c. Inability to work extended hours most days
    d. Support of family members and friends

2. The network effect is the:

    a. Advantage of being the first to market.
    b. Increased value to each participant as the number of participants increases.
    c. Sending of unsolicited messages or advertisments.
    d. Ability of an e-business idea to function well regardless of growth.

3. Which of the following factors is not important when evaluating an e-business idea?

    a. The idea provides a new, unique product.
    b. The idea exploits the network effect.
    c. The idea has a first-mover advantage.
    d. Your Aunt Harriet hates the idea.

4. Which of the following is not a useful trait for an entrepreneur?

    a. Determination to succeed
    b. Great organizational skills
    c. Hesitant to take risks
    d. Ability to act quickly and decisively

5. Which of the following innovative marketing ideas secured early success for Hotmail?

    a. Virtual marketing
    b. Spam
    c. Viral marketing
    d. Scalable marketing

6. Low-cost Internet technology discourages the growth of e-businesses by denying easy entry into a market. **True or False?**

7. First-mover advantage alone guarantees the success of an e-business. **True or False?**

8. Many traditional businesses are expanding into e-businesses where they are far less restricted by the constraints of physical space and time. **True or False?**

9. Two advantages of an auction e-business are the low barrier to entry and no need to maintain an inventory. **True or False?**

10. The network effect is the ability of an e-business to continue to function successfully regardless of how large the e-business grows. **True or False?**

Exercises

1. Using Internet search tools or other relevant sources, research current trends in e-business. Select two e-businesses that are pursuing new ideas and write a one-page paper describing these ideas, including how each idea exploits the built-in advantages of doing business on the Web.

2. Using Internet search tools or other relevant sources, research the current status of two e-business examples from this chapter. Then write a one-page paper describing the current status of each e-business.

3. Review Tom Ashbrook's book The Leap: A Memoir of Love and Madness in the Internet Gold Rush or other books or magazine articles that discuss the effects that starting a new business (preferably an e-business) has on an entrepreneur's personal and professional life. Then select one entrepreneur and write a one-page paper summarizing those experiences.

4. Using information in this chapter or other relevant sources, evaluate whether or not you believe you are an entrepreneur. Then write a one-page paper explaining why you would or would not start your own e-business.

5. Using Internet search tools or other relevant sources, identify several e-businesses in different markets that use Internet technology to maximize their customer relationships. Then select one of the e-businesses and write a one-page paper describing the e-business and the way it maximizes its customer relationships.

CASE PROJECTS

◆ 1 ◆

You have an idea for an online business but are concerned that you may not have the entrepreneurial abilities to develop the idea into a successful e-business. You would like to know more about how Jeff Bezos of Amazon.com transformed his idea for an online bookstore into a successful e-business. Using the Internet, the library, and any other helpful resources, write a one-page report describing how Jeff Bezos used his entrepreneurial abilities to create Amazon.com.

◆ 2 ◆

You have decided that you are an entrepreneur and that you want to create a new e-business. Considering the current trends in e-business, the current status of the economy, recent changes in technology, and other forces, create of list of five new e-business ideas you might like to pursue.

◆ 3 ◆

As the owner of Suzie's PartyTown, a small brick-and-mortar party supply store, you are considering selling party supplies online. Write a one-page paper describing how you could exploit the inherent advantages of an online business to expand your party supply business.

Useful Links

AllBusiness.com – Business Resources
http://www.allbusiness.com/

Bigstep.com – Business Resources
http://bigstep.com/

Business N@tion – Small Business Resources
http://www.businessnation.com/smallbiz.html

Center for Entrepreneurial Leadership and Clearinghouse on Entrepreneurship Education
http://www.celcee.edu/

ENTERWeb – The Enterprise Development Web Site
http://www.enterweb.org/

Entrepreneur.com
http://www.entrepreneur.com/

Entrepreneurs' Help Page
http://www.tannedfeet.com/

EntreWorld – Resources for Entrepreneurs
http://www.entreworld.com/

eWeb – The Resource for Entrepreneurship Education
http://www.slu.edu/eweb/

Microsoft bCentral – Business Resources
http://www.bcentral.com/

Links to Web Sites Noted in This Chapter

Afternic.com
http://www.afternic.com/

AFundRaiser.Com
http://www.afundraiser.com/

AllAdvantage.com
http://www.alladvantage.com/home.asp

Amazon.com
http://www.amazon.com/

America Online (AOL)
http://www.aol.com/

auctions.eders.com
http://auctions.eders.com/

beenz
http://www.beenz.com/splash.html

Butterfields
http://www.butterfields.com/index2.html

Charles Schwab & Co., Inc.
http://www.schwab.com/

ClickRewards
http://www.clickrewards.com/

Cisco Connection Online
http://www.cisco.com/

CNET Auctions
http://auctions.cnet.com/

Costco Online
http://www.costco.com/home.asp

Cybergold
http://www.cybergold.com/

DanceAuction
http://www.danceauction.com/

Dell.com
http://www.dell.com/us/en/gen/default.htm

E*TRADE
http://www.etrade.com/

eBay
http://www.ebay.com/

Egghead.com
http://www.egghead.com/category/inv/00388140/
    03454130.htm

ePhysician
http://www.ephysician.com/

eToys
http://www.etoys.com/

Failure Magazine
http://failuremag.com/

Fidelity Investments
http://www100.fidelity.com/

Fingerhut
http://www.fingerhut.com/

FreeRide.com
http://www.freeride.com/

foodlocker.com
http://www.foodlocker.com

GroceryWorks.com
http://www.groceryworks2.com/

Half.com
http://www.half.com/

Half.com, Oregon
http://town.half.com/

iScribe
http://www.iscribe.com/

J.C. Penney
http://www1.jcpenney.com/jcp/default.asp

Jotter
http://referral.jotter.com/download.jsp?refid=money
    forsurfing

LoveThatLook.com
http://www.4myspecialday.com/

MoneyForMail
http://www.moneyformail.com/home.asp

MSN Hotmail
http://lc2.law5.hotmail.passport.com/cgi-bin/login

MyPoints
http://www.mypoints.com/

NetMind
http://www.netmind.com/

NetZero
http://www.netzero.com/

Pier 1 Imports
http://www.pier1.com/

Southwest Airlines
http://www.southwest.com/

Spedia Network
http://www.spedia.net/

Sun Microsystems
http://www.sun.com/

Third Voice
http://www.thirdvoice.com/

Ticketmaster.com
http://www.ticketmaster.com/

TrailBreaker
http://www.trailbreaker.com/trailbreaker/

W.W. Grainger, Inc.
http://www.grainger.com/index.htm

Yahoo!
http://www.yahoo.com/

yesmail.com
http://www.yesmail.com/

For Additional Review

Ackerman, Jerry. 2000. "Boston-Area Business Personalities Prepare to Launch Internet Start-Up," *The Boston Globe*, July 10.

All Things Considered. 1999. "ThirdVoice.com," October 18. Audio available online at: http://search.npr.org/cf/cmn/cmnpd01fm.cfm?PrgDate=10%2F18%2F1999&PrgID=2.

All Things Considered. 2000. "Failure," National Public Radio, July 17. Audio available online at: http://search.npr.org/cf/cmn/cmnps05fm.cfm?SegID=79633.

Allis, Sam. 2000. "Making a Success Out of Failure," *The Boston Globe*, July 14. Available online at: http://failuremag.com/news_boston_globe.html.

Andres, Clay. 2000. "Born Loser, and Proud of It," *The Westchester County Times*, July. Available online at: http://failuremag.com/news_westchester.html.

Arlington Morning News. 2000. "GroceryWorks Plans to Serve Tarrant County," March 16. Available online at: http://www.groceryworks2.com/Help/CommunityCafe/PressReleases/arlington_morning_news.html.

Asbrand, Deborah. 2000. "Taking Stock in Trading Exchanges," 26(4), *Electronic Business,* April, 2.

Ashbrook, Tom. 2000. *The Leap: A Memoir of Love and Madness in the Internet Gold Rush* New York, NY: Houghton Mifflin Company.

Business2.0. 2000. "Small Biz Busts into B2B," July 17. Available online at: http://www.business20.com/content/research/numbers/2000/07/17/14433.

Calem, Robert E. 2000. "The Next eBay? After Fending Off Big-Name Competition, eBay Remains the Undisputed King of the Auction World." *Ziff Davis Smart Business for the New Economy,* July 1, 52.

Church, Allan H. 1998. "From Both Sides Now: Leadership—So Close and Yet So Far," *The Society for Industrial and Organizational Psychology*, January. Available online at: http://www.siop.org/tip/backissues/tipjan98/church.html.

Cisco-Eagle Case Studies. 2000. "Click! Shop! GroceryWorks.com Delivers Fast, Efficient, Online Grocery Shopping with Cisco-Eagle Order Picking System." Available online at: http://www.cisco-eagle.com/Case Studies/index.htm.

D Magazine. 2000. "Who's Who in Dallas High Tech," July. Available online at: http://www.groceryworks2.com/Help/CommunityCafe/PressReleases/dmagazine2.html.

Davis, Jillian J. 2000. "TrailBreaker Sells Industry Advice," *dbusiness.com*. March 20. Available online at: www.dbusiness.com/Story/0,1118,NOCITY_11785,00.html.

Deise, Martin V. et al. 2000. *Executive's Guide to E-Business: From Tactics to Strategy*. New York: John Wiley & Sons, Inc.

Detroit News. 1996. "Meg Whitman," January 22. Available online at: http://www.detnews.com/menu/stories/32861.htm.

Elder, Laura. 2000. "Online Grocers Check Out City with Designs on Delivery," *Houston Business Journal*. March 27. Available online at: http://www.groceryworks2.com/Help/CommunityCafe/PressReleases/houston_business_journal.html.

Fitzpatrick, Michele. 2000. "Diary of a Web Start-Up: Chicago-Based Yesmail Endures Whirlwinds," *Chicago Tribune*, July 11.

Fortune. 2000. "Dot-Coms: What Have We Learned?" 142(10), October 30, 82.

Fortune. 2000. "Starting a Business? Read This First," 142(2), July 10, 274.

Fortune. 1999. "10 Companies That Get It," 140(9), November 8, 115.

Foster, Ed. 2000. "THE GRIPE LINE: Viral marketing Goes One Step Too Far—To a Place Where Friends Spam Friends," *InfoWorld*, 22(6), February 7, 93.

Feuerstein, Adam. 2000. "Safeway Invests $30M in Dotcom Grocer," *Upside.com*, April 17. Available online at: http://www.groceryworks2.com/Help/CommunityCafe/PressReleases/upsidecom.html.

Gimein, Mark. 2000. "Meet the Dumbest Dot-Com in the World," *Fortune*, 142(2), July 10, 46.

GroceryWorks.com. 2000. "Kelby Hagar: A Founder's Vision," Press Kit. Available online at: http://www.groceryworks2.com/Help/CommunityCafe/kit.zip.

Gunn, Bob. 1999. "Leadership from Within (part 2)," *Institute of Management Accountants: Strategic Finance*, October 1, 81(4), 20–22.

Halkias, Maria. 2000. "GroceryWorks Names New CEO," *Dallas Morning News*, January 7. Available online at: http://www.groceryworks2.com/ Help/CommunityCafe/PressReleases/ dallas_morning_news2.html.

Hartman, Amir, Sifonis, John, and Kador, John. 2000. *NetReady: Strategies for Success in the E-conomy*. New York. McGraw Hill.

Henry, Holly. 1999. "Groceries Are Just the Beginning," *San Angelo Standard-Times*, November 7. Available online at:http:// www.groceryworks2.com/Help/CommunityCafe/ PressReleases/san_angelo.html.

Herbeck, Nancy. 1999. "The New Online Grocery Services Are Full of Whizbangs…and Pitfalls," *D Magazine*, December. Available online at: http://www.groceryworks2.com/Help/ CommunityCafe/PressReleases/youve_got_food.html.

Ignatius, David. 2000. "Revenge of the Dinosaurs," *Washington Post*, April 19. Available online at: http://www.groceryworks2.com/Help/Community Cafe/PressReleases/washington_post.html.

InfoWorld. 2000. "Sites to Watch," 22(42), October 16, 65.

Jurvetson, Steve and Stavropoulos, Andreas. 2000. "Does Your Idea Make Sense?" *Business 2.0*, March, 138–139.

Kirby, Carrie. 2000. "Pay-to-Surf Not Paying Off for Web Sites," *San Francisco Chronicle*, July 12, FINAL, PSA-2638, Business Section. Available online at: Northern Light Special Collection Documents, http://library.northernlight.com/.

Koenig, David. 2000. "Safeway Agrees to Buy up to Half of Texas Online Grocery Company," April 17. Available online at: http://www.groceryworks2.com/Help/Community Cafe/PressReleases/associated_press.html.

Melymuka, Kathleen. 2000. "Internet Intuition: CEO Meg Whitman Powers eBay on Instinct and Experience and Gains an Education in Technology," *Computerworld*, January 10, 48(1).

Mowbray, Rebecca. 2000. "Grocery Will Open Online in Houston," *Houston Chronicle*, March 16. Available online at: http://www.groceryworks2.com/Help/Community Cafe/PressReleases/houston_chronicle.html.

Murphy, Chris. 2000. "A Financial Advantage?" *Information Week*, ISSN: 8750-6874, June 12, 18.

O'Brien, Chris. 2000. "eBay Buys Half.com," *San Jose Mercury News*, June 14.

Ransdell, Eric. 1999. "Network Effects," *Fast Company*, September, 27, 208. Available online at: http://www.fastcompany.com/online/27/ neteffects.html.

Roth, Daniel. 2000. "Meet eBay's Worst Nightmare: Bidder's Edge Stormed the Online Auction Industry with Napster-like Force. Then It Met eBay. Their Legal Clash May Change the Way the Web Works Forever," *Fortune*, 142(1), June 26, 1999.

San Francisco Women on the Web. 1999. "Meg Whitman." Available online at: http://www.top25.org/99_bios/ mw.shtml.

Schneider, Gary P. and Perry, James T. 2000. *Electronic Commerce*. Cambridge, MA: Course Technology.

Segaller, Stephen. 1999. *Nerds 2.0.1: A Brief History of the Internet*. New York: TV Books, L.L.C.

Stevens, Larry. 2000. "MDs Welcome E-Prescriptions— New Web Services Let Physicians Prescribe Drugs Using PDAs," *InternetWeek*, March 27, 31.

Stone, Martin. 2000. "AllAdvantage.com Alienates Ad-Watchers," *Newsbytes*, June 5. Available online at: http://www.newsbytes.com.

Stross, Randall E. 2000. "The Auction Economy," *U.S. News & World Report*, 128(25), June 26, 44.

Sullivan, Robert. 1998. *The Small Business Start-up Guide*. Second Edition. Great Falls, VA: Information International.

Timmers, Paul. 1999. *Electronic Commerce: Strategies and Models for Business-to-Business Trading*. Chichester, England. John Wiley & Sons, Ltd.

U. S. Department of Commerce. 2000. "Digital Economy 2000," June. Available online at: http://www.esa.doc.gov/.

Ward, Leah Beth. 2000. "GroceryWorks, Safeway to Form Alliance Deal Gives Dallas Company Storefronts, Provides Chain With Online Presence," *Dallas Morning News*, April 18. Available online at: http://www.groceryworks2.com/Help/Community Cafe/PressReleases/dallas_morning_news6.html.

Williams, E. E. 2000. *Rice University, MGMT 520*. Fall 1999, Figure 2-2.

Wirth, Greg. 2000. "Online Brokers Slug It Out In a Sagging Market….. With the Top Four Taking Half the Trades, What's Left? *Investment Dealers Digest*. June 19. Available online at: Northern Light Special Documents, http://www.northernlight.com/.

Young, Steve and Jorgensen, Jim. 1999. "CEO AllAdvantage.com Interview on Digital Jam," *Cable News Network (CNNfn)*, July 1, 19:30:00 (Transcript prepared by *Federal Document Clearing House, Inc.*). Available online at: Northern Light Special Collection Document, http://library.northernlight.com/.

CHAPTER **3**

Traditional and Electronic Payment Methods

In this chapter, you will learn to:

Identify traditional payment methods

Discuss electronic credit, debit, and charge card payments

Describe the electronic card payment setup process

Discuss an e-business's exposure to credit card fraud and ways to prevent it

Describe electronic check payment processing

Identify electronic cash payment methods

Traditional Payment Methods

Thousands of years ago nomadic hunter-gatherers lived in a world where there was little need for money, and most valuable goods were perishable. Owning too many goods could be a liability for nomadic tribes who were constantly on the move looking for food and shelter. The advent of agricultural communities around 8,000 B.C. led first to producing enough grain and other agricultural products for the needs of individual families, and then to surplus production. Trading systems developed as families exchanged their surplus grain or agricultural products or bartered them for other household needs. As the number of goods available grew and families became more selective in satisfying their household needs, the barter system became increasingly inefficient. Adoption of a monetary commodity, or "money," that had value independent of other goods simplified the trading process.

> **TIP**
>
> The inefficiencies of the barter system were not the only factors, and may not have been the primary factors, in the origins and earliest development of money. Many early societies maintained social or religious laws requiring compensation or payment in some form for crimes of violence or the loss of services to a family when a daughter married. Rulers also imposed taxes or tribute on their subjects that required payment in some form, and religious obligations often entailed payment of tribute or sacrifices of some kind. Therefore the need in many ancient societies to pay "blood money," "bride money," taxes, and tributes may well have influenced the spread of money.

A brief history of money and banking

People have used all kinds of objects as money at different times and in different places. For example, amber, beads, cowrie shells, drums, feathers, ivory, jade, leather, rice, salt, oxen, pigs, stones, gems, gold and other precious metals, and other items have been used as money throughout history. Because of this diversity, it is easier to define money by its functions than by its physical form.

Money functions as a unit of account, a common measure of value, a medium of exchange, and a means of payment.

Banking originated in ancient Mesopotamia and Egypt, where the royal palaces, temples, and granaries served as secure places for the safekeeping of grain and other commodities. Receipts for stored commodities were used for transfers not only to the original depositors but also to third parties, including tax gatherers, priests, and traders.

The Lydians in Asia Minor were the first to develop **coinage**, stamping small round pieces of precious metals as a symbol of purity and weight, and the practice quickly spread to Greece and Persia. Because of the variety of coins in circulation during this time, money changing became a common form of banking. Early Greek bankers also supported shipping, mining, and construction of public buildings. Early coins of bronze or copper were also used in China.

The importance of banking decreased with the Romans, who preferred to conduct cash transactions with coins. After the fall of the Roman Empire, banking was largely forgotten in Europe. Banking reemerged in Europe around the time of the Crusades. The Crusades stimulated the need for a safe and speedy means of transferring cash to pay for supplies, equipment, soldiers, and ransom; this led to the development of banking services provided by others such as the Knights Templar and Hospitallers in addition to Italian city states. In the Italian city states of Rome, Venice, and Genoa, and in medieval France, bills of exchange were developed to satisfy the need to transfer sums of money for trading purposes.

The development of movable type in 1456 led to the production of improved printing presses and the production of fine coins called "milled" money—rather than to the development of paper money as might be expected. In England in the 1600s, goldsmiths were moneylenders or bankers. In 1694 the Bank of England was founded as a central banking system. The Bank of England began issuing bank notes backed by gold bullion reserves. With the emergence of bank notes, paper money became widely circulated.

Paper money was also widely used in the American colonies. The Massachusetts Bay colony issued the first paper money in the colonies in 1690. In 1775 the Continental Congress issued paper money to finance the Revolutionary War. The new nation's first real bank, the Bank of North America in Philadelphia, was chartered in 1781 by the Continental Congress to provide further support for the Revolutionary War effort, and in 1785 the Continental Congress adopted the dollar as the unit for national currency. The 1800s and early 1900s saw the issuance of state bank notes, treasury notes, gold certificates, and silver certificates. In 1913 the Federal Reserve Act created the Federal Reserve System as the nation's central bank. The Federal Reserve System issues Federal Reserve notes—the only U.S. currency now produced, which represents 99 percent of U.S. currency in circulation.

> **TIP**
>
> The first American coins were struck in 1793.
>
> In addition to coins and paper money, "near cash" forms such as money orders, checks, credit cards, charge cards, and debit cards are important payment methods today.

Checks and money orders

A **check** is a written order on a bank or other financial institution to pay money belonging to the signer of the check to the check's presenter. A personal check is drawn on an individual's bank account. A cashier's check is a check drawn on the financial institution's own funds and signed by the financial institution's cashier. Checks came into use in the late 1800s, and by 1890 90 percent of the value of payment transactions was being carried out by check. Today, payment by check remains one of the most popular offline payment methods. Figure 3-1 illustrates the traditional check-processing system.

A **money order** is an order for the payment of a specified amount of money, usually issued and payable at a bank or post office. Money orders are convenient for people who do not have bank accounts, or for circumstances where checks are not accepted in payment.

Consumer credit, debit, and charge cards

While it is likely that credit (or money lending) originated much earlier, some of the earliest recorded laws concerned the issue of credit and interest, the price of credit. Around 1800 B.C., Hammurabi, king of the first dynasty of Babylon, authored the earliest known formal laws, including the first recorded attempts to regulate interest rates. One of the first advertisements for consumer credit occurred in 1730. Christopher Thornton, who sold furniture, advertised that a furniture purchase could be paid off on a weekly basis. Also, from the eighteenth to the early part of the twentieth century, merchants called "tallymen" sold clothes in return for small weekly payments. They tallied a customer's credit sales and payments on a wooden stick: one side for sales and one side for payments.

In the early part of the twentieth century, individual merchants began to extend credit to their customers. By the 1950s retail stores began issuing credit cards as an extension of the credit accounts they had long offered their customers.

Figure 3-1
Check
processing

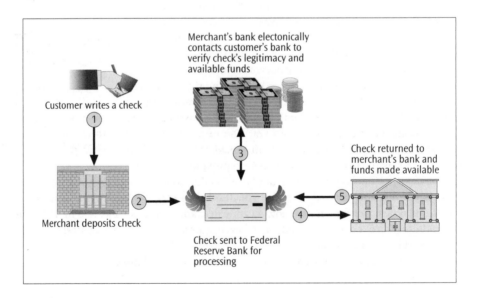

A **credit card** is a rectangular piece of plastic used instead of cash or checks to pay for goods and services. Credit card issuers send their users a monthly statement of charges made by credit card. Users can pay their monthly balance in full or make a payment on their credit card balance, generally paying interest on the unpaid balance. Oil companies followed retailers with their own credit cards. Next, banks began issuing cards and forming associations to act as clearinghouses for their credit card transactions. These clearinghouses are now consolidated into two associations for MasterCard and Visa card issuers.

A **charge card** is similar to a credit card and is used instead of cash or checks to pay for goods or services. However, users must pay their charge card balances in full upon the receipt of the statement. In the 1950s, Diners Club, American Express, and Carte Blanche launched charge cards that became known as T & E, or travel and entertainment, cards.

A debit card is also a rectangular piece of plastic used instead of cash or checks to pay for goods and services. Banks issue **debit cards** in connection with a bank account and deduct all debit card transactions directly from the account.

Cash, coins, traditional paper checks, and money orders are neither efficient nor safe payment methods for an e-business to employ. Instead, there are several electronic payment methods that are more appropriate.

Electronic Credit, Charge, and Debit Card Payments

Although there are many different types of electronic payment methods for online consumers, the most common and currently the consumer-preferred method is the use of credit, debit, and charge cards. In order to accept these cards, an e-business must have a merchant account, payment-processing software, and procedures to protect its customers and itself against fraud. Figure 3-2 illustrates a credit, debit, or charge card online payment-processing procedure.

Merchant accounts

To accept credit, debit, or charge cards, such as Visa, MasterCard, Discover, and American Express, an e-business must first set up a **merchant account** at a financial institution. To do this, an e-business must apply for an account, much in the same way an individual applies for a personal credit card account, by supplying requested financial information. The kinds of information requested may vary by financial institution but will likely include:

- The size of the e-business
- How long the e-business has been in business
- The kind of products and services the e-business offers
- The anticipated average size of each transaction
- The kinds of cards to be accepted
- The volume of anticipated transactions

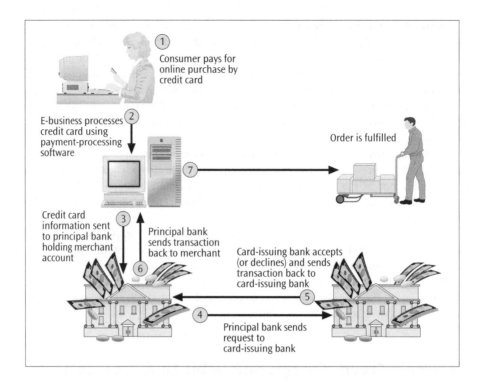

Figure 3-2
Online payment processing

The fees charged to an e-business for its merchant account might be a fixed monthly fee, a monthly minimum transaction fee, a one-time setup fee, or a percentage of each transaction, called a **discount rate**. In establishing its fees for a merchant account, the financial institution assesses its level of risk in providing the e-business with a merchant account, which depends on the e-business's credit worthiness and the financial institution's experience with similar merchant accounts.

A financial institution's risk of issuing a merchant account is based on the rate of chargebacks it may experience on the account. A **chargeback** results from a consumer's refusal to pay a charge on his or her credit card account for a variety of reasons, including returned products, billing errors, and fraudulent charges. To protect itself against excessive chargebacks, a financial institution may require an e-business to pay higher fees or may hold back a percentage of an e-business's money each month as a hedge against future chargebacks.

Certain other "high-risk" factors also influence the fees charged for a merchant account:

- *Cardholder not present:* Brick-and-mortar businesses, which can get immediate authorization or declination for a credit card purchase at the time the card is presented by the cardholder, generally pay more favorable fees than e-businesses, which must accept credit cards over the Internet or via telephone, with the cardholder not present.

- *High-risk products or services:* Financial institutions maintain chargeback risk statistics for specific types of businesses. If an e-business falls into the "high-risk" category, such as gambling and other politically incorrect and nontraditional products or services, it will likely pay higher fees for its merchant account.

An e-business also wants to be certain that the financial institution selected to provide the merchant account is knowledgeable about doing business on the Web. Reviewing the Web sites of major payment-processing service providers is a good way to identify financial institutions that offer e-business merchant accounts.

TIP

There are a number of e-businesses that offer a payment-processing system and access to a merchant account as part of their bundled services. Using Internet search tools and the keywords "online store" and "merchant account" is a good way to identify these e-businesses.

Payment-processing software

In addition to securing a merchant account, an e-business must also have in place a process for getting card transactions authorized and getting the credit card transaction processed. A third party typically provides payment-processing software for a monthly fee. The software may be downloaded to an e-business's site and the payments processed at the site.

Another alternative is for an e-business to outsource payment processing to a third-party provider. An e-business passes the card data to the third-party service provider for authorization and transaction processing. The e-business typically pays a monthly fee for the processing services. Well-known payment-processing service providers include Authorize.Net, CyberCash, and ICVERIFY (Figure 3-3).

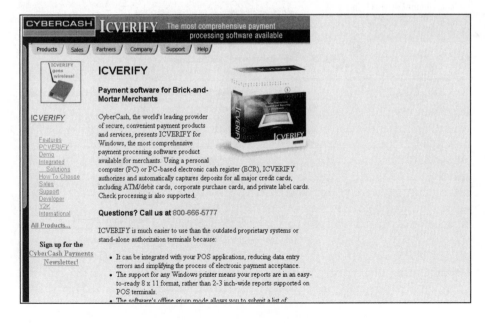

Figure 3-3
ICVERIFY

It is important that an e-business accepting credit, debit, or charge card payments provide adequate security for the card information being transmitted.

Security for credit, debit, and charge card processing

Security for online card transactions requires some type of encryption and digital authentication to provide secure transaction processing. **Cryptography** is the art of protecting information by encrypting it. **Encryption** is the translation of data into a secret code called **ciphertext**. Ciphertext that is transmitted to its destination and then **decrypted**, or returned to its unencrypted format, is called **plaintext**. A **protocol** is a standard or agreed-upon format for electronically transmitting data. One of the earliest Internet security protocols, **Secure Sockets Layer** (**SSL**), provides server-side encrypted transactions for electronic payments and other forms of secure Internet communications. To use the SSL protocol, an e-business's ISP or Web hosting company places the e-business's online order form Web page on a secure server. The URL for the SSL-secured online order form then begins with https:// instead of just http://, indicating that information is transmitted using the SSL protocol. Additionally, most browsers indicate by a warning message or some type of icon, such as a lock or key, that information being transmitted is or is not secure.

Figure 3-4 illustrates the process of encrypting and decrypting credit card transaction information.

Figure 3-4

Encrypting and decrypting transaction information

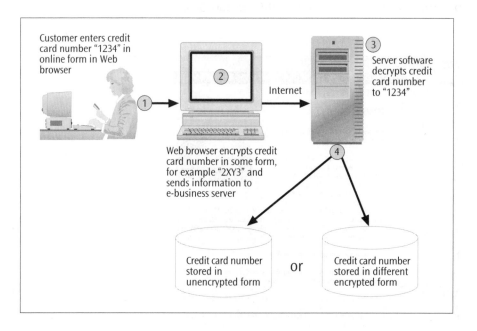

Authentication is the process of identifying an individual or e-tailer and is usually based on a combination of username and password. Digital signatures and digital certificates are often used for authentication. A **digital signature** is a unique code attached to an electronically transmitted message that identifies the sender. A **digital certificate** is an electronic message attachment that verifies the sender's identity. A **certificate authority**, a trusted third-party organization that guarantees the identity of the sender, issues digital certificates. An e-business using SSL must get a digital certificate. Again, an e-business's ISP or Web site hosting company can help by generating a Certificate Signing Request. The e-business then submits the request to a certificate authority, such as VeriSign, that for a fee authorizes the digital certificate. The ISP or Web hosting company then installs the digital certificate at the e-business's Web site. Figure 3-5 illustrates a digital certificate.

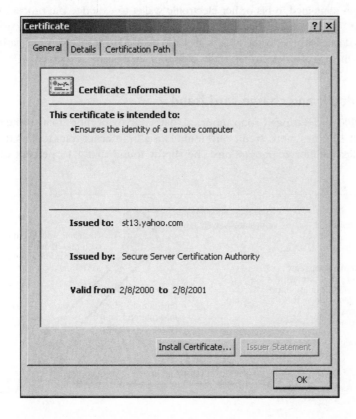

Figure 3-5
Digital certificate

While many B2C transactions are paid for by credit or debit card using SSL to transmit the credit or debit card information, SSL is not the only protocol for transmitting credit or debit card information. In 1996, Visa and MasterCard, with the backing of Microsoft, IBM, Netscape, and others, announced support for a protocol called **Secure Electronic Transactions** (**SET**). SET is designed to be the standard protocol for presenting credit card transactions on the Internet. SET works in real time from Web sites and in a "store and forward" mode such as e-mail. SET is an **open standard** protocol, which means that companies that write buyer, e-tail, and banking software can develop software for their clients that gathers and transmits financial data such as credit card information independently and have competing software products work together successfully. SET uses digital certificates to authenticate both the credit card holder and the e-tailer. The buyer's digital certificate, name and address, and credit card information are maintained in an electronic or digital wallet.

An **electronic** or **digital wallet**, sometimes called an **e-wallet**, is encryption software that stores payment information much like a physical wallet. An electronic wallet may reside on the buyer's PC or may reside on the credit card issuer's or e-business's server. Instead of completing a long, complicated Web-based form during the checkout process, a buyer can use the payment information contained in his or her electronic wallet to expedite the process. Companies such as Q*Wallet, InstaBuy, Gator, Jotter, and Microsoft Passport provide electronic wallets users can download. Electronic wallets may also be used to store electronic cash. Figure 3-6 illustrates the use of an electronic wallet stored on an e-business's server.

Credit, debit, and charge card fraud

In July 2000, the Gartner Group reported that a survey of more than 160 e-tailers indicated that 12 times more credit card fraud exists in Internet transactions than in traditional brick-and-mortar transactions. The survey found that 1.15 percent of all online

Figure 3-6
Electronic wallet

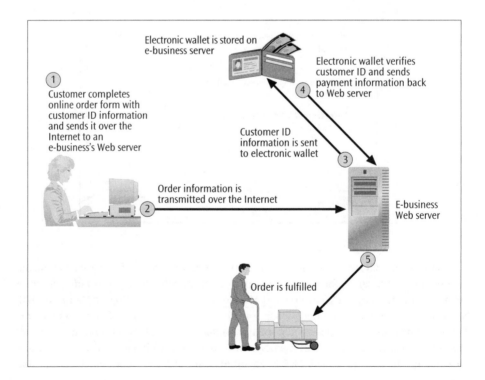

Electronic wallet is stored on e-business server

Electronic wallet verifies customer ID and sends payment information back to Web server

① Customer completes online order form with customer ID information and sends it over the Internet to an e-business's Web server

④

Customer ID information is sent to electronic wallet

③

Order information is transmitted over the Internet

②

E-business Web server

⑤

Order is fulfilled

purchases were fraudulent, while only .06 percent to .09 percent of offline purchases were fraudulent. The survey also found that 64 percent of online chargebacks were the result of fraud, while only 44 percent of offline chargebacks were the result of fraud.

Additionally, the survey indicated that e-tailers are bearing the liability and costs of electronic credit card fraud, instead of the credit card companies, which generally absorb the fraud for traditional retailers (who follow rules regarding documentation that include the cardholder's physical signature). E-tailers are also more likely to pay discount rates to credit card companies that are 66 percent higher than fees for brick-and-mortar retailers. Finally, e-tailers spend about four times what brick-and-mortar retailers spend to resolve and process chargebacks.

Because a cardholder is not responsible for more than $50 of fraudulent charges, and the bank issuing the card simply charges back an e-business for fraudulent charges (and usually adds a chargeback fee), an e-business must consider all possible measures to reduce its exposure to card fraud. Some of those measures include:

- Stating a clear return policy on the e-business's Web site
- Displaying a disclaimer on the e-business's Web site holding the buyer responsible for all charges
- Sending orders via national shipping companies to avoid having customers claim that the products were not received
- Not accepting orders unless complete billing and shipping information, including complete address and phone number information, is provided
- Accepting orders only from ISP-based or domain-name-based e-mail addresses, not from free, Web-based, or e-mail forwarding addresses, so that transactions can be traced to an actual person
- Calling the phone number on a questionable order to verify the order
- Waiting to ship items until authorization and verification are ensured, especially for orders being shipped out of the country
- Carefully checking very large orders and overnight delivery requests
- Double-checking international orders where the shipping address is outside the U.S. but the charge, credit, or debit card is issued by a U.S. bank; for example, double-checking an order to be shipped to central Europe paid for by a credit card issued by a Denver, CO, bank.

In the fall of 2000, American Express and MBNA, the world's largest independent Visa and MasterCard credit card issuers, both announced plans to provide consumers and e-businesses with a new tool to fight online credit card fraud—disposable credit card numbers. American Express's Private Payments system allows registered cardholders to download software that generates a unique, disposable credit card number for each online transaction. Registered users can also log on to the American Express Private Payments Web page and request a disposable credit card number. MBNA cardholders can download third-party software that generates a unique disposable credit card number to be used for a single transaction. MBNA cardholders can also set a dollar limit on the transaction that used the disposable card number.

Using disposable credit card numbers reduces an e-business's risk of failing to protect customer's credit card information from theft. Each number becomes worthless after the transaction is completed. However, disposable credit card numbers may not be suitable for all online transactions. For example:

- One-click ordering such as that offered by Amazon.com, which uses stored credit card numbers
- Grocery or pet supplies replenished at preset times
- Payment of automatic charges such as ISP monthly charges
- Guaranteed payment for auto or hotel reservations

E-CASE foodlocker.com

foodlocker.com accepts American Express, MasterCard, Visa, and money orders for mail orders. The management at foodlocker.com chose to accept credit cards and not other forms of electronic payments because they believe that online customers are still more comfortable using credit cards than other forms of online payment, because a majority of potential customers have credit cards, and because of the consumer protections credit cards provide. Additionally, it was easier for foodlocker.com to set up their e-business Web site to accept credit cards than to accept other forms of electronic payments.

Another alternative for an online consumer, to reduce his or her anxiety about purchasing products online, is to have his or her credit card funds held in escrow until the products are delivered and approved. The advantage to the consumer is that he or she can inspect the product and return it if the product turns out to be not as described. There are also advantages to an e-business of accepting payment from a third-party escrow service such as i-Escrow (Figure 3-7). Escrowed transactions do not require a merchant account or the purchase and installation of payment-processing hardware or software. Escrow payments also ensure against claims of undelivered products. Additionally, offering escrowed payments can inspire trust and confidence in buyers afraid of doing business with unknown sellers.

Although credit, debit, and charge cards are the preferred form of electronic payment today, several other electronic payment forms are gaining in popularity.

Figure 3-7

i-Escrow

Other Electronic Payment Forms

Other electronic payment forms include electronic cash, electronic checks, smart cards, and person-to-person payment systems.

Early electronic cash

Electronic cash, sometimes called **digital cash**, **cybercash**, or **e-cash**, is a method that allows buyers to pay for goods or services online by transmitting over the Internet a unique electronic number or other identifier that carries a specific value. There are two primary advantages of using electronic cash instead of a credit card: lower processing costs for the seller, and no special credit card type authorization for the buyer. Electronic cash is now a popular online payment method in Europe and Japan. To date, electronic cash has not been as widely accepted by U.S. consumers, who still prefer to use credit or debit cards to pay for their online purchases.

Early electronic cash issuers were not successful partly because of the complexity of setting up and funding an electronic wallet, and partly because the concept of electronic cash did not catch on with consumers. Electronic cash pioneer DigiCash filed for bankruptcy in 1998, First Virtual became MessageMedia, and CyberCash now concentrates on its online credit card verification and processing products.

TIP

Some analysts predict electronic cash will only chip away at the dominance of the credit card as the preferred online payment method. Credit cards now account for approximately 95 percent of online purchases. Analysts predict that number will drop to the 75 percent to 81 percent range by 2003, as other online payment methods, such as electronic cash, become more popular.

David Chaum is a renowned cryptographer and mathematician. He began his work in cryptography in the 1980s as a graduate student in computer science at Berkeley, where he wrote his master's thesis on an electronic mail system so private it could be used for voting. After receiving a doctorate at Berkeley, Chaum went to the Center for Mathematics and Computer Science in Amsterdam, where his interest in privacy and the Internet led to his invention of a form of electronic money whose legitimacy could be verified, while its source could not. Chaum named his electronic cash "eCash" and founded DigiCash in 1990 to market eCash to banks.

DigiCash's eCash product gained widespread attention in 1994 and 1995 as banks and other businesses began to look at the commerce opportunities the Internet provided. DigiCash arranged licensing with several banks, including Deutsche Bank, Den Norske Bank, Bank Austria, Advance Bank of Australia, and Mark Twain Bank of St. Louis. Unfortunately, DigiCash's eCash product was ahead of its time. In 1995 commercial Web browser technology was in its infancy, and the idea of using the Internet to purchase goods and services was so new that the term "e-commerce" had yet to be invented. Consumers who were beginning to buy items online turned out to be less resistant to providing credit card information over the Internet than industry analysts had originally thought.

Lack of success with eCash pilot programs, the complexity of the eCash system, and internal management problems led to DigiCash filing for bankruptcy in November 1998. In May 1999 a group of entrepreneurs bought software and patents from DigiCash and formed a company called eCash Technologies, Inc., known as eCash. eCash (Figure 3-8) hopes to revive Chaum's electronic cash idea. Meanwhile, Chaum, who has no involvement in the new company, continues to pursue his interest in cryptography and privacy issues.

Figure 3-8
eCash

Today there are several competing forms of "second-generation" electronic cash available to consumers.

Second-generation electronic cash products

The growth of e-business has also prompted the growth of newer, second-generation forms of electronic cash. E-tailers are using these electronic cash products in customer rewards systems and promoting them as convenient online payment methods for children and teenagers, who likely do not have access to credit cards. Three of these more modern electronic cash products are beenz, Cybergold, and Flooz.

beenz (Figure 3-9), launched in 1999, is electronic cash used by e-tailers in consumer rewards programs. Consumers earn beenz from a participating Web site by visiting the Web site, by purchasing items from a Web site, or by accessing the Internet via a participating ISP. beenz.com signed up 500,000 users in 1999, an additional 250,000 users in January and February 2000, and another 250,000 in the first two weeks in March 2000. In July 2000 beenz.com announced a new program called beenzBack Shopping that enables consumers to earn beenz by shopping at such well-known e-tailers as ibeauty.com, mySimon.com, Fossil.com, and Gear.com. Consumers shopping at beenzBack Shopping sites receive up to 40 beenz for every dollar spent at the participating e-tailers. beenz.com plans to move beyond the Web into mobile phone, interactive television, smart card, PDA, and video game platforms.

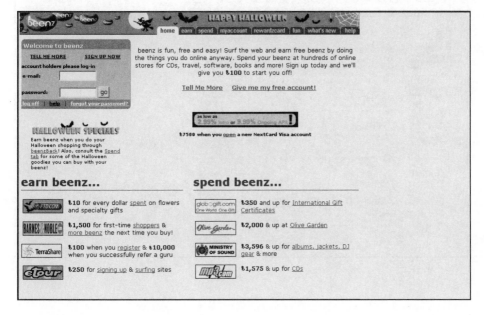

Figure 3-9
beenz

Cybergold (Figure 3-10), like beenz, is electronic cash used by e-tailers in consumer rewards programs. Cybergold pays users to look at offers by participating e-tailers. Users can spend their Cybergold electronic cash by donating it to one of Cybergold's participating nonprofit organizations or by converting the electronic cash to dollars and depositing the dollars to a Visa card account or bank account.

Flooz (Figure 3-11) is gift currency sent to recipients via e-mail. Flooz is funded by credit card, and recipients can spend Flooz at more than 60 popular e-tailers such as Starbucks, Tower Records, Toys 'R' Us, and Godiva Chocolates. As of May 2000, more than 600,000 people sent or received Flooz gift certificates worth about $5 million. Half of Flooz's sales came from 75 corporate clients such as Bell Atlantic, Eastman Kodak, Sun Microsystems, and Cisco Systems, which use Flooz as employee and customer incentives.

Figure 3-10
Cybergold

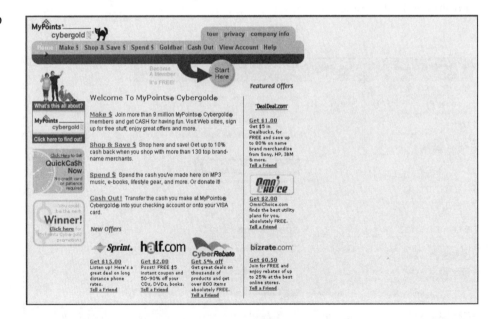

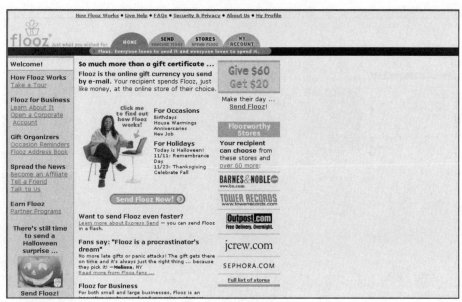

Figure 3-11

Flooz

E-CASE The Gold Standard

Two of the properties of gold that led to its use for monetary payments for thousands of years are its aesthetic attractiveness and its resistance to corrosion. While gold is rarely used for monetary payments today, some people still prefer to keep some of their wealth in gold and other precious metals.

e-gold (Figure 3-12) is an Internet monetary payment method that enables participants to use gold (or other precious metals) as money. An e-gold account holder can fund his or her account by (1) exchanging government-issued money for gold, silver, platinum, or palladium, (2) transferring metal into his or her account from another source, or (3) accepting transfer of metal from another e-gold account holder. Metals in an e-gold account can be transferred to another account holder or exchanged for government-backed currencies to make any payment ordinarily made by cash, check, or credit card. As of June 2000, e-gold reported over 10,000 active accounts.

(Continued on the next page)

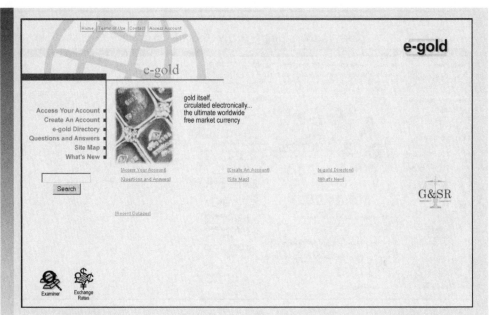

Figure 3-12
e-gold

Smart cards

A **smart card** is a small electronic device approximately the size of a credit card that contains electronic memory. Smart cards are used for a variety of purposes, including storing medical records, generating network identification, and storing electronic cash. Mondex, a popular smart card in Europe and Asia, holds and dispenses electronic cash. Mondex electronic cash can be transferred over a telephone line or over the Internet. Users can download up to five different currencies from their bank account straight to their Mondex card. Figure 3-13 illustrates the acquisition and use of a smart card to store electronic cash.

Mondex cards support micropayments as small as three cents. One disadvantage of the Mondex card is the need for special equipment. Internet users must attach a special card reader to their PCs to read the card, and offline merchants must also have a specific card reader to read a Mondex card. Another disadvantage is the risk of theft, which may deter users from loading a Mondex card with very much money. Also, Mondex cards dispense their cash immediately; there is no deferred payment, as there is with some credit cards that allow almost a month before the users incur interest charges or have to pay the credit card bill. In the U.S., Wells Fargo is now using over 250 Mondex (Figure 3-14) cards for an electronic cash Internet pilot program. Six merchants, ATMs, and twenty three POS (point of sale) devices are available for the pilot program.

Smart card activities are growing at 30 percent a year, predominantly outside the U.S. Although there are fewer than one billion smart cards now in use, analysts predicted that 3.4 billion smart cards would be used worldwide by 2001. For example, MasterCard International, Inc. is running a pilot program in France, testing the use of smart cards and Web-enabled cell phones to make Internet payments.

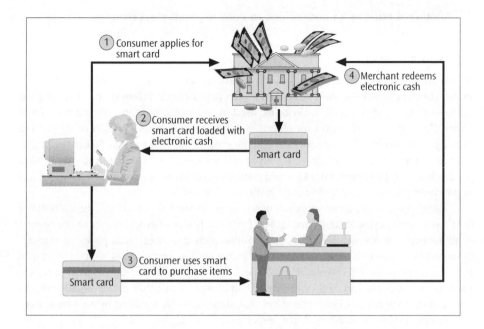

Figure 3-13
Smart card usage

Figure 3-14
Mondex USA

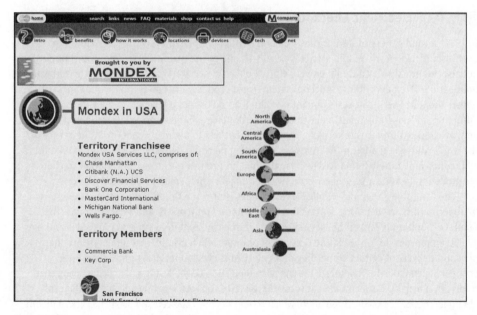

Another alternative for both offline and online payments is electronic checks.

Electronic checks

An **electronic check**, or **e-check**, is the electronic version of a paper check. An electronic check contains the same information as a paper check, is based on the same legal framework as a paper check, and can be used in any transaction in which a traditional paper check is used. However, an electronic check is much less expensive to process than a paper check. Costs for processing an electronic check are about 50 cents a check instead of the 75 cent to $3 range for a paper check. Analysts estimate that converting paper checks to electronic checks could reduce or eliminate the estimated $76.7 billion in processing costs faced by financial institutions in 2000.

While many B2B payment transactions are processed via electronic funds transfer (EFT), whereby payment funds are moved from the buyer's bank account to the seller's bank account electronically, the study also indicated that many B2B payment transactions will continue to be conducted at least partly offline, because of the volume of B2B services that cannot effectively be purchased via an online form. Professional services, such as engineering services, custom product design, and other professional services transactions may begin online, but their final steps must be handled by an e-business's employees in direct contact with the buyer.

E-CASE Electronic Checkout

In 1932, second-generation Americans Ben and Bill Golub opened Public Service Market in Green Island, New York. Public Service Market was one of the region's first "one stop shopping" outlets. Not exactly a traditional grocery store, Public Service Market was located in an old warehouse and offered everything: groceries, meat, fresh produce, dry goods, home appliances, a barber shop, a shoe repair store, an antique shop, a book shop, a jeweler, and a cafeteria. Customers walked from department to department, placing bargain-priced items in wicker baskets. At checkout time customers would take their baskets to a grocery checker in the center of the store, who would add up the items on an adding machine. Customers would then take the adding machine tape bill to a cashier, pay the bill, and return to the grocery checker to claim the groceries. In 1932 a $5 grocery order was too big for one person to carry out alone!

The "one stop shopping" Public Service Market is now the Golub Corporation, which owns 97 supermarkets in New York, Massachusetts, Vermont, Connecticut, and Pennsylvania. The Golub Corporation is ranked between 43rd and 45th in the nation in supermarket sales and has 17,500 employees. In July 2000 the Golub Corporation, which also offers online grocery shopping and delivery at certain stores, began an electronic check pilot program at their Price Chopper Supermarket in Dunmore, PA. If the pilot, thought to be the first test of digital imaging check readers by a major U.S. supermarket, is successful, all Price Chopper Supermarket stores will accept electronic checks. The Golub Corporation, which accepts between 800,000 and 900,000 checks a month at its Price Chopper supermarkets (Figure 3-15), hopes that accepting electronic checks will make the checkout process more efficient for its customers and payment processing less costly for its stores.

Figure 3-15
Price Chopper

Electronic checks are processed at the point of sale much like paper checks. For a purchase at a brick-and-mortar merchant, the buyer writes out a check and gives it to a cashier at the point of sale. The cashier then scans the check through a MICR (magnetic ink character recognition) device that captures the account number, check serial number, and financial institution routing number. The scanned data moves electronically to a check authorization service such as TeleCheck, which verifies that the check is drawn on an open account and that the buyer does not have a record of writing "hot" or insufficient funds checks. The paper check is then voided and may be returned to the buyer or maintained by the seller. The electronic check data then goes to the ACH (Automated ClearingHouse) network, where the amount is credited to the merchant's account and the data is forwarded to the buyer's financial institution for debiting the appropriate account. Just as with a paper check, it takes approximately two to three days for the electronic check to clear the buyer's bank. Figure 3-16 illustrates the electronic check process.

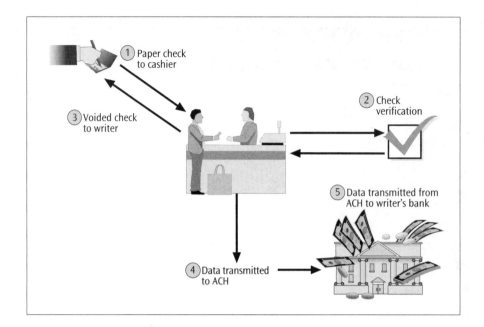

Figure 3-16
Electronic
check process

Buyers using electronic checks at an online store fill out a Web-based form at check-out to authorize the electronic payment and provide the account number and bank routing number for the check. Then the data is submitted to an approval agency and on through the ACH system just like an electronic check presented at a brick-and-mortar store.

In order to reduce costs and speed up the processing associated with paper checks, U.S. banks are leading the effort to replace paper checks with electronic checks. In June 2000 Wells Fargo and 10 other banks, including Bank of America, Chase Manhattan, and Citibank, joined together in a pilot program to replace paper checks with electronic checks at the point of sale. The goal is to stop merchants and banks from moving paper checks through their systems. The banks involved in the pilot program represent more than $2 trillion in deposits, or about 51 percent of U.S. bank deposits.

In June 2000 Wells Fargo and eBay, the Internet auction site, launched a pilot electronic check program using Billpoint, eBay's preferred payment system. A limited number of U.S.-based eBay sellers were chosen to accept electronic checks. During the pilot program, all U.S.-based eBay buyers who have a U.S. checking account can use electronic checks on any auction offered by one of the eligible Billpoint pilot electronic check sellers. There is no fee to the buyer to use an electronic check. During the pilot program, sellers pay 35 cents for each electronic check payment.

E-businesses such as PayByCheck.com (formerly i-check.com), OnlineCheck Systems, and CyberSource (Figure 3-17) provide other e-businesses and government agencies with the software and processing support they need to accept electronic checks for payment of magazine and newsletter subscription services; gas, electricity, and water utility services; cable, telephone, and Internet connectivity services; and taxes, business licenses, traffic fines, park use fees, and so forth.

Transferring dollars from person to person to make payments electronically is now easy, quick, and fun.

Figure 3-17
CyberSource

Person-to-person systems

Person-to-person (or P2P) Internet payment systems are becoming increasingly popular for online auction buyers and sellers, small business owners, and individuals who need to transfer money to family and friends. The primary reason P2P payment systems work so well is that they use e-mail, the most popular activity on the Internet. Like e-mail, P2P payment systems such as eMoneyMail, Western Union MoneyZap, eCash P2P, ProPay.com, Billpoint, and PayMe.com are cheaper, faster, and easier to use than traditional alternatives. Figure 3-18 illustrates the process of person-to-person payment systems.

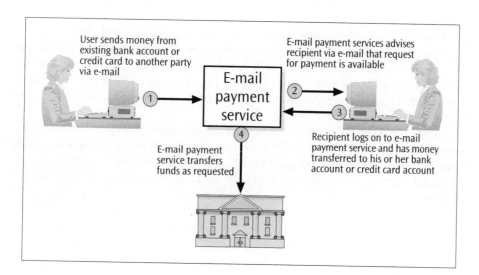

Figure 3-18
Person-to-person payment systems

eMoneyMail (Figure 3-19) is an e-mail money transfer service of Bank One, the fourth largest U.S. bank. For a $1 transaction fee, a user can send money from a Visa credit card, Visa debit card, or checking account to someone via e-mail. The recipient can have the funds transferred to his or her Visa credit or debit card or, for an additional $1 fee, receive a check. If the sender uses a checking account to transfer funds, it may take three to four business days for the funds to clear. The cleared funds then become available to the recipient in one or two business days if transferred into a checking account, or three to four business days if transferred to a Visa credit card account.

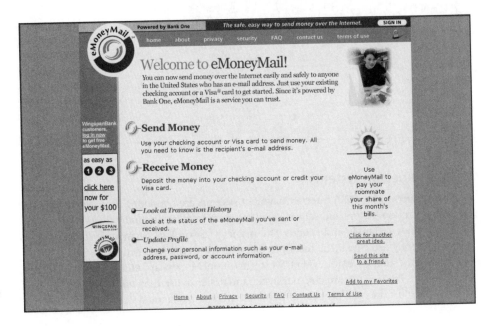

Figure 3-19
eMoneyMail

Users register at the Western Union MoneyZap (Figure 3-20) Web site to create an account. Then users can send money by specifying the recipient's name, e-mail address, and a personal message if desired. The user specifies a checking account or bankcard payment method to complete the transaction. The recipient receives an e-mail notice when the funds are received, usually within four to six business days, and identifies how the funds are to be disbursed.

eCash P2P (Figure 3-21) allows eCash customers to send eCash from their accounts to other eCash accounts, using e-mail. Users can fund their eCash accounts using credit cards and paper checks. Recipients of the eCash transfers can convert their eCash into traditional currency by transferring the eCash from their eCash accounts to traditional bank accounts.

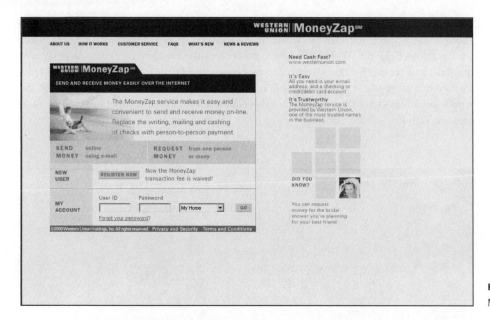

Figure 3-20
MoneyZap

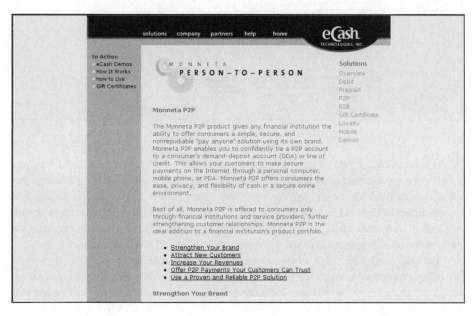

Figure 3-21
eCash P2P

ProPay.com (Figure 3-22) allows individuals to accept credit card payments as though they were merchants. With an established ProPay account, a seller visits ProPay.com, clicks the "Web Pay" button, and enters the customer's e-mail address, sales amount, and reason for the charge. The customer immediately receives an e-mail from ProPay itemizing the seller's request. When the customer clicks a URL provided in the e-mail and enters his or her credit card information, ProPay promptly transfers the sales amount into the seller's account.

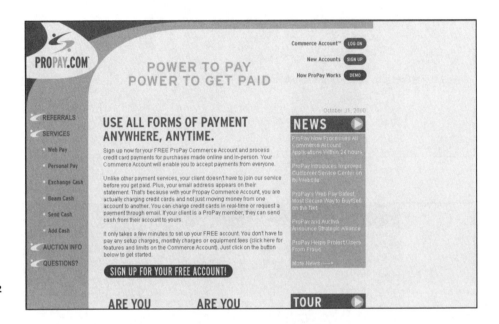

Figure 3-22
ProPay.com

Billpoint (Figure 3-23), founded in 1998 to facilitate person-to-person credit card payments over the Internet, was acquired by eBay in 1999 and became eBay's preferred payment system. Billpoint, free to buyers, processes international credit card payments from buyers around the world. Sellers register with Billpoint to accept credit card payments and pay a small transaction fee on each payment processed. Wells Fargo, a strategic partner and part owner of Billpoint, handles the payment processing and customer service.

PayMe.com (Figure 3-24), a free service of PayMyBills.com, an online bill management provider, is typical of many of the P2P systems. PayMe.com allows registered users to transfer money from credit cards to pay for items bought at online auctions or from online classified ads. Registered users can also send money to any individual who has an e-mail account by entering the recipient's e-mail address, the dollar amount, and subject and message information, and then sending the e-mail. The recipient receives an e-mail notifying him or her of the payment, which is then deposited directly into the recipient's designated account.

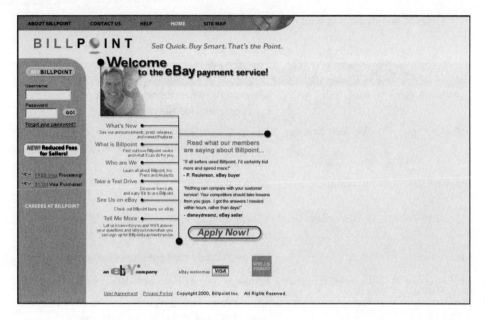

Figure 3-23
Billpoint

PayMe.com registered users can also request that money be sent to their account. Registered users log in to their secure PayMe.com account and select "Request Money." They then enter the recipient's e-mail address, the dollar amount they are requesting, and subject and message information, and then send the e-mail. When the e-mail arrives, the recipient clicks a special link to make the payment. PayMe.com notifies the person requesting payment via e-mail and deposits the payment into the designated account.

Figure 3-24
PayMe.com

Table 3-1 summarizes several of the most popular person-to-person payment systems.

Table 3-1
Person-to-person payment systems

System	Sponsor	Service
eMoneyMail	Bank One	Allows you to send money from credit or debit cards or checking account
MoneyZap	Western Union	Allows you to send money from Western Union account to checking account or bank card account
eCash P2P	eCash	Allows you to send electronic cash from one eCash account to another
ProPay	ProPay.com	Allows individuals to accept credit cards as though they were merchants
Billpoint	eBay	Processes credit card payments for auction sellers
PayMe.com	PayMyBills.com	Transfers money between credit card accounts
PayPal	X.com	Allows you to send money via e-mail or PDA from checking account or credit card

There are thousands of potential consumers who cannot purchase items online because they do not have access to bank accounts or credit cards. Using prepaid cards and prepaid accounts may be one solution for these consumers.

Prepaid cards and prepaid accounts

Thousands of teens and children 12 and under would like to spend their allowances or part-time job income shopping online for clothes, music CDs, and other popular items. Unfortunately, most young shoppers lack access to the credit cards accepted by e-tailers, and parents are often reluctant to allow children to use their credit cards. Additionally, there are thousands of potential e-business consumers who do not have or do not want access to traditional credit cards. One way e-tailers are tapping into this lucrative market is by accepting prepaid shopping cards and prepaid accounts offered by e-businesses such as Cobaltcard, RocketCash, DoughNET, and Cybermoola. Figure 3-25 illustrates how consumers and e-businesses use prepaid cards and accounts.

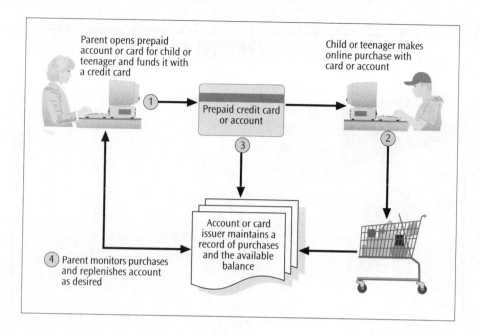

Figure 3-25
Prepaid cards
and accounts

E-CASE **Emergency Cash to the Rescue**

When William Scheurer, a father of four, grew weary of driving to suburban Chicago ATMs to get emergency cash for his teenage daughter, he began to think that a debit card for teenagers that could be controlled by parents was a good business idea. Scheurer, president and owner of Welcome America, a credit card services company, pitched his debit card idea to David Allen, a managing partner in the Chicago firm Divine InterVentures, and the sponsored payment card financial service named PocketCard was born.

PocketCard, which earns revenue by charging fees to activate an account, a per-transaction fee, and a merchant fee, is a Web-based Visa debit card and can be used anywhere Visa is accepted. A parent or adult sponsor enrolls at PocketCard to link a checking or other bank account with the Visa PocketCard account. Funds can be transferred quickly on the Web or by touchtone phone to a PocketCard account from a checking or other bank account. Each PocketCard account has a preset spending limit and, unlike a credit card, does not have interest charges, late payment fees, or end-of-month balances. PocketCard (Figure 3-26) accounts are targeted to teenagers 13 years or older and other sponsored individuals who do not have access to credit cards. PocketCards can also be used at ATMs.

(Continued on the next page)

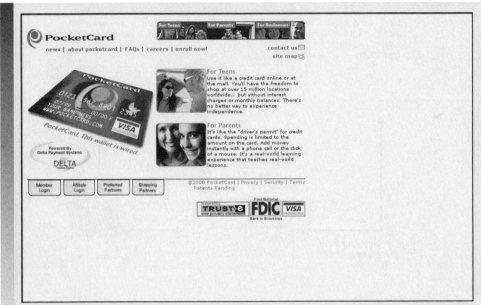

Figure 3-26
PocketCard

Cobaltcard (Figure 3-27), targeted to young adults and other persons without access to credit cards, is a free prepaid shopping card funded by transfers from a bank account. A Cobaltcard carries no fees or interest charges. Cobaltcard members register themselves, fund the Cobaltcard from their bank account, and then monitor spending through a monthly statement. A Cobaltcard is good everywhere American Express is accepted.

Figure 3-27
Cobaltcard

RocketCash (Figure 3-28), a free service targeting 10–18 year olds, provides prepaid accounts for teens and kids that can be used to pay for items purchased from e-tailers advertising on the RocketCash site. Money is added to a RocketCash account by check, money order, or credit card. Account holders can also redeem electronic cash such as beenz or Cybergold to add to their accounts. Family or friends can purchase RocketCash gift certificates for an account holder. When an account holder links to a participating e-tailer's site, makes a purchase, and goes to the checkout page, RocketCash takes over the checkout process by displaying a Checkout Wizard that helps the account holder pay for his or her purchases.

TIP

Children and teens make up one of the largest growth markets on the Internet. Analysts expect 38.5 million teens and children to be online by 2002. Analysts also predict that by 2002 teens will account for $1.2 billion in e-business sales and that children 12 and under will account for $100 million in e-business sales.

DoughNET (Figure 3-29) takes a more full-service approach. DoughNET wants to educate children and teens to make them smarter about money and how to incorporate it into their everyday lives. DoughNET offers banking services, investment kits, and a donation section. Parents can control transactions by setting controls on spending limits, blocking purchases at individual stores, and blocking donations at nonprofit organizations. Parents can also choose to be notified by e-mail when certain kinds of transactions are processed.

Cybermoola (Figure 3-30) targets teenagers and children who want to shop online much as they do offline. Cybermoola offers a prepaid shopping card much like prepaid phone cards. Cybermoola cards and certificates are purchased with cash at brick-and-mortar stores such as Footaction USA, and then "deposited" in the owner's Cybermoola online account by entering the card or certificate's unique serial number. Cybermoola online accounts can also be funded from a credit card or by check. A buyer selects Cybermoola as his or her payment option at participating e-tailers who deduct purchases from the buyer's Cybermoola account.

Figure 3-28
RocketCash

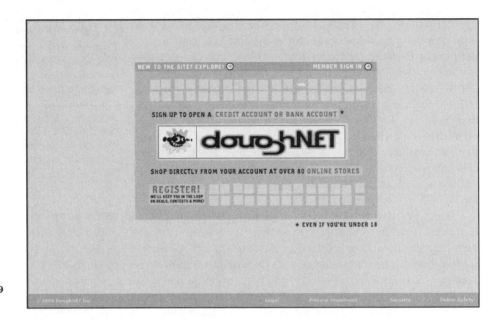

Figure 3-29
DoughNET

Figure 3-30
Cybermoola

Visa Buxx (Figure 3-31), a venture from Visa International, is a prepaid card for teenagers 13 years old and older that parents fund and monitor. Visa Buxx cards are accepted anywhere Visa is accepted, including brick-and-mortar stores, Web sites, and ATM machines. Parents fund the card via credit card, debit card, or bank account. The teenaged user cannot add money to the card, and only parents can order a new card or reactivate an existing card.

Figure 3-31
Visa Buxx

Table 3-2 summarizes a number of popular prepaid card and account methods.

Card/Account Name	Description
PocketCard	Web-based Visa debit card targeted to teenagers and sponsored by parent or other adult sponsor, funded by checking account or credit card
Cobaltcard	Prepaid shopping card funded by checking account transfers, targeted to young adults and others without access to credit cards
RocketCash	Prepaid accounts for 10–18 year olds funded by check, money order, or credit card
DoughNET	Offers banking services, investment kits, and donation opportunities for children and teens
Cybermoola	Prepaid shopping card for children and teenagers similar to prepaid telephone card; buyer purchases cards and gift certificates at participating brick-and-mortar stores
Visa Buxx	Prepaid Visa card for teenagers 13 years and older, funded and monitored by parents

TABLE 3-2
Prepaid cards and accounts

Another important new technological advance for e-business may be consumers' ability to use wireless devices to conduct everyday transactions such as vending machine purchases.

M-commerce

Some analysts think that **mobile commerce**, or **m-commerce**, has the potential to turn a cell phone or PDA into an electronic wallet. With a growing number of Internet-enabled cell phones and PDAs able to access weather reports, e-mail, stock quotes, movie listings, and other information, some proponents think the next logical step is to use these Internet-enabled devices to purchase items on the fly, such as movie tickets, soft drinks, and fast food. Most of the advances in m-commerce are now taking place in Europe and Asia, where consumers are more comfortable with the idea and where companies have adopted a single standard for digital cellular devices. Phone manufacturers such as Nokia and QUALCOMM and carriers such as AT&T Wireless and Sprint are complying with the **Wireless Application Protocol (WAP)**, which allows delivery of Internet content via cellular networks to small-screen devices.

One company that has its own solution for the American market is TeleVend Inc. (Figure 3-32), a small startup company based in Israel. TeleVend's m-ABLE™ technology is designed to work with any existing mobile phone and does not use Internet access. With TeleVend's technology, a consumer could dial the number of a vending machine, purchase an item like a soft drink, and have the charge added to his or her phone bill. Other companies experimenting with m-commerce include Delta Airlines, which is partnering with IBM and Modem Media to develop and test a wireless system that provides flight and gate information. Visa is developing a system that allows users to transfer money from their debit or credit cards to their mobile phones.

Figure 3-32
TeleVend Inc.

Selecting an electronic payment method or methods

A startup e-business should select appropriate electronic payment method(s) based on the kinds of products and services the e-business proposes to offer and the type of customers who will buy those products and services. For example, a B2C e-business selling products to adults would likely select electronic credit, debit, or charge card payments. However, a B2C e-business that targets teenagers and young adults should remember that many potential customers in this category do not have access to traditional credit, debit, or charge cards; therefore, prepaid debit cards or prepaid accounts might be the best payment approach. A C2C e-business might find it more efficient and cost-effective to use one or more person-to-person electronic payment methods. A B2B e-business may well use a combination of electronic payment methods, such as credit cards and EFT, with traditional payment methods, such as paper checks.

An e-business should evaluate the different electronic payment methods—credit, debit, charge cards, electronic checks, and electronic cash—and then determine which will best suit the e-business idea.

. . . WILL THAT BE CASH, CHECK, CREDIT CARD, E-MAIL, OR PDA?

PayPal, a free service from X.com, was one of the first e-mail and PDA payment methods. Peter Thiel, who received his B.A. and J.D. from Stanford University and Stanford Law School, had stints as a derivatives trader, securities lawyer, and head of Thiel Capital Management, LLC. In December 1998, Thiel launched a company called Confinity with the idea of doing something with finance, Palm Pilot PDAs, and security. This idea evolved into PayPal, a service that sends U.S. dollars via the Internet, using Web-enabled devices and e-mail.

PayPal allows users to send money from an existing bank account or credit card to anyone via e-mail. The money is debited from the payer's credit card or bank account and credited to the recipient's PayPal account. When notified via e-mail that money is in his or her account at PayPal, the recipient logs on to PayPal and transfers the money to a bank account or another individual, or requests a check from PayPal. If the recipient does not already have a PayPal account, one is created when he or she logs on to request disbursement. Because PayPal is a free service to consumers, it plans to make money by investing the "float," or cash held before it is transferred out of the recipients' accounts, and by charging a service fee to auction sites and e-tailers.

In July 1999, Confinity secured $3 million in first-round financing from Nokia Ventures, the investment arm of the Finnish wireless company Nokia. To publicize PayPal, Confinity held a press conference where a representative from Nokia Ventures "beamed" the $3 million to Peter Thiel using infrared technology and a Palm Pilot. Thiel received notification of the transfer on his Palm Pilot. In December 1999, PayPal was officially launched by having celebrity spokesperson James Doohan, aka "Scotty" on the original Star Trek™ television program, use his PDA to beam money to selected e-mail users.

By February 2000, Confinity's PayPal was adding 9,000 new users each day. In March 2000, PayPal, now a leading payment system for eBay customers, merged with X.com, an online banking service. The company, named X.com Corporation, quickly announced partnerships with

(*Continued on the next page*)

14 additional auction services. In April 2000, X.com announced that it was growing at a rate of 10,000 new customers a day and had signed up over one million customers. Additional partnerships with the online communities eGroups and eCircles were announced.

In May 2000, X.com had 1.5 million customers and 25 percent of eBay's auction payments. By June 2000, X.com's CEO, Elon Musk, announced that the customer base had increased to about two million and its percentage of eBay's auction payments to 39 percent, far exceeding eBay's in-house payment system, called Billpoint. He also noted that X.com's PayPal services were processing transactions in the amount of $4 to $4.5 million each day. In June 2000, X.com also announced the launch of its PayPal services for mobile devices. By July 2000, X.com had to upgrade its hardware and add 200 customer service reps to keep its PayPal.com (Figure 3-33) site running smoothly. PayPal's competitors include Western Union MoneyZap, eCash P2P, ProPay.com, Billpoint, and PayMe.com.

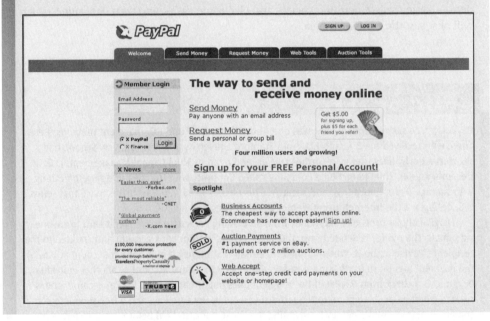

Figure 3-33
PayPal

Summary

- The inefficiencies of the barter system, together with social, political, and religious factors, led to the development of money, which acts as a unit of account or measure of value.
- Amber, beads, cowrie shells, drums, feathers, ivory, jade, gems, gold and other precious metals, oxen, pigs, rice, and salt have all been used as money at different times and in different places.
- Checks, money orders, credit cards, charge cards, and debit cards are "near cash" forms popular today.
- Electronic payment methods usually rely on encryption and digital authentication to provide secure transaction processing.
- Although early electronic cash vendors were not successful, consumers may begin to accept "second-generation" electronic cash such as beenz, Flooz, and Cybergold.

- Smart cards, which can store electronic cash, are popular in Europe and Japan and are expected to gain popularity in the U.S.
- An electronic check is the electronic equivalent of a paper check and can be used at both brick-and-mortar stores and online stores.
- Person-to-person electronic payment methods allow family, friends, and business associates to transfer real dollars to others, using e-mail.
- Teenagers, children, and others who do not have access to credit or debit cards, can now use prepaid cards or prepaid accounts for their online purchases.
- M-commerce, using mobile wireless devices to pay for goods or services, is an emerging technology.

CHECKLIST Payment Methods for Your E-Business

❑ Evaluate the different electronic payment methods, and determine which will best suit your e-business idea on the basis of kinds of products and services your e-business proposes to offer, and the type of customers who will buy those products and services.

❑ Determine how your e-business idea can allow customers to make micropayments, if necessary.

❑ Consider how your e-business idea can use electronic cash to create customer reward or incentive programs.

❑ Evaluate the advantages and disadvantages of accepting electronic checks from your online customers.

❑ Study how you can effectively use person-to-person payment methods at your e-business.

❑ Consider providing prepaid cards or accounts for customers without access to other online payment methods.

❑ Make sure your e-business idea can adapt, if necessary, to m-commerce and the use of mobile wireless devices for payment.

Key Terms

authentication
certificate authority
chargeback
charge card
check

ciphertext
coinage
credit card
cryptography
cybercash

debit card
decryption (decrypted)
digital cash
digital certificate
digital signature
digital wallet
discount rate
electronic cash (e-cash)
electronic checks (e-checks)
electronic wallet (e-wallet)
encryption
merchant account

micropayments
mobile commerce (m-commerce)
money
money order
open standard
person-to-person (P2P) payment systems
plaintext
protocol
Secure Electronic Transactions (SET)
Secure Sockets Layer (SSL)
smart card
Wireless Application Protocol (WAP)

Review Questions

1. The most common method(s) of payment for online consumers is:
   a. Electronic cash.
   b. Credit, debit, and charge cards.
   c. Electronic checks.
   d. E-mail.

2. Which of the following conditions is likely to increase merchant account fees?
   a. Low chargeback rate
   b. Cardholder present
   c. Several years in business
   d. Nontraditional products or services

3. A protocol is:
   a. A standard format for electronically transmitting data.
   b. An electronic message attachment that verifies the sender's identity.
   c. A unique electronic number that carries a specific value.
   d. A small electronic device that contains electronic memory.

4. A chargeback results from:
   a. Failure to code a credit card number.
   b. An increased discount rate.

   c. Sending money via e-mail.
   d. A consumer's refusal to pay a charge on his or her credit card.

5. Which of the following is a suitable payment method for a PDA?
   a. Paper check
   b. Cash
   c. Money sent via e-mail
   d. Coinage

6. A charge card is similar to a credit card except that its balance must be paid in full upon receipt of a statement. **True or false?**

7. SSL and SET are competing methods of securing credit card transactions. **True or false?**

8. A digital signature is a trusted third party that guarantees the identity of the owner of a digital certificate. **True or false?**

9. beenz, Cybergold, and Flooz are P2P payment systems. **True or false?**

10. An e-business shouldn't bother to double-check international orders paid for with a U.S.-issued credit card. **True or false?**

Exercises

1. Using Internet search tools or other relevant resources, review the current status of three P2P e-businesses mentioned in this chapter. Write a one-page paper describing their current operational status and programs.

2. Using Internet search tools or other relevant sources, research current B2B payment methods. Write a one-page paper giving examples.

3. Using Internet search tools or other relevant sources, research the current status of micropayments. Write a one-page paper that describes that current status, including, if possible, at least one e-business that currently accepts micropayments.

4. Using Internet search tools or other relevant sources, locate three e-businesses not listed in this chapter that offer payment-processing services. Write a one-page paper describing the three e-businesses.

5. Use the www.echeck.org Web site and other relevant sources to learn more about electronic checks, and then write a one-page paper describing the advantages of accepting electronic checks for both e-tailer and buyer.

CASE PROJECTS

◆ 1 ◆

You are planning your startup B2C e-business and are evaluating various traditional and electronic payment methods. Using Internet search tools or other relevant sources, research security or potential fraud issues with traditional paper checks, credit, debit, or charge cards, and various forms of electronic cash. Then write a one- to two-page paper ranking each method, from most secure to least secure, and give your reasons for the ranking.

◆ 2 ◆

You and Liz, your best friend, are planning an e-business that sells equipment and uniforms for women's sports and provides auctions for women's sports memorabilia. Based on research and personal experience, you believe your e-business idea targets young adults 19–30 years old, teenagers 13–18 years old, and children 10–12 years old. You and Liz want to include payment methods appropriate for sales and auction participants and for all age groups. Write a one-page paper describing the different payment systems you plan to offer, including the reasons for selecting each payment system.

◆ 3 ◆

As the program director for a local association of online retailers, you must plan a 15-minute presentation on the current status of mobile commerce, or m-commerce. Using Web search tools and other relevant resources, create an outline for a presentation entitled "M-Commerce Today" that includes relevant topics, statistics, and Web site examples.

You and three classmates are starting a B2C e-business and are considering which payment methods to offer customers. Working together, define the e-business, its products and services, and the type of customer who will buy those products and services. Then, using Internet search tools or other relevant sources, identify several electronic payment methods appropriate for the proposed e-business.

Using Microsoft PowerPoint or other presentation tools, create a 5–10 slide presentation defining the e-business idea and its customers. Then list the electronic payment method or methods your team has selected. Give the reasons for your selection. If you select a method that is vulnerable to fraud, list the ways in which the e-business will protect against potential fraud.

Present your e-business idea and payment selections to a group of classmates, selected by your instructor, who will critique the e-business idea, your payment selections, and, if necessary, your fraud-prevention procedures.

Useful Links

4B2B – Business to Business Guides
http://4b2b.4anything.com/

Advancing Women – B2B Hub
http://www.advancingwomen.com/

Credit Cards 101
http://www.paypros.com/creditcards.htm

Cryptography Research, Inc.
http://www.cryptography.com/

didyouknow.com – Credit Cards
http://didyouknow.com/creditcards.htm

eCheck.org
http://www.echeck.org/

E-Commerce Webopedia
http://e-comm.webopedia.com/

Electronic Benefits Transfer
http://www.nacha.org/ebt/

Electronic Check Clearing House Organization
http://www.eccho.com/

Electronic Funds Transfer
http://www.fms.treas.gov/eft/INDEX.html

Electronic Payment Systems
http://theweb.badm.sc.edu/701fstu/risley/EPS.htm

FierceWireless.com
http://www.fiercewireless.com/

Financial Services Technology Consortium
http://www.fstc.org/

Gomez.com
http://www.gomezadvisors.com/

History of the Credit Card
http://www.triumphant.com/history.htm

History of Money from Ancient Times to the Present Day
http://www.ex.ac.uk/~RDavies/arian/llyfr.html

History of Paper Money
http://www.bep.treas.gov/currency/facts6.cfm

Institute for eCommerce – Epayment Links
http://www.ecom.cmu.edu/resources/elibrary/
    epaylinks.shtml

internet.com – E-commerce Payment Solution Reviews
http://ecommerce.internet.com/reviews/glance/
    0,,3691_5,00.html

MasterCard Shop Online: SET
http://www.mastercard.com/shoponline/set/

NACHA – The Electronic Payments Association
http://www.nacha.org/default.htm

Online Fraud Prevention Tips
http://www.antifraud.com/tips.htm

Ovum – E-commerce Information
http://www.ovum.com/ecommerce

Payment mechanisms designed for the Internet
http://ganges.cs.tcd.ie/mepeirce/Project/oninternet.html

Secure Online Payment – Vanderbilt Student Projects
http://www.ecommerce.vanderbilt.edu/Student.Projects/
secure.payment.systems/Introduction.html

Smart Card Forum
http://www.smartcardforum.org/

SSL – How SSL Works
http://developer.netscape.com/tech/security/ssl/
howitworks.html

The Journal of Internet Banking and Commerce
http://www.arraydev.com/commerce/JIBC/

Links to Web Sites Noted in This Chapter

American Express – Private Payments
http://www26.americanexpress.com/
privatepayments/info_page.jsp

Authorize.Net
http://www.authorizenet.com/

beenz
http://www.beenz.com/us/home.ihtml

Billpoint
http://www.billpoint.com/index.html

Cobaltcard
http://www.cobaltcard.com/index.html

CyberCash
http://www.cybercash.com/

Cybergold
http://company.cybergold.com/index.html

Cybermoola
http://www.cybermoola.com/index.htm

CyberSource
http://www.cybersource.com/services/payment/echeck/

DoughNET.com
http://www.doughnet.com/

eCash
http://www.ecash.net/

eCash P2P
http://www.ecash.net/Solutions/p2p.asp

eCircles
http://www.eCircles.com/

e-gold
http://www.e-gold.com/e-gold.asp?cid=100142

eGroups
http://www.egroups.com/

eMoneyMail
https://www.emoneymail.com/
Default.asp?WCI=Home&cell=

Flooz
http://www.flooz.com/

Fossil
http://www.fossil.com/retail/MainFrameSet.asp

Gator
http://www.gator.com/ads.html

Gear.com
http://www.gear.com/

Godiva.com
http://www.godiva.com/

foodlocker.com
http://www.foodlocker.com

ibeauty.com
http://www10.ibeauty.com/index.jsp

ICVERIFY
http://www.cybercash.com/icverify/

i-Escrow
http://www.iescrow.com/whatisescrow.html

InstaBuy
http://www.instabuy.com/

Jotter
http://www.jotter.com/

MBNA
http://www.mbnainternational.com/

MessageMedia
http://www.messagemedia.com/

Microsoft Passport
http://www.passport.com/

Mondex USA
http://mondexusa.com/

mySimon.com
http://www.mysimon.com/

OnLineCheck Systems
http://www.onlinecheck.com/main.html

PayByCheck.com
http://www.i-check.net/

PayMe.com
https://www.payme.com/

PayMyBills.com
http://www.paymybills.com/index.html

PayPal
http://www.x.com/

PocketCard
http://www.pocketcard.com/

Price Chopper Supermarkets
http://www.pricechopper.com/orderonline/index.shtml

ProPay.com
http://www.propay.com/

Q*Wallet
http://qwallet.com/index.shtml

RocketCash
http://www.rocketcash.com/

Secure Electronic Transaction LLC (SETCo)
http://www.setco.org/

Starbucks Coffee Company
http://www.starbucks.com/

TeleCheck
http://www.telecheck.com/home/home.html

TeleVend Inc.
http://www.televend.com/

TowerRecords.com
http://www.towerrecords.com/

Toysrus.com
http://www.toysrus.com/index.cfm?sc=rnhome

VeriSign
http://www.verisign.com/

Visa Buxx
https://www.visabuxx.com/index.cfm?

Western Union MoneyZap
https://www.moneyzap.com/main.asp

For Additional Review

Aberdeen Group. 2000. "Mobile Electronic Commerce: The New Economy on the Move," October 4. Webcast available at: http://www.Aberdeen.com/ab_company/hottopics/webcast/mcom.htm.

Abramson, Ronna. 2000. "The Throwaway Credit Card," *TheStandard*, October 18. Available online at: http://www.thestandard.com/article/display/0,1151,19505,00.html.

Alaimo, Dan and Morrow, Elizabeth. 2000. "Price Chopper Set to Test Electronic Checks," *Supermarket News*, July 3, 15.

Andress, Mandy. 2000. "Smart Is Not Enough: Cards Must Also Be Easy and Useful – With a Little More Refinement and Development, Smart Cards Just May Save Businesses a Ton of Cash," *InfoWorld*, 22(42), October 16, 94.

Aragon, Lawrence. 1999. "Cybermoola Wants Your Kids' Money," *Redherring.com*, September 17. Available online at: http://www.redherring.com/insider/1999/0917/news-moola.html?cod-fd.

Bekker, Scott. 1999. "Boom Then Bust: How Electronic Cash Faltered," *ENT*, 4(5), March 10, 40(1).

Bodnar, Janet. 2000. "Pocket Power," *Kiplinger's Personal Finance Magazine*, 54(6), June, 98.

Brown, Duncan. 2000. "Mobile E-Commerce's Dark Side," *TheStandard.com*, April 17. Available online at: http://www.thestandard.com/ article/display/1,1151,14042,00.html.

Campbell-Holt, Moriah. 2000. "Dissecting Disposable Credit Cards," *E-Commerce News Opinion*, October 4. Available online at: http://www.zdnet.com/ecommerce/stories/main/0,10475,2636916,00.html.

Carpenter, Dave. 2000. "E-Mail Your Check," *ABCNEWS.com and The Associated Press*, March 8. Available online at:http://www.abcnews.go.com/sections/tech/DailyNews/ecash000307.html.

Caufield, Brian. 2000. "Micropayments: Turning Pennies Into Big Money – Small Fortunes: The Technology's Here – But Are Net Businesses Ready to Use It?" *Internet World*, October 1. http://www.internetworld.com/article_bot.asp?inc=100100/10.01.00Cover1&issue=10.01.00

CCH Business Owners Toolkit. 2000. "Accepting Credit Cards." http://www.toolkit.cch.com/Text/P06_2500.asp.

Choi, Soon-Yong and Whinston, Andrew B. 1998. "Smart Cards: Enabling Smart Commerce in the Digital Age," *KPMG and Center for Research in Electronic Commerce, The University of Texas at Austin*. May. Available online at: http://cism.bus.utexas.edu/works/articles/smartcardswp.html.

Conley, Jim. 2000. "Small Change, Big Bucks – Your Pocket Change Is About to Disappear Into Your PDA and Cell Phone," *PC/Computing*, March, 52.

Davies, Glyn. 1996. *A History of Money from Ancient Times to the Present Day, Revised Edition*. Cardiff, Wales: University of Wales Press.

Dugas, Christine. 2000. "Services Let Consumers E-mail Cash," *USA TODAY*, February 1. Available online at: http://www.usatoday.com/life/cyber/tech/review/crg862.htm.

Elmore, Darrel Ray. 2000. "Free Money, Part Two," *Goodauthority*, March 15. Available online at: http://goodauthority.org/buzz/0003/db00315/db00315.htm.

Elmore, Darrell Ray. 1999. "Hey Buddy, Could You Beam Me a Dime?" *Goodauthority*, September 14. Available online at: http://goodauthority.org/buzz/9909/db99914/db99914.htm.

Enos, Lori. 2000. "Report: Net Payment Options Taking Hold," *E-Commerce Times*, October 19. Available online at: http://www.ecommerce-times.com/news/articles2000/001019-7.shtml.

Espe, Erik. 1999. "Confinity Hopes to Change the Way We Pay Each Other," *The Business Journal*, 17(14), July 30, 5(1).

FierceWireless.com. 2000. "Elon Musk + Your Mobile Phone = Your New Wallet," June 9. Available online at: http://www.paypal.x.com/html/fw-060900.html.

Financial Times. 2000. "Financial Express: Computers: Shopping Online? Buy It With beenz," July 15. http://news.ft.com/.

Francisco, Bambi. 2000. "X.com Launching Service for Wireless," *CBS MarketWatch*, June 11. http://www.cbsmarketwatch.com/archive/.

Frezza, Bill. 1999. "Cell Phones As Global Currency? It's Just an IP Address Away," *InternetWeek*, November 22, 27.

Gartner Survey: 2000. "Retail Internet Fraud Is Twelve Times Higher Than Offline Fraud," Gartner Interactive, July 17. Available online at: http://gartner6.gartnerweb.com/public/static/aboutgg/pressrel/pr20000717a.html.

Hafner, Katie. 2000. "Will That Be Cash or Cell Phone?" *The New York Times*, March 2. Available online at: http://www.paypal.x.com/html/nyt-030200.html.

Hicks, Matt. 2000. "Shopping Online With No Wires Attached – MasterCard Explores a Number of Secure Ways to Make Payments Over the Wireless Internet," *eWeek*, July 3, 27.

Hoy, Richard. 2000. "Your Merchant Account and 'Puffy Director Pants' Taxes," *ClickZ*, May 5. Available online at: http://www.clickz.com/cgi-bin/gt/print.html?article=1678.

Johnson, Patrice D. 2000. "Plastic With Parental Guidance," *Money*, 29(4), April 1, 156.

Kahan, Stuart. 2000. "Sign in, Please!" *The Practical Accountant*, 33(10), October, 64.

Kandra, Anne. 2000, "Heads Up," *PC World*, 18(8), August, 35.

Kosiur, David. 1997. *Understanding Electronic Commerce*. Redmond, WA: Microsoft Press.

Latour, Almar. 1999. "PayPal Electronic Plan May Be On the Money in Years to Come," *The Wall Street Journal Interactive Edition*, November 15. Available online at: http://www.paypal.x.com/html/wsj.html.

Lidsky, David. 2000. "The FSB 25…Continued," *Fortune Small Business*, April 24. Available online at: http://www.fsb.com/fortunesb/articles/0,2227,704,00.html.

Lillington, Karlin. 1999. "PayPal Puts Dough in Your Palm," *Wired News*, July 27. Available online at: http://www.wired.com/news/technology/0,1282,20958,00.html.

Manzetti, Amy. 2000. "Electronic Checks: A New Way to Pay," *Credit Union National Association, Inc.*, May. Available online at: http://hc-fcu.org/hffo98/0500_a.htm.

Morris, Charlie. 1999. "Accepting Credit Cards: Getting a Merchant Account," *Web Developer's Virtual Library*, April 29. Available online at: http://wdvl.internet.com/Internet/Commerce/MerchantAccounts/.

Nickell, Joe Ashbrook. 1999. "Are E-Wallets More Trouble Than They're Worth?" *webreview.com*, November 12. Available online at: http://www.webreview.com/pub/1999/11/12/feature/index2.html.

Nobel, Carmen. 1999. "Putting Money in Your Palm," *PC Week*, November 15, 49.

Nocera, Joseph. 2000. "Easy Money: E-cash Can Solve the MP3 Dilemma and Make E-commerce Simpler," *Money*, 29(8), August 1, 71.

Pitta, Julie. 1997. "David Chaum," *Forbes*, 159(14), July 7, 320(2).

Power, Carol. 2000. "E-Sign Law Gives Equality to E-Signatures," *American Banker*, 165(96), October 12, 14A.

Power, Carol and Kutler, Jeffrey. 1998. "Bankrupt DigiCash to Seek Financing, New Allies," *American Banker*, 163(216), November 10.

PRNewswire. 1999. "Beaming Money by Email Is Web's Next Killer App," November 16. Available online at: http://www.paypal.x.com/html/pr-111699.html.

PRNewswire. 1999. "PayPal.com and Star Trek's 'Scotty' Put the Power to Beam Money in the Palm of Your Hand," December 17. Available online at: http://www.paypal.x.com/html/pr-121799.html.

PRNewswire. 2000. "X.com Chosen by Auction Sites to Boost E-Commerce," March 14. Available online at: http://www.paypal.x.com/html/pr-031400.html.

PRNewswire. 2000. "X.com Surpasses One Million Customer Mark," April 5. http://www.paypal.x.com/html/pr-0405b.html.

Revolution Magazine. 2000. "Kid Commerce – Hey Little Spender," April. Available online at: http://www.revolutionmagazine.com/archive/Magazine_Story.asp?Article=6117.

Roberts, C. R. 2000. "Old World Meets New at Paybycheck.com," *Tacoma News Tribune*, May 21. Available online at: http://www.i-check.net/press1.html.

Rosen, Cheryl. 2000. "Major E-Payment Initiatives Combat Online Fraud – Visa and Citigroup Roll Out Payment Infrastructure Designed to Meet Customer Demand," *InformationWeek*, July 24, 38.

Ross, Robyn. 1999. "e-Commerce Means e-Allowance for Kids," *Money.net*, October 28. Available online at: http://www.money.net/scripts/readcontent?941143644-3079.

Saliba, Clare. 2000. "Credit Cards Losing Grip on E-Commerce," *E-Commerce Times*, September 28. Available online at: http://www.ecommerce-times.com/news/articles2000/000928-1.shtml.

Sanghera, Sathnam. 2000. "Teenage Revolution on the Cards," *Financial Times*, May 17. http://www.ft.com/

Sapsford, Jathon. 2000. "PayPal Sees Torrid Growth With Money-Sending Service," *The Wall Street Journal Interactive Edition*, February 16. Available online at: http://www.paypal.x.com/html/wsj3.html.

Schneider, Gary P. and Perry, James T. 2000. *Electronic Commerce*. Cambridge, MA: Course Technology.

Schoenberger, Chana. 2000. "Big Brother (And Sister)," *Forbes*, January 24, 142.

Schu, Jennifer. 2000. "E-Paying Your Way," *CNNfn*, May 17. Available online at: http://cnnfn.cnn.com/2000/05/17/electronic/q_wc_epay/.

Schu, Jennifer. 2000. "The Check's in the E-Mail: The Growing Popularity of Person-to-Person Payment Systems," *womenCONNECT.com*, May. http://www.womenconnect.com/LinkTo/may1600_biz.htm.

Sirbu, Marvin. 1997. "Credits and Debits on the Internet," *Electronic Payments*. Available online at: http://www.gsia.cmu.edu/afs/andrew/gsia/45-871/Readings/Spectrum/netp.html.

Slatalla, Michelle. 2000. "Easy Payments Put Hole in the Pocketbook," *The New York Times*, March 2. Available online at: http://www.paypal.x.com/html/nyt-062900.html.

Soloman, Melissa. 2000. "Micropayments," *Computerworld*, May 1, 62(1).

Stevens, Curtis. 2000. "Tenfold: Ten Tips to Help You Beat Credit Card Fraud," *Sell It!* Available online at: http://sellitontheweb.com/ezine/tentips010.shtml.

Stock, Helen. 2000. "DigiCash Idea Finds New Life in More Flexible eCash," *American Banker*, 165(67), April 6, 9.

The Economist. 2000. "E-Cash 2.0," February 18. Available online at: http://www.paypal.x.com/html/ecom.html.

The History of Paper Money. 2000. Currency Facts: The U.S. Treasury Department. Available online at: http://www.bep.treas.gov/currency/facts6.cfm.

Trager, Louis. 2000. "Smart Card, E-wallet Transactions to Increase," *E-Commerce News*, October 2. Available online at: http://www.zdnet.com/ecommerce/stories/main/0,10475,2635258,00.html.

Trombly, Maria. 2000. "Banks Launch Effort to Purge Paper Checks: Electronic Funds Transfer Systems Piloted," *Computerworld*, June 19, 6(1).

Weitzman, Jennifer. 1999. "Star Trek Promise Fulfilled: Wireless Cash Transfer," *American Banker*, 164(235), December 9, 22.

Westland, J. Christopher and Clark, Theodore H. K. 1999. *Global Electronic Commerce: Theory and Case Studies*. Cambridge, MA: Massachusetts Institute of Technology.

Yakal, Kathy. 2000. "Sign on the Digital Line," *PC Magazine*, September 19, 32.

CHAPTER **4**

Creating an
E-Business Plan

In this chapter, you will learn to:

Organize an e-business plan

Prepare an executive summary

Create vision and mission statements

Create marketing, operations, and financial plans

Understand legal forms of businesses

Describe e-business partnerships

Jenny Lefcourt and Jessica DiLullo met early in their first year at Stanford Business School, where they were M.B.A. candidates. Lefcourt, who received her B.S.E. from the University of Pennsylvania's Wharton School, was a senior product manager at MySoftware Company, which provides productivity software and Internet services for small and midsized companies. DiLullo, who graduated with a B.A. in Economics from Stanford University, was a senior product manager at NetRatings, a leader in the online audience measurement market.

Lefcourt and DiLullo got acquainted by attending the same entrepreneurial meetings, and soon discovered that they both wanted to create a startup company. They also discovered that they shared the same idea: finding an easy way for people to register for a broad range of gifts and for givers to send those gifts. Lefcourt's original idea wasn't for a Web-based registry. She wanted to find a way that people could register for gifts that were in keeping with their lifestyle. DiLullo came at the idea from an e-business perspective. She wanted to aggregate gift information—including price and availability—in one place, using the convenience of the Internet. Combining the idea of a large choice of gift products and the convenience of the Internet was the basis for a startup company idea: an online bridal registry.

Lefcourt and DiLullo decided to create an e-business plan for their online bridal registry idea and enter the plan in an entrepreneurship contest at Stanford. In the contest, a panel of judges reviewed business plans through several rounds that could ultimately result in $25,000 in funding. Lefcourt and DiLullo relied on their real-world experience as consumers to prepare their plan. After all, both had been bridesmaids many times, and DiLullo was getting married. In the Stanford library, Lefcourt and DiLullo worked on their business plan, always asking each other, "Would you register (for gifts) that way?" If the answer was "No!" then they went back to the drawing board. Their hard work in creating, reviewing, and revising their e-business plan paid off. After the first round of judges' reviews, Lefcourt and DiLullo pulled their e-business plan out of the competition, and for good reason!

E-Business Plan Organization

An **e-business plan** is used to seek funding for a new or existing e-business and also serves as a design for operating an e-business after it is funded. Creating an e-business plan forces you to take a critical, objective, and unemotional look at the e-business idea.

Developing an e-business plan takes thought, effort, and time. However, a thorough e-business plan can help you find and exploit hidden strengths as well as identify and fix hidden weaknesses in an e-business idea. For some time, a five-year business plan based on older business cycles and proven business models has been standard. Because flexibility is critical in today's quickly changing e-business environment, three-year e-business plans are also gaining in popularity.

There are a number of nonprofit, governmental, or educational Web sites that provide information about creating a business plan, such as the Service Corps of Retired Executives (SCORE), the Small Business Administration (SBA) (Figure 4-1),

TIP

Strategic business planning involves determining a company's long-term goals and then identifying the actions required to reach those goals. The strategic planning process is ongoing for any existing business. For a startup e-business, creating a business plan is also part of the strategic planning process. Therefore, business plans are sometimes called strategic plans.

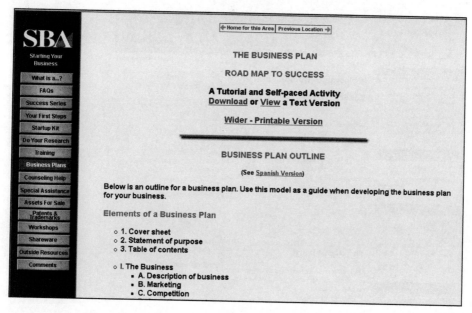

Figure 4-1

Small Business Administration

and the Small Business Advancement National Center (SBANC) at the University of Central Arkansas. While the tips and examples provided are generally based on traditional business plans, they do provide information and insight on how to structure any business plan and the kinds of information a plan may contain.

There are also several e-business Web sites such as BizPlanIt.com and Bplans.com (Figure 4-2) that sell business plan preparation services or desktop application software containing business plan tips and templates. Many of these e-businesses also offer free tips and business plan examples at their Web sites.

Although there are no hard and fast rules for the arrangement of plan components or for content, an e-business plan should include some or all of the following items:

- A cover sheet and a title page
- A table of contents
- An executive summary
- A description of the e-business idea
- Information on products or services to be offered
- Analyses of the e-business's overall industry, target market, and competition
- Marketing, operational, financial, and managerial plans
- Identification of critical risks
- An exit strategy

Headings and subheadings such as Introduction, or Business Description, may also vary from plan to plan as a matter of professional style. Although paying attention to style in developing a professional-looking e-business plan is important, the real goal is to prepare an easy-to-read and fact-filled plan with clearly stated objectives.

Figure 4-2
Bplans.com

When you prepare multiple copies of an e-business plan, it is a good idea to number each copy and to keep a list of who has each copy. Another consideration is confidentiality. Each plan should be clearly marked as confidential, with a notation that additional copies should not be made.

E-CASE IN PROGRESS foodlocker.com

Michael Bates and Stuart Wagner began working on their e-business plan almost as soon as they conceived the idea for foodlocker.com. They began by researching specialty food sites on the Web and the specialty food industry to determine the viability of the foodlocker.com concept. Part of their research included attending specialty food shows in New York, San Francisco, and Chicago, where they started talking with vendors and arranging vendor partnerships. From these meetings Bates and Wagner determined that foodlocker.com needed a strategic food industry partner to add credibility and provide access to important resources. After several discussions and negotiations, they were able to develop a strategic partnership with a local specialty food store chain.

The foodlocker.com business plan continued to evolve with the growth of the company, the need for additional funding, and other marketplace changes. The foodlocker.com e-business plan examples used in this chapter are excerpted from or based on the complete plan, and sensitive or confidential information has been modified or deleted from the illustrations.

Cover sheet and title page

The **cover sheet** identifies the e-business, displaying the title of the document, the preparer's name, the plan copy number, and a "Confidential" notation. A custom cover for a bound e-business plan is appropriate if expense is not an issue. An inside cover sheet, or **title page**, repeats the information from the cover sheet and adds the contact numbers for the preparer and any associates. It may also include the name of the person to whom the copy of the plan is assigned. Figure 4-3 illustrates a sample cover sheet, and Figure 4-4 illustrates a sample title page for foodlocker.com's e-business plan.

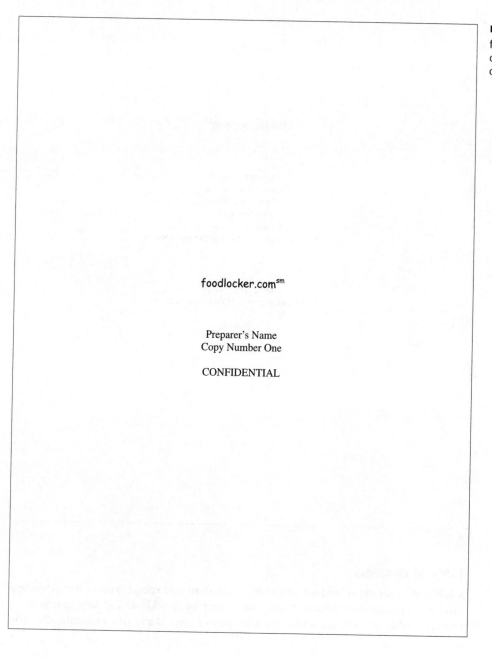

Figure 4-3
foodlocker.com sample cover sheet

*foodlocker.com*sm

Preparer's Name
Copy Number One

CONFIDENTIAL

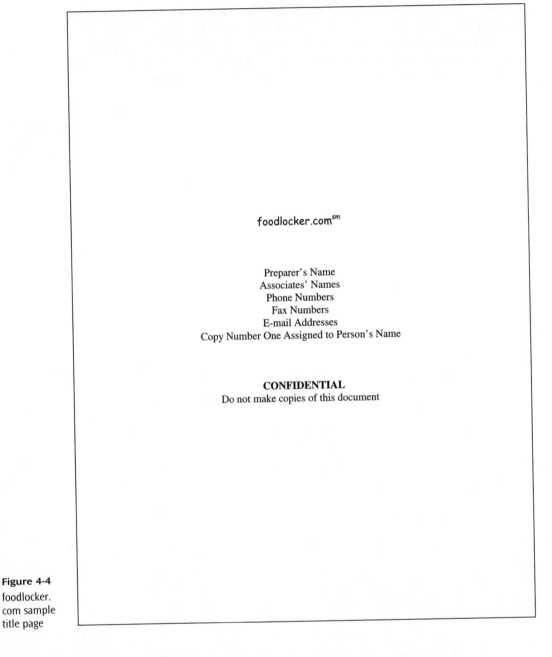

foodlocker.com^{sm}

Preparer's Name
Associates' Names
Phone Numbers
Fax Numbers
E-mail Addresses
Copy Number One Assigned to Person's Name

CONFIDENTIAL
Do not make copies of this document

Figure 4-4
foodlocker.
com sample
title page

Table of contents

A **table of contents** should list all the major sections and subsections of the e-business plan by page number, which allows the reader to quickly locate key information. Prepare a table of contents, which must be well organized and free of errors, after you have finalized the remainder of the business plan. However, it may be helpful to maintain a table of contents checklist as you prepare different sections of the plan. You can then review the checklist when preparing the final table of contents. When preparing the final table of contents, try to avoid these common mistakes: missing sections or subsections, cluttered presentation with too much detail, sloppy layout, and incorrect page numbering. Figure 4-5 illustrates an example of a table of contents for the foodlocker.com e-business plan.

Executive summary

The executive summary may be the most important section of your business plan. An **executive summary** is a miniature version of a complete e-business plan. An executive summary allows readers such as investors, bankers, and managers to quickly understand and get excited about the unique e-business opportunity being offered, without having to wade through the entire plan. Because investors, bankers, and other sources of funding usually have more business plans and proposals than they have time to evaluate, they can use the executive summary to quickly eliminate an e-business plan that does not generate serious interest.

When writing an executive summary, you should use clear and concise language and words that grasp the reader's attention and generate excitement and interest. An executive summary should be no more than two or three pages in length, although one page is even better. A reader should be able to read through an executive summary in five minutes or less and come away with knowledge about how the e-business concept works. In clear, concise, and convincing language, an executive summary should focus on the unique e-business opportunity and use concrete facts and figures to explain the e-business concept. It should highlight the reasons why the e-business concept will be successful, and it can include brief information on:

- The staff and management team
- A definable market
- Any competitive advantages
- Financial projections

Although you can write an executive summary at any time in the planning process, writing it after the remainder of the plan is complete often allows for more effective summation. Figure 4-6 illustrates an executive summary.

TIP

Some e-business plan styles place the executive summary before the table of contents, and some place it after the table of contents. Both the bulleted list and narrative styles are acceptable for an executive summary. No matter which style you use or where you position it in relation to the table of contents, don't forget that the executive summary is normally the first section of an e-business plan that potential investors read—and, if it is poorly prepared, it may well be the last!

TABLE OF CONTENTS

Figure 4-5
foodlocker. com sample table of contents

Executive Summary

Wilderness Treks, Inc. (WT) begins online operations in January, 2002 to offer wilderness adventure travel packages to customers worldwide. WT also offers wilderness survival training to individuals or groups sponsored by corporations and government agencies. WT tour and training team members are all experienced wilderness tour professionals or wilderness survival trainers and are excited about the travel and training services WT provides.

WT's projected total sales for the first year of operations are $5,600,000 increasing 10% annually for the following two years.

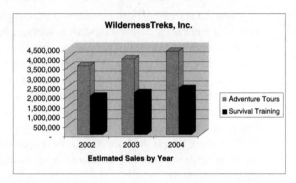

The market for wilderness adventure travel is growing at 15% annually. Additionally, the market for wilderness survival training is growing at 20% annually. WT is ready to exploit this growth and a lack of competition with its experienced tour and training team, online availability, and effective management and marketing.

WT's target markets are 1) adventure-oriented individuals, couples, and 2) groups interested in backpacking to rugged, out-of-the-way, unspoiled locations and sponsored individuals or groups for whom wilderness survival preparation is critical. WT provides specialized and differentiated services to its target markets, both of which are exploitable niches. WT's prices for wilderness adventures and wilderness survival training are competitive with the remainder of the market.

Figure 4-6
Executive summary example

Vision and mission statements

Some e-business plans include a vision and/or mission statement. Although closely related, a vision statement and a mission statement are often defined differently. A **vision statement** can be defined as a business's statement of long-term dreams and goals. A business's vision is its reason for being and a picture of what it is striving to become. A vision statement should inspire, guide, and encourage people toward achieving goals. A **mission statement** is a statement of challenging but achievable actions that a business will take to realize its vision. A mission statement can be a living, working document and a reference point for everything an organization does. Many businesses have vision and/or mission statements but define them in other ways, such as corporate identity statements or statements of goals, values, and objectives. The terms "vision" and "mission" are not universally used and are often interchangeable with similar terms such as "goals," "aims," "philosophy," and "direction." In this text we use the term "mission statement" to broadly define similar statements of company philosophy.

When defining a mission statement, an e-business should consider:

- The purpose of the statement
- The audience for which it is designed
- Who benefits from the statement

TIP

Critics of mission statements argue that many mission statements are just "pretty words" that look good framed on the wall but have little basis in reality. Some critics suggest that many companies waste resources in developing mission statements, and create them simply because no major company can afford to be seen not to have one. Critics also single out over-generalized and wordy mission statements as meaningless, unbelievable, and lacking in an audience focus. For funny examples of how *not* to write a mission statement, check out Catbert's Mission Statement Generator at Comics.com.

Mission statements can be targeted to one group or a combination of audiences: customers, shareholders, company staff, competitors, the government, the press, and the general public. Mission statements often focus on themes of customer care, product quality, market leadership, staff motivation, and innovation.

The length of a mission statement depends on its purpose and defined audience. The length of some real-world mission statements varies greatly from seven or fewer words to a thousand or more words. Johnson Controls' (automotive systems and building controls) mission statement is an example of a shorter mission statement:

> *Continually to exceed our customers' increasing expectations.*
>
> *Johnson Controls*

Again, depending on its purpose and audience, a mission statement's tone can be informative and straightforward or rousing and inspirational. AMR Corporation's (American Airlines and related companies) mission statement, for example, is of medium length, and is written in a straightforward style.

> *We will be the global market leader in air transportation and related information services. That leadership will be attained by:*
>
> ◆ *Setting the industry standard for safety and security.*
>
> ◆ *Providing world-class customer service.*
>
> ◆ *Creating an open and participative work environment which seeks positive change, rewards innovation, and provides growth, security, and opportunity to all employees.*
>
> ◆ *Producing consistently superior financial returns for shareholders.*
>
> *AMR Corporation*

Many e-businesses such as Cisco Systems include their mission statement on their Web site. Cisco Systems has a vision statement and different mission statements for primary business segments, each of which supports Cisco's vision. Figure 4-7 illustrates Cisco's vision statement, and Figure 4-8 illustrates the mission statement for Cisco's Federal Business Development Operations segment.

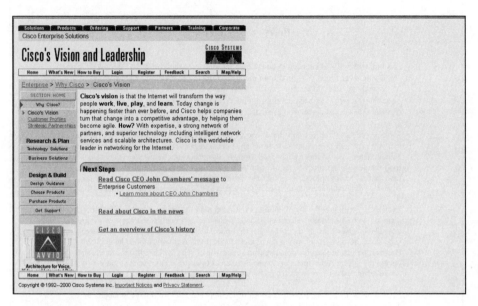

Figure 4-7
Cisco Systems vision statement

Business description

The company or **business description** provides the reader with an outline of the e-business's background and business concept. It may include information about the legal form of the business, when and where it was formed, its history and its current status,

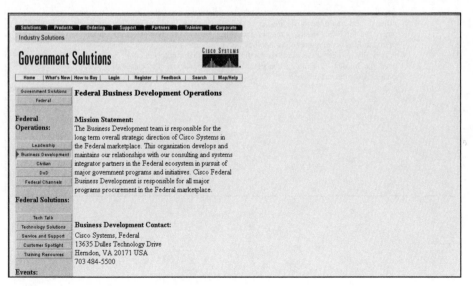

Figure 4-8
Cisco Systems mission statement example

and its future goals. The business description should be brief and include only pertinent information. Figure 4-9 illustrates a sample portion of foodlocker.com's business description.

Figure 4-9
foodlocker.
com sample
business
description

BUSINESS DESCRIPTION

foodlocker.com^{sm} is intended to be the Internet source for food people want most – the "food you never forget"^{sm} – food from the town or region in which they grew up, lived, or visited at one time but which they cannot find in their local markets. foodlocker.com^{sm} has been developed to market, on the Internet, the best specialty and regional foods to customers who desire these foods and cannot buy them locally.

foodlocker.com^{sm} offers only the highest quality specialty and regional foods, foods that will eventually represent most of the large cities, regions, and distinct cultures of the United States. The ultimate goal of the Company is to be a superlative specialty/regional food retailer on the Internet, to offer an overwhelmingly impressive array of regional food and specialty products within the scope of a single transaction, and to take advantage of 'first mover' dominance in this aspect of e-commerce on the Internet.

Even today, with grocery distribution in the United States at its most sophisticated level, many of the best specialty and regional foods, which are very popular in their locality, are unavailable elsewhere in the country. In some areas of the nation, newer grocery superstores have begun to carry more of these foods from other parts of the United States. However, selection remains limited and an abundance of such merchandise is often found only in the largest stores in major metropolitan areas (except for cities such as New York, where urban real estate is at such a premium that such large superstores are not viable).

Several traits distinguish the expectations of foodlocker.com^{sm} from the commodity and deflationary pricing model that dominates Web commerce:

- the familiarity of customers with specific products

- a local scarcity or unavailability of the products that customers seek

- a desire for specific foods that is bred from deprivation

foodlocker.com^{sm}'s customer, one who is relatively computer literate, somewhat affluent, and interested in buying specific foods, is often living away from the locale where the food they desire is readily available. As a result, they are willing to pay a premium for obtaining the foods that they desire but cannot easily obtain. Because there is typically no alternative, local source for the foods that foodlocker.com^{sm} will feature, even the burden of shipping charges becomes secondary to the customer's goal of obtaining the desired product.

2

Products and services

The **products and services section** of an e-business plan is one of the most important sections in the plan. It provides a description of the products and/or services offered and the sales that each product or service group is expected to generate. It is important that the products and services section provide the reader with enough information to understand the e-business, but not too much detailed information, which might confuse the reader. It is also important that all products and services be described from the customers' perspective and that all e-business advantages are highlighted.

If the e-business is selling products, the products section describes what the products are and what the products do, and highlights each product's customer benefits. It is helpful to include in the appendices high-quality photocopies, pictures, or graphics of each product listed in the narrative. If the product line is large, include only a few major products, and place information about the complete product line in the appendices, or add a statement offering a complete catalog listing upon request. Supporting documents such as patents, trademarks, and copyrights belong in the appendices. If the e-business is selling a service, the services section describes what the service is, how the service works, what makes the service different, what materials or equipment are needed to use the service, and what marketplace need the service addresses.

If you have multiple products or services, you can categorize them into product or service groups. You can provide analyses of past and/or projected sales by product or service group, and a graph illustrating product or service group mix can also be included. Common mistakes made in the products and services section include failure to identify benefits to the customer for products or services, using language that is too technical and industry-specific, failure to provide a third-party analysis of products or services, using brochures or catalog sheets in place of text and product pictures or illustrations, and failure to mention the specific need a product or service addresses. Figure 4-10 illustrates a sample portion of foodlocker.com's e-business plan products and services section.

Industry analysis

An **industry** consists of businesses that make or sell similar, complementary, or supplementary products or services. For example airlines, hotels, and travel agencies are all part of the travel industry. Auto, tire, and auto maintenance businesses are all part of the auto industry. Computer hardware manufacturers, software developers, and silicon chip manufacturers are all part of the computer industry. An **industry analysis** can describe industry size, characteristics, trends, growth factors, barriers to entry, government regulations, distribution systems, competitors, the effects of technology on the industry, and other relevant topics. An industry analysis should be based on verifiable data gathered from recognized sources such as government agencies, industry trade associations, or qualified studies from reliable organizations.

PRODUCTS AND SERVICES

A key to success for foodlocker.com℠ is the nature, quality, and quantity of the goods that are offered. Only premium examples of foods from notable producers will be incorporated into the Web site's offerings. The Company's management believes that, based upon their personal experience with foods and vendors that they know or recognize the Company's customers would be willing to presume that all of the foods offered on the Company's Web site are exemplary examples within their respective categories, thus increasing the propensity of cross-category and multi-product sales.

The Company believes there are three categories of products that customers might desire to buy:

> **LEGENDARY FAVORITES**: Foods from producers recognized as exemplary in their category and perhaps already known nationally or internationally for the quality of their products.
>
> **REGIONAL FAVORITES**: Foods exceedingly popular in their locales (top quality examples of what they are) but not widely known beyond the particular city or region in which they are produced.
>
> **NEW CLASSICS**: Foods that deserve inclusion on the site, that merit wider recognition than they have yet achieved, and that represent well the efforts of the producer and the character of the product.

Examples of Legendary Favorites are pasta sauces from Patsy's Restaurant in Manhattan, Virginia ham from S. W. Edwards, and onion products from Vidalia Onion Brothers in Georgia. In Regional Favorites, one might find Utz Snacks from Pennsylvania and bagels from H & H in New York City. New Classics may include corn chowder from Turtle Island Soups of Illinois or preserves and condiments from Wild Thymes of California.

A customer who finds the specific item he or she seeks on foodlocker.com℠ is able to purchase it by way of a direct transaction with the foodlocker.com℠ store. Having found something keenly desired and long sought, a customer will be tempted and favorably inclined to try other items offered on the site. After all, one transaction provides the customer an opportunity to obtain many products, unlike a visit to each vendor's proprietary site where only one type of goods may be available and a separate transaction required at each vendor's site. If a customer wants more selection from a particular vendor than is offered through foodlocker.com℠, links to sites of certain vendors may be available (wherein the Company will receive a fee based upon the amount of any sale).

4

Figure 4-10 foodlocker.com sample products and services section

Emerging trends such as changes in consumer preferences, shifts in consumer demographics, and new technologies that affect the e-business belong in this section. You can include industry characteristics, including product consumption patterns. Product pricing trends and any pricing advantages or disadvantages the e-business enjoys are noted in this section, as well as industry trends the e-business can exploit to attract new customers.

An industry analysis also addresses supplier issues such as labor shortages or any legal, technical, or personnel issue that might affect the e-business's ability to deliver products or services. Major competitors can be identified, including market share. Use charts and graphs when appropriate to enhance the reader's understanding of the data presented. Charts and graphs should be large enough to be readable, but not so large that they interfere with the business plan format. Figure 4-11 illustrates a sample portion of foodlocker.com's e-business plan industry analysis.

Figure 4-11
foodlocker.
com sample
industry
analysis

INDUSTRY ANALYSIS

E-Commerce

The United States currently leads the world in technology, sophistication of Web sites and Internet savvy of consumers. The number of Americans who shop on the Web far exceeds those of any other country, and the Internet will be an integral part of every business retail strategy and customer's shopping list in the very near future *("STORES: Global Online Retailing An Ernst & Young Special Report" [January 2000] Section 2).*

The U.S. e-tail marketplace appears to be entering a period of hyper-growth. Online sales tripled from 1998 to 1999, and 2000 could be a breakout year. Despite this growth, however, a significant portion of the population still does not own a personal computer and remains unconnected to the Internet. This opportunity gap could signal enormous potential for online shopping in coming years.

Ernst & Young compares the Internet today with the advent of the shopping mall in the 1960's. The Company's management believes that this is an appropriate comparison but it is potentially short of the long-term impact that the Internet will have on society and commerce worldwide. The ultimate ease and benefits of shopping online can almost not be overestimated. In addition, the Internet brings a scope of goods, a worldwide mall, if you will, to anyone with a computer, modem, and online access. People in locations never before adequately served by traditional, sticks-and-bricks retailers (specifically small vendors or specialty merchants), will eventually be on par with shoppers located in the largest of metropolitan areas.

If the explosion of online buying marked 1998 as a watershed year for the U.S. Internet retailing business, it was anything but an anomaly. Rather, it was a clear signal of things to come. In October and November 1999, Ernst & Young surveyed consumers and online retail companies in six countries. The result was near universal approval and optimism. According to this Ernst & Young survey, consumers reported that they were generally pleased with the online shopping experience and planned to do much more of it in the future. Companies reported that they had been successful in launching their online businesses, and they were bullish about the future. Online shopping could grow to 10% of U.S. retailing by 2002 if the current growth trajectory continues. Industry leader Bill Gates predicts that one-third of all food sales will be handled electronically by 2005.

Several factors are driving the online purchase decision. First – by a wide margin – is item selection, followed by competitive prices, ease of use, and availability of product. Online buyers reported that 40% of their Internet purchases were made because of product availability on the Internet and would not have been made otherwise. Categories of items purchased by at least 10% of Internet shoppers included hotel reservations, flowers, event tickets, sporting goods, artwork, car rentals, and food and drink (the food and drink category included 13% of Internet shoppers).

7

Marketing plan

A **marketing plan**, which is a critical component of an e-business plan, helps establish, direct, and coordinate marketing efforts for an e-business. It describes all the marketing efforts for an e-business for a specific period of time (usually one year). A marketing plan includes background information and the research results used to select marketing efforts. It also includes the costs of those efforts. Finally, a marketing plan provides a way to measure the success (or failure) of the selected marketing efforts. Marketing plan components can vary by industry, the size of the e-business, and the stage of the e-business's growth. However, many marketing plans include a definition of the target market, a competitor analysis, marketing objectives, marketing strategies, action programs, budgets, the method for measuring results, and supporting documents.

A typical marketing plan contains information about an e-business's **target market**, the group of potential customers that shares a common set of traits that sets them apart from other groups. Defining a target market and understanding its characteristics enable an e-business to more accurately determine which marketing efforts will be most effective. A target market can be defined by several characteristics. **Demographic characteristics** are specific, observable, and objective identifiers shared by those in a target market. General demographic characteristics include age, gender, income level, family life cycle, occupation, education, and race/ethnic group. **Geographic characteristics** such as country/region, state, city/town, size of population, climate, and population density are based on the location of the target market. **Psychographic characteristics** relate to attitudes, beliefs, hopes, fears, prejudices, needs, and desires. Psychographic characteristics include social class, lifestyle, and other attributes such as extroversion or introversion, conservatism or liberalism, and socially conscious or self-centered behaviors. **Consumer characteristics** involve customer purchasing and usage traits, such as frequency of use, method of use, and frequency of purchase of products that are the same as or similar to those offered by the e-business.

After identifying the e-business target market and defining its demographic, geographic, psychographic, and consumer characteristics, it is important to determine its size. It is difficult and expensive to try to reach a too large target market; targeting a niche market, a smaller segment of the larger market, might be preferable. If the target market is too small, it may be difficult to capture enough customers to be successful. Figure 4-12 illustrates a sample portion of foodlocker.com's target market analysis.

A marketing plan can present an analysis of competitive issues that affect the e-business's potential success. A **competitive analysis** may include both direct and indirect competitors' names and a summary of each competitor's products and market strengths, weaknesses, objectives, and strategies. It may also provide charts or graphs indicating various competitors' market share. A competitive analysis may also highlight the e-business's specific competitive advantages and strengths and include a discussion of the strength of the overall market. Figure 4-13 illustrates a sample portion of foodlocker.com's e-business plan competitive analysis.

Marketing Plan

Primary Market Segment

foodlocker.com[sm] will focus on fulfilling the needs and desires of a customer base that could not adequately be served prior to both the existence of the Internet and a willingness of the public to transact business via e-commerce. A principal characteristic of the Company's customers is wide-ranging geographic distribution. In order to build a physical store incorporating the Company's concept, one would have to bear all of the costs of a traditional retailer, including stocking a large selection of goods with potentially low rates of turnover, plus the added task of attracting customers desiring foods from precise regions of the country. The strain on both the advertising budget and inventory controls would be unrealistic.

On the Internet, goods may be added to, or deleted from, "virtual" inventory at practically no cost. If vendors complete order fulfillment, there is no real inventory at all. In instances where vendors are not able to ship directly to customers, the Company will bear the cost and risk of carrying a sufficient quantity of items in its own inventory that particularly suit the needs of the Company's Web store. As demand increases and becomes more predictable, the Company believes that virtually all vendors will accommodate the Company as direct shippers of their goods.

The size of the market for the Company's goods is hard to estimate because it is a new market segment that overlaps previously established market segments. This opportunity for customers was not previously available and was therefore neither measured nor measurable. The Company believes that the opportunity could amount to a sizeable portion of the specialty food market, plus a smaller portion of both the grocery and restaurant markets. The Company further believes that foodlocker.com[sm] will build a significant business, and become a market leader within this new segment of the food distribution industry.

11

Figure 4-12 foodlocker.com sample target market analysis

Marketing objectives should be clearly stated, be measurable, have a time frame, and lead to sales. Multiple marketing objectives should be consistent and should not conflict with each other. Marketing strategies, budgets, action plans, and measurements must support the stated marketing objectives. **Marketing strategies**, often defined as the 4Ps of marketing—product, price, promotion, and place (distribution)—describe in detail the features and benefits of the products and services offered, prices and pricing

Competitive Analysis

Competitors

OnlineStoreA.com, based in Florida, began as a small retail store that located and sold unusual food items to retirees and out-of-state visitors (particularly those from the Northeast). The retail establishment has purportedly been closed so that the owner can focus on online sales. The Web store is apparently based out of the owner's residential garage and reportedly achieved approximately one half million dollars in sales in 1999 with only a minimal public relations campaign for promotion. This site tends toward a selection of what were once commonplace candies and confections but is increasing its offerings in other lines of goods. It is the belief of the Company that the small-scale success of OnlineStoreA.com serves only to prove the viability of our concept and demonstrate the potential for our business model. The Company believes that all of the items featured on this potentially competitive site would be available to be added to the Company's inventory. Further, the Company believes that this site has demonstrated little in the way of sophisticated merchandising, marketing, or finance acumen. Notwithstanding the existence of OnlineStoreA.com, the Company does not believe that there is an identifiable, distinct, 'first mover' extant in the category segment of e-commerce that the Company is targeting. An opportunity exists to seize that dominant position.

Another Web site, OnlineStoreB.com, provides information on regional food products but offers few products for direct sale, and, so far as the Company can determine, does not advertise its existence. Customers at OnlineStoreB.com must generally link from it to vendor sites and enter into a separate transaction with each vendor.

There are a number of specialty food sources on the Internet. SpecialtyA.com, SpecialtyB.com, SpecialtyC.com, SpecialtyD.com, SpecialtyE.com, and others offer the lines of product typically associated with 'brick and mortar', high-end, specialty food stores found in most cities and large towns. SpecialtyA.com, SpecialtyC.com, and SpecalityE.com offer international or ethnic foods. On the whole, these sites confine themselves either to a very high demographic slice of what these companies believe 'is' the e-commerce customer or to a narrow element of the population demarqued by ethnicity or foreign origin.

There are Web stores, such as WebStoreA.com, that offer food items alongside multiple other types of goods. However, their food selection is limited and their site is trying to be all things to all customers, a strategy that the Company believes will prove to be ineffective.

There are many sites devoted to the promotion of a specific genre of food, such as barbecue, hot sauce, and seafood, or of products from a specific locale, such as New York City, Philadelphia, or Chicago. Several of these sites have been successful, but only to a limited extent because, by the nature of their overly focused or localized business model, they have intentionally restricted the size of their customer pool.

12

Figure 4-13 foodlocker.com sample competitive analysis

strategies, how the products and services will be promoted, and how the products and services will be distributed.

A **marketing budget** is an estimate of the costs for all the activities described in the marketing strategies portion of the marketing plan. Typical marketing budget cost or expense categories include communications, research, promotions, advertising, special events, and public relations. An **action plan** is the "to do" list for promoting products and services. It can be in a chart, table, time-line, or narrative format. However it is formatted, an action plan describes specific promotional tasks, when each will start and finish, and who will do each one.

The measurements section of a marketing plan details the numerical targets and time limits for the marketing efforts. This section is used to regularly assess the company's progress toward achieving its marketing objectives. Common mistakes made in preparation of a marketing plan include defining the target market too widely, not specifically identifying advertising and promotion activities, and omitting the details about when, where, and how to reach the target market and what it will cost. Figure 4-14 illustrates portions of foodlocker.com's marketing strategies section.

Operations plan

A traditional **operations plan** describes the business location, necessary equipment, required labor, and any manufacturing and/or service processes related to the products and services to be sold. Information about business location can include headquarters location, branch office locations, warehouse space, and manufacturing space. Any advantages inherent in the specified business locations may be noted, and a layout of the facilities provided in the appendices. An outline of any significant vehicle, computer, office equipment, or manufacturing equipment needs, including purchase or lease costs, may be included.

E-business operational plans should also incorporate these traditional elements when applicable. For example, GroceryWorks.com's e-business model relies on strategically located grocery distribution centers, a sophisticated order-picking system in each distribution center, and a fleet of specially designed delivery trucks. For an e-business such as GroceryWorks.com, specifying the strategic distribution center locations, describing the order-picking system, identifying labor requirements, and illustrating the delivery process would be critical portions of an operations plan. The operations plan for other e-business models would focus on the operations critical to that model. For example, the operations plan for an e-business that requires a large customer support staff might focus on call center locations and the available labor pool for customer support personnel near those locations.

An e-business operations plan may also include information on how the company's Web site is integrated into the company operations. A diagram of the Web site elements and a description of each element in accompanying notes can help readers better understand the Web site operations. If the Web site is already developed, a picture of the Web site may also be provided. It would also be appropriate to include summary information about the staff needed to create and/or maintain the Web site and whether these staffing requirements will be met internally or outsourced. Figure 4-15 illustrates a portion of the foodlocker.com operations plan.

MARKETING STRATEGIES

Low Cost to Promote Products

Due to the excessive cost and widespread difficulty of obtaining space on grocery and market shelves for regional, specialty foods, the Company believes that the Internet will become a significant channel for these types of products. It may also evolve as the customary proving ground for new or improved food products or product variants that producers know will sell, but not in quantities sufficient to successfully compete for, or to justify paying for, space on grocers' shelves.

Advertising

The Company plans to purchase advertising space, primarily in print media and via "opt-in" direct e-mail. Opt-in e-mail provides access to consumers who have given permission to be sent e-mail offers, generally for a specific area of interest. This method will allow the Company to send an e-mail offer or announcement to consumers who have expressed interest in specialty foods and are receptive to receiving the Company's e-mail. This method is different from "spam" e-mail, where no permission has been obtained and e-mails are sent indiscriminately. Evidence suggests that opt-in e-mail generates a much higher response and conversion rate because the Company would be sending the offer to a consumer who is already interested in specialty foods and who can immediately click on the offer and be taken directly to a page on the foodlocker.com^sm Web site that corresponds to that offer.

Traditional venues with a high likelihood of being seen by regular travelers or more affluent persons appeal most. These may include:

- USA Today
- The Wall Street Journal
- New York Times' Sunday Magazine
- Airline in-flight magazines
- Selected travel publications (Traveler)
- Selected regional publications (Southern Living, Texas Monthly)
- Selected lifestyle publications (*Bon Appétit*, *Better Homes & Gardens*, *Martha Stewart Living*)

Targeted, online banner advertisements will also be placed to reinforce the foodlocker.com^sm brand and bring traffic to the site. These banner ads will be highly specific to appropriate keywords or key phrases (such as New England clam chowder or New Orleans-style coffee). These ads will be more cost effective in generating traffic to the Web site than random banners, especially in conjunction with the print ad campaign and public relations initiatives.

15

Figure 4-14
foodlocker.com sample marketing strategies

Order and Fulfillment

The actual sales process at the Company works as follows:

• Customer places order on Web site. • Company e-mails receipt and link to check order status to customer.	• Company downloads order from server. • Company confirms credit card authorization. • Company separates orders and faxes, e-mails or phones orders to vendors. • When failed authorizations occur, they are referred to foodlocker.com[sm] Customer Care.	• Vendor prepares and ships order. In some cases, orders will be filled from inventory of Rice Epicurean Markets. • Vendor sends confirmation of shipping to foodlocker.com[sm], including tracking numbers. • Company charges authorized credit card for order. • Company updates order status on Web site and e-mails customer that order has shipped. • Problems or delays are referred to foodlocker.com[sm] Customer Care.

All products on foodlocker.com[sm] can be easily located by searching one of three categories: geographic location, food type, or product brand name. After a customer identifies the product(s) in which he or she is interested, the customer may click on the 'Add to Shopping Cart' button in order to add the item to an online "shopping cart". All items are priced to include (per the customer's selection) standard, 2-day, or overnight shipping. Only premium shipping increases the base price, but even that is apparent to the customer at the time the item is placed into the shopping cart.

Shipping is included in the posted price of the product for two reasons. First, many products ship directly from the vendor, each incurring separate shipping costs. Second, evidence indicates that e-commerce suffers a high rate of shopping cart abandonment because of customer shock upon commencing checkout before becoming fully aware of actual, total cost. The Ernst & Young survey data confirms that shipping cost is far and away the leading deterrent to Internet purchasing and is also the biggest reason for abandoned shopping carts.

Cited by 60% of those surveyed by Ernst & Young, the key improvements online buyers are looking for are lower shipping costs and lower/more competitive prices. The Company estimates shipping as closely as possible to average actual cost, anticipating that a mixture of shorter and longer distances should average out. A close eye will be kept on this cost sector and pricing adjusted accordingly. Anticipated system upgrades should eventually reduce or eliminate the risk associated with variations in "freight inclusive" pricing.

19

Figure 4-15 foodlocker.com sample operations plan

Financial plan

The financial plan section of an e-business plan shows the reader how all the ideas, concepts, and strategies discussed elsewhere in the plan can come together in a profitable way. At minimum, a financial plan should include a pro forma balance sheet (Figure 4-16), income statement (Figure 4-17), and cash flow statement (Figure 4-18).

Figure 4-16

Balance sheet format example

Comparative Pro Forma Balance Sheet

Assets	FY 2002	FY 2003	FY 2004
Short-term Assets			
Cash	$ -	$ -	$ -
Accounts Receivable	-	-	-
Other Short-term Assets	-	-	-
Total Short-term Assets	$ -	$ -	$ -
Capital Assets	$ -	$ -	$ -
Less Accumulated Depreciation	-	-	-
Total Long-term Assets	-	-	-
Total Assets	$ -	$ -	$ -
Liabilities and Equity			
Accounts Payable	$ -	$ -	$ -
Short-term Notes	-	-	-
Other Short-term Liabilities	-	-	-
Long Term Liabilities	-	-	-
Total Liabilities	-	-	-
Capital Account	-	-	-
Current Period Net Income (Loss)	-	-	-
Total Liabilities and Equity	$ -	$ -	$ -

Figure 4-17

Income statement format example

Projected Profit and Loss

	2002	2003	2004
Sales			
Adventure Travel	$ -	$ -	$ -
Wilderness Training	-	-	-
Total Sales	$ -	$ -	$ -
Direct Sales Costs	-	-	-
Gross Margin	$ -	$ -	$ -
Gross Margin %	%	%	%
Advertising	$ -	$ -	$ -
Depreciation			
Insurance			
Leased Equipment			
Miscellaneous			
Payroll Expense			
Payroll Taxes			
Rent			
Travel	-	-	-
Utilities	-	-	-
Total Operating Expenses	$ -	$ -	$ -
Profit before Interest and Taxes	$ -	$ -	$ -
Interest Expense Short-term	-	-	-
Interest Expense Long-term	-	-	-
Taxes	-	-	-
Net Profit	$ -	$ -	$ -
% of Sales	%	%	%

Planned Monthly Cash Flow

	12/31/02	12/31/03	12/31/04
CASH IN			
Cash on Hand	$ -	$ -	$ -
Gross Receipts	-	-	-
Loan Proceeds	-	-	-
Interest Income	-	-	-
Sales Proceeds			
Other	-	-	-
Cash Inflow	$ -	$ -	$ -
CASH OUT			
Cash on Hand 12/31/xx	$ -	$ -	$ -
Expenses	-	-	-
Asset Purchases	-	-	-
Estimated Tax Payments	-	-	-
Other	-	-	-
Cash Outflow	$ -	$ -	$ -

Figure 4-18
Cash flow statement format example

A financial plan may also include a statement of the financial assumptions used to generate the plan numbers, a break-even analysis (Figure 4-19) that identifies the amount of sales needed to cover fixed and variable expenses, a statement of the sources and uses of funds that explains how the e-business expects to secure capital and how it will spend it, a standard financial ratios analysis (Figure 4-20) illustrating how well the e-business will perform compared to other companies in the same industry, and an investment proposal outlining who owns the e-business, its financial partners, how much capital is needed, and the expected return on investment. If the projected e-business has a financial history, historical financial statements may be included. Some common mistakes to avoid when preparing a financial plan include failing to include projected financial statements, presenting unrealistic sales and profit projections, failure to include a statement of financial assumptions, underestimating, and failure to plan for unexpected costs.

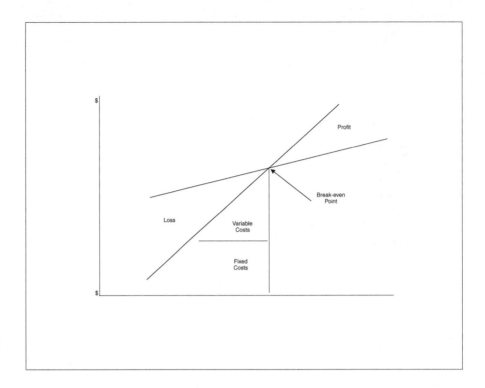

Figure 4-19

Break-even
analysis chart
format
example

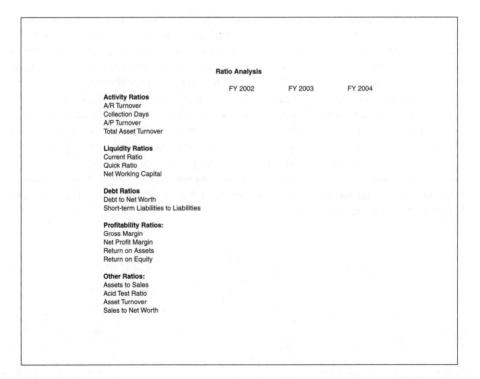

Figure 4-20

Ratio analysis
format
example

Management plan

A strong management team is critical, especially for a startup e-business, because many investors base their investment decisions on the strength of the management team behind a new venture. The management plan section of an e-business plan focuses on the management team, including outside advisors and consultants. The management team described in the plan should be three to five people involved in day-to-day operations and in those that are most critical to the future success of the e-business. The management plan can contain a *brief* narrative description of each team member, including title, duties and responsibilities, previous related experience, previous successes, and education. A detailed resume can be provided in the e-business plan appendices, if necessary.

A new e-business should also have a **board of advisors** that consists of individuals with valuable industry or business experience, who act as advisors to the business. These outside advisors can be very important to a new venture by bringing in experience and background that might be missing in the management team. The management plan section should *briefly* list the name, background, and expected contribution for each advisory board member. An impressive board of advisors can add credibility to a new e-business venture.

The management team planning a new e-business must make a number of insourcing and outsourcing decisions as part of the planning process. **Outsourcing** refers to work done for an e-business by people or organizations other than the e-business's employees. **Insourcing** refers to work done by the e-business's employees. A new e-business's management team must decide which business activities are best done in-house and which can be outsourced to consultants to save time, cut costs, or take advantage of skills not available in-house. Outsourcing allows the management team and personnel of a new e-business to concentrate on the e-business's core activities, while allowing experienced consultants to focus on areas of the new e-business in which the management team may lack experience or skills. New e-businesses outsource many activities to a variety of consultants such as accountants, attorneys, bankers, insurance agents, and technology experts. For example, it is important to get advice from an attorney on the legal form of the e-business, the content of the business plan, the wording of confidentiality agreements, and contractual obligations. Also, the financial plan should be prepared by or reviewed by professional accountants. Many e-business startups outsource the business's Web site development and Web site operations to technology firms. Critical outsourcing arrangements, such as Web site development and management, should appear in the management plan. Other outside consultants such as attorneys and accountants can also be *briefly* listed in the management plan (Figures 4-21 and 4-22) to add to the plan's credibility.

Office and Staffing

Initially, the Company will lease office space from Rice Food Markets in a building adjacent to the corporate office of Rice Food Markets on a month-to-month basis. In addition, the Company anticipates the following hiring schedule:

IMMEDIATE HIRES:	HIRE BY OCT. 2000:	HIRE IN 2001:
Marketing Director Webmaster (or Technology Manager) Controller* Vendor Relations Associate Customer Care Associate Administrative Assistant	Director of Operations Marketing Associate Customer Care Associates Programmer/Graphic Designer Accounting Associate*	Vendor Relations Associate Programmer/IT Support Marketing/Sales Administrative Assistant Additional Vendor Relations or Customer Care Associates (as required by the volume of business)

* *MAY BE OUTSOURCED*

The Company will hire labor for packing and shipping orders as needed to complete the volume of orders filled directly by the Company. The Company will hire temporary personnel to help with customer service during the anticipated peak months of November and December.

26

Figure 4-21
foodlocker. com sample management plan – office and staffing

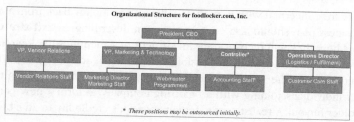

Organizational Structure

The following is a chart of Company roles. The organization is very organic, without extensive hierarchy, due to the rapidly changing needs of an Internet start-up. During initial operations, it is expected that employees will perform multiple functions across the Company and a team-based culture will be encouraged. From time to time, the Company may elect to hire the services of outside consultants to fill specific roles when it is not cost effective to employ such persons within the Company.

Organizational Structure for foodlocker.com, Inc.

President, CEO

VP, Vendor Relations — VP, Marketing & Technology — Controller* — Operations Director (Logistics / Fulfillment)

Vendor Relations Staff — Marketing Director Marketing Staff — Webmaster Programmers — Accounting Staff* — Customer Care Staff

* These positions may be outsourced initially.

27

Figure 4-22 foodlocker. com sample management plan – organizational structure

Issues analysis and critical risks

An **issues analysis** or **risk assessment** identifies threats or opportunities the e-business faces from outside influences such as the overall economic outlook, impending product innovations, technological advancements, environmental issues, government regulations, and any barriers to entry into the market. The plan should take into account all potential losses and any legal factors, staffing concerns, competitor information, technological factors, or other information that could have a negative impact on the e-business. An issues analysis also identifies internal strengths or weaknesses, such as lack of managerial expertise by the e-business principals. After you have determined which issues are most significant, potential problems and any contingency plans to resolve them can be integrated into a one-page issues statement.

Exit strategies

Investors are concerned with how they will get their money back from a new venture. An e-business plan should also contain a section describing an **exit strategy** that tells investors how they will recover their investment, and identifies the long-term plans for the e-business and its principals. Possible exit strategies include going public with an IPO (initial public offering), being acquired by another company, selling the company to other individuals, joining with an existing company to form a new company, and a stockholder buyout. A realistic exit strategy also adds credibility to an e-business plan.

Appendices

The appendices to an e-business plan can include items that provide additional details to the plan—for example, pictures of products, marketing materials, management resumes, details of order fulfillment locations and equipment, and any other supporting documentation.

An e-business plan should also identify the e-business's legal form of organization.

Legal Forms of Organization for an E-Business

One of the issues in the planning process is determining the legal form of organization the new e-business should adopt. This legal form is noted in an e-business plan, generally in the business description section. The most common legal forms of organization include:

- ◆ Sole proprietorship
- ◆ Partnership
- ◆ Corporation

The oldest and simplest form of organization is a sole proprietorship. A **sole proprietorship** is a business started and operated by an individual, without the formalities associated with other legal organization forms. In a sole proprietorship, the individual (proprietor) and the business are one and the same for tax and legal liability purposes. The proprietorship does not pay taxes as a separate entity, and legal claimants can pursue the personal property of the proprietor, not just the assets used in the business.

A **partnership** is a legal business form or entity that consists of two or more owners. A **general partnership** consists of multiple co-owners of a for-profit business. Because partnerships can be complicated, there must be a written partnership agreement that details fundamentals of the partnership. These fundamentals include the amount and nature of each partner's capital contribution to the partnership, the allocation of the partnership's profits and losses, salaries and drawings against profits, and management responsibilities. The partnership agreement should also indicate what happens if a partner withdraws, retires, becomes disabled, or dies. The agreement must also state how the partnership can be dissolved. As with a proprietorship, tax and legal liabilities flow through to the individual partners. Earnings are distributed according to the partnership agreement, and each individual partner pays taxes. Partners are legally liable jointly and severally. This means that a claimant may pursue any of the partners for any amount, regardless of the partner's proportion of invested capital or percentage of distributed earnings. A **limited partnership** has both general partners and limited partners. In a limited partnership, the general partners assume management responsibility and unlimited liability for the partnership. The limited partners have no management participation and are legally liable only for the amount of their capital contribution plus any specifically accepted debt.

A common legal organization form for a business is a regular, or "C," **corporation**. The owners of a C corporation are its shareholders. Unlike a sole proprietor or partner, shareholders of a corporation are protected from liability. Their liability is limited to the extent of their investment in the business. A corporation is a taxpaying entity. Earnings of a corporation are taxed twice: once as income to the corporation and again when shareholders pay taxes on dividends issued by the corporation. Two other corporate forms are the "S" corporation and the limited liability company (LLC). An S corporation affords owners the tax status of a partnership with the protection from liability of a corporation. An organization must meet very restrictive conditions to qualify for S corporation status. Although state laws differ somewhat, a limited liability company generally provides owners with the same tax advantages and limited liability protection of an S corporation, but without some of the S corporation restrictions. A limited liability company is similar to a partnership in that the operating agreement may distribute profits and losses in many ways, not just in proportion to the owners' capital contributions.

TIP

A board of directors can play an important role in guiding an e-business to success. A board of directors is often made up of key internal team members and outside members such as investors, advisors, and strategic partners. If the e-business is a corporation and has a board of directors, the management plan section of its e-business plan also briefly lists the name and background of each board member.

When planning the new e-business's legal form, it is important to consider who the investors and owners will be. A partnership, S corporation, or LLC might be appropriate for a small group of individual owners and investors. If venture capital or other professional investors are solicited, if there is a need for the flexibility of stock options in incentive stock plans, or if there is a plan to take the business public, a corporation form is more suitable. It is important to consult with attorneys and accountants, so as to fully understand the details and ramifications of state laws and different legal business forms, when preparing an e-business plan.

Relationships with strategic business partners, which can also be described in an e-business plan, are often at the core of an e-business's future success.

E-Business Partnerships

Any business's success strongly depends on its relationships with its business partners, such as its suppliers and distributors. Partnering, building alliances, and collaborating with others are the order of the day for the Internet, because they offer a faster way to get a business's products and services sold and distributed. Brick-and-click businesses are finding new and different ways to use the Internet to partner with key allies and others.

One example of brick-and-click businesses forming an e-partnership is PartsAmerica.com (Figure 4-23), formed to provide broader and more efficient distribution of auto parts and accessories. Advance Auto Parts and CSK Auto announced that they would become e-partners in PartsAmerica.com early in 2000, to offer automotive parts and delivery options to both consumers and the commercial market. Advance Auto Parts, ranked second in the Aftermarket Business Auto Chain Report Top 50, has more than 1,600 stores in 37 states, primarily in the East, Midwest, Puerto Rico, and the Virgin Islands. CSK Auto, ranked third in the same report, has 1,120 stores in 17 states, primarily in the western United States. Together, the companies cover the U.S. and have only approximately 30 stores in overlapping areas in their primary sales regions. By partnering as PartsAmerica.com, the two companies benefit from a 50-state coverage of 2,700 stores, 3,000 delivery vehicles, 59 distribution facilities, and more than $1 billion in inventory. PartsAmerica.com customers benefit from broader product options, same-day or overnight delivery, and in-store order pickup. PartsAmerica.com customers can also return or exchange merchandise at their local CSK Auto or Advanced Auto Parts stores.

An e-business may have a variety of e-partners: for marketing, sales assistance, technology support, or financial services. As an e-business grows and the e-business environment evolves, old partnerships may disappear and new partnerships may take their place. For startup e-businesses there are two general categories of partnerships:

- Partnerships that help the e-business build market awareness
- Partnerships that assist in the actual e-business operations, such as customer service or shipping and transportation

Figure 4-23
PartsAmerica.com

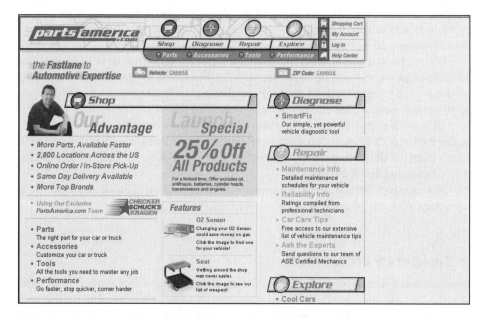

As part of the planning process for a new e-business, appropriate partnerships for the start-up phase of the e-business can be defined and noted in the e-business plan.

The first step in defining appropriate e-partnerships is to be clear on what the e-business wants to provide to customers and how the potential customers want to buy the e-business's products and services. This analysis can illustrate areas in which the e-business needs help from various partners. The next step is to identify potential partners by asking for recommendations from advisors and by monitoring with whom the e-business's competition is partnering. Then prepare a mutually advantageous e-partnership proposal, and deliver it to potential partners. Upon agreement, the e-partnership should be formalized in a contract containing clearly identified expectations for both organizations, a specific period of time for the contract obligations, and an exit clause.

Two examples of e-businesses forming partnerships to share business improvements are Amazon.com's partnerships with Toys 'R' Us and Greenlight.com. Undaunted by the failure of its online furniture sales partnership with Living.com, which filed for bankruptcy in August 2000, Amazon.com announced two new partnerships that same month. First, Amazon.com and Toys 'R' Us entered into a partnership to sell toys. Toys 'R' Us experienced many operational problems with its online Christmas 1999 sales, as did Amazon.com. The new partnership was designed to allow both companies to do what they do best: Amazon to warehouse and ship toy merchandise, while Toys 'R' Us handles the buying and marketing of the toy merchandise. Customers who go to the Toys 'R' Us Web site are directed to the toy section at Amazon.com, where they use their Amazon account to purchase the toys along with books and other merchandise that can be shipped together in one box. Toys 'R' Us buys the inventory, sets the prices, writes the item descriptions, and gets credit for any revenue and potential profits or losses on the toy sales. Toys 'R' Us pays Amazon.com to use its Internet site and warehouses. This partnership allows Toys 'R' Us to solve its order fulfillment problems, while Amazon.com is relieved of the costs associated with buying and maintaining a toy inventory.

Amazon.com's second announced partnership was with car seller Greenlight.com (Figure 4-24). Greenlight.com allows its customers to browse its online showroom, get access to offline dealers, and review trade-in and financing options. Greenlight.com agreed to pay Amazon.com $15.25 million over two years as part of the deal. This second partnership provides additional revenue for Amazon.com and gives Greenlight.com access to Amazon.com's 23 million+ online shoppers.

> **TIP**
>
> An affiliate program is a revenue-sharing partnership program in which e-partners provide links at their Web site to an e-tailer's site. When a user at the e-partner's site clicks through to the e-tailer's site and buys something or signs up for something, the e-partner is paid a commission. Many e-tailers, such as Amazon.com, look to affiliate programs to drive customers to their Web site. It is estimated that Amazon derives 11 percent of its revenues from its pioneering affiliate program and has over 100,000 affiliate e-partners linking to its site and selling its products. Other e-businesses participating in affiliate e-partner programs include eToys, CDNow, BarnesandNoble.com, and Autoweb.com.

A credible, professional business plan that describes an e-business and its business concept, illustrates its products and services and how they will be marketed, shows how the e-business operations can make money, and lists its strategic partners, advisors, and management team can be a map to success for a new e-business.

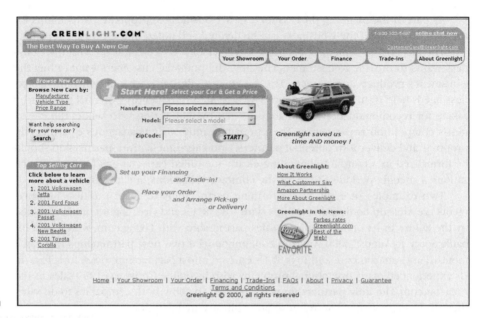

Figure 4-24
Greenlight.com

One of the judges during the first round of the entrepreneurial contest was Dave Wharton, an associate at Kleiner Perkins Caufield & Byers, a venture capitalist firm. Wharton thought the online bridal registry was such a good idea that he contacted Lefcourt and DiLullo after the first round. Following a series of conversations, Lefcourt and DiLullo decided to pull their plan out of the contest. They were getting serious about it and didn't want to circulate their idea too widely. By the end of their first B-school year (May 1998), Kleiner Perkins funded their online gift registry idea, and Della & James, named for the two characters in O. Henry's story "The Gift of the Magi," was born. Lefcourt and DiLullo left Stanford at the end of that first year to enter Della & James in the estimated $17 billion-per-year bridal registry market.

During the first year after funding, Lefcourt and DiLullo concentrated on creating partnerships with retailers and hiring staff. By the time the Della & James Web site was launched in June 1999, there were 50 employees, including CEO Rebecca Patton, formerly a senior vice president with E*TRADE Group. Della & James retail partners included Crate & Barrel, Neiman Marcus Group, Williams-Sonoma, Dillard's, Recreational Equipment, and Gump's, a well-known San Francisco specialty retailer.

In September 1999, Della & James announced $45 million in equity financing from Amazon.com, Neiman Marcus Group, Inc., Crate & Barrel Inc., Williams-Sonoma, Inc., Trinity Ventures, and Kleiner Perkins. In February 2000, Tiffany & Co. acquired a 5 percent stake in Della & James, now Della.com. Della.com became the exclusive wedding registry partner for Tiffany's. In April 2000, Della.com merged with WeddingChannel.com (Figure 4-25) to form an e-business offering a complete array of services and products for engaged couples and their guests.

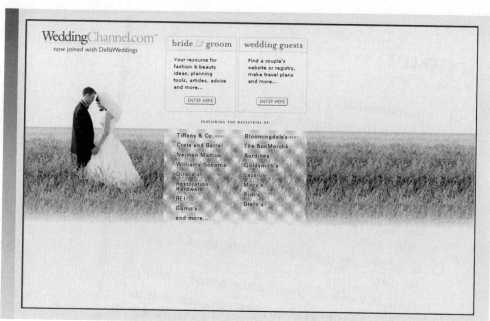

Figure 4-25
Wedding
Channel.com

Summary

- An e-business plan is used to seek funding for a new or existing e-business and as a design for operating the e-business.
- An e-business plan should include some or all of the following components: a cover sheet and title page, a table of contents, an executive summary, a business description, information on products and services, and industry, target market, and competitive analyses.
- An e-business plan should also include some or all of the following components: a marketing plan, an operations plan, a financial plan, an analysis of critical risks, and a formalized exit strategy.

- The appendices to an e-business plan include items such as resumes, pictures of products, or other supporting documentation, which provide additional detailed information for the plan.
- The e-business plan business description can also include a notation about the legal form of organization of the e-business: a sole proprietorship, a partnership, or a corporation.
- An e-business plan can also note any key business partnerships and alliances, such as those developed with suppliers or distributors.

CHECKLIST Your E-Business Plan

- ❏ Demonstrates overall professionalism
- ❏ Is neatly prepared, assembled, and bound
- ❏ Is clearly and concisely written
- ❏ Has been professionally edited for organization, clarity, spelling, and grammar
- ❏ Begins with a cover sheet and a title page with contact information and a proprietary caution against reproducing the plan
- ❏ Is numbered so that you can account for all copies of the plan
- ❏ Contains a table of contents and executive summary
- ❏ Contains some or all of the following components, as appropriate: a business description, a summary of products and services, an industry analysis, a marketing plan, an operational plan, a financial plan, a management plan, a survey of critical risks, and an exit strategy
- ❏ Uses verifiable data from reliable sources and substantiates all data sources with footnotes or endnotes
- ❏ Has been reviewed by attorneys and accountants as necessary and contains any appropriate legal notations relating to the plan's contents and financial projections

Key Terms

action plan
board of advisors
business description
competitive analysis
consumer characteristics
corporation
cover sheet
demographic characteristics
e-business plan
executive summary
exit strategy
general partnership

geographic characteristics
industry
industry analysis
insourcing
issues analysis
limited partnership
marketing budget
marketing objectives
marketing plan
marketing strategies
mission statement
operations plan

outsourcing
partnership
products and services section
psychographic characteristics
risk assessment
sole proprietorship
table of contents
target market
title page
vision statement

Review Questions

1. An e-business plan executive summary:
   a. Allows plan readers to locate specific plan contents.
   b. Provides a miniature version of the complete plan.
   c. Describes the products and services offered.
   d. Is a "to do" list for promoting products and services offered.

2. An operations plan:
   a. Identifies the e-business's niche market.
   b. Emphasizes the customer benefits inherent in the e-business's products and services.
   c. Identifies emerging trends in the e-business's industry.
   d. Describes any sales and services processes related to the products and services offered.

3. An issues analysis identifies:
   a. Outside threats or opportunities.
   b. Marketing strategies for pricing, promoting, and distributing the e-business's products and services.
   c. The e-business's idea or concept.
   d. Customer purchasing or usage traits.

4. A financial plan:
   a. Describes business activities that will be outsourced to consultants.
   b. Identifies outside advisors who bring additional background and experience to the e-business.

   c. Provides supporting documents such as patents, trademarks, and copyrights.
   d. Explains how the e-business ideas, concepts, and strategies can be profitable.

5. A marketing plan:
   a. Identifies threats or opportunities an e-business faces from outside sources.
   b. Identifies the outside advisors with valuable industry experience who are assisting in the startup process.
   c. Helps establish, direct, and coordinate marketing efforts for an e-business.
   d. Outlines the e-business's owners, its financial partners, and its return on investment information.

6. An e-business plan is only used to solicit funding. **True or false?**

7. It is not necessary to number multiple copies of an e-business plan. **True or false?**

8. Demographic characteristics are specific, observable, and objective identifiers shared by a target market. **True or false?**

9. A general partnership is a taxable entity. **True or false?**

10. E-partnerships allow e-businesses to share business improvements and enjoy joint rewards. **True or false?**

Exercises

1. Using Alta Vista or another search tool, search the Web for two e-business sites that include a company mission statement. Review each business's Web site and mission statement. Write a short paper describing each business and quote its mission statement. Include a brief analysis of each mission statement, identifying the mission statement's theme or focus and its target audiences. Based on your analysis, explain why you think the mission statement is or is not effective and useful.

2. Using Bplans.com or similar e-business Web sites that provide examples of business plans, review at least three sample business plans. Then create the cover sheet, title page, and table of contents for an e-business plan whose e-business idea is based on the e-tailing model.

3. Using the Cisco-Eagle GroceryWorks case study located at www.cisco-eagle.com/CaseStudies/index.htm, review the case study and then write a one-page operations plan section describing the order-picking process for GroceryWorks.com.

4. Using Internet search tools or other relevant sources, identify three recent e-business partnerships. Describe each partnership and the shared business improvements or joint rewards to be achieved by each e-partnership.

5. Using the Online Women's Business Center Web site marketing plan information located at www.onlinewbc.org/DOCS/market/mk_plan_why.html or other appropriate resources, create an outline for a marketing plan for an e-business based on the B2B model.

CASE PROJECTS

◆ 1 ◆

You have been thinking about creating a new e-business to sell your custom needlework designs, needlework supplies, and the completed handcrafted needlework projects you accept on consignment. You decide it's time to get serious about your idea. Write a two-page paper describing your e-business idea and identifying the types of e-partnerships that would benefit your e-business.

◆ 2 ◆

As the director of a nationwide youth sports organization, you are considering ways to increase traffic to your organization's Web site and generate revenue. Currently the organization's Web site receives moderate traffic from both the young participants in the sports activities and their adult sponsors. After reading a newspaper article about e-partnerships and Web-based affiliate revenue sharing programs, you think that establishing one or more e-partnerships might be a good way to boost traffic and generate some revenue. Using Internet search tools and other appropriate resources, research different types of e-partnerships, including affiliate programs. Then write a short paper describing three types of e-partnerships or affiliate programs that would be suitable for your organization.

◆ 3 ◆

You are the product manager for an exclusive line of cosmetics and fragrances targeted to teenage girls and sold in shopping mall kiosks. You are considering submitting a proposal to your management, encouraging the company to begin selling its products online. Before you begin to prepare a proposal, you want to know more about the market for online cosmetic and fragrance sales. Using Web sites and other relevant sources, prepare a two-page summary of a cosmetic industry analysis. Briefly describe any trends, industry characteristics, growth factors, barriers to entry, and other relevant factors related to selling cosmetics and fragrances online. Add a third page listing your research sources.

TEAM PROJECT

You and several associates are planning a new e-business based on the expert information exchange e-business model, and you need to draft the e-business's description, business concept, and mission statement. Working together with your assigned team members, define the e-business's concept and description, and then draft its mission statement. Consider the following points when drafting the mission statement text:

1. The purpose of the mission statement is clear.
2. The mission statement reflects the e-business's objectives.
3. The mission statement's language is meaningful and understandable.
4. The target audiences are clearly defined.

As a group, present the e-business concept and mission statement to an audience selected by your instructor. Have the audience evaluate the mission statement's clarity of wording, identify its target audiences, and evaluate its effectiveness in conveying its message to those target audiences.

After evaluation of the mission statement, at the direction of your instructor, prepare a complete business plan for the e-business, including (but not limited to) cover page, title page, executive summary, products and services section, industry analysis, and marketing, operations, financial, and management plans. Limit the business plan to 20 pages or less.

Useful Links

American Success Institute
http://www.success.org/

BASES – Business Association of Stanford Engineering Students – Entrepreneurial Challenge Links
http://bases.stanford.edu/

Biz/ed – General Information on Financial Ratios
http://bized.ac.uk/stafsup/options/ratio/ragi.htm

Center for Business Planning
http://www.businessplans.org/

BusinessTown.com – Creating Business Plans
http://www.businesstown.com/mindspring/
     planning/creating.asp

Canada Business Service Centres – Interactive Business Planner
http://www.cbsc.org/ibp/

CCH Business Owner's Toolkit—Guide to Financial Ratios
http://csi.toolkit.cch.com/text/p02_5651.asp

CCH Business Owner's Toolkit—Guide to Preparing Financial Statements
http://aol.toolkit.cch.com/text/P06_1570.asp

CCIPS – Department of Justice
http://www.cybercrime.gov/

Earthlink Biz Resource Center
http://www.earthlink.net/business/

eBusiness Advisor – Online Magazine
http://www.advisor.com/MEB

Entrepreneurial Edge
http://209.241.14.8/fmpro?-db=homepage.fp5&-
format=indexscream.htm&homepage=yes&-find

eWeb - Business Planning
http://www.slu.edu/eweb/businessplan.htm

Garage.com
http://www.garage.com/

Idea Café
http://www.ideacafe.com/

The Small Business Knowledge Base
http://www.bizmove.com/index.htm

Women's Wire
http://www.womenswire.com/

Links to Web Sites Noted in This Chapter

Amazon.com
http://www.amazon.com/

AMR Corporation
http://www.amrcorp.com/index.html

Autoweb.com
http://www.autoweb.com/

BarnesandNoble.com
http://www.barnesandnoble.com

BizPlanIt.Com
http://www.bizplanit.com/

Bplans.com
http://www.bplans.com/

CDNow.com
http://cdnow.com/

Cisco Systems – Federal Business Development Operations
http://www.cisco.com/warp/public/779/gov/fed_bus_
devlpmnt.html

Cisco Systems – Cisco's Vision and Leadership
http://www.cisco.com/warp/public/779/largeent/why
_cisco/vision.html

Comics.com – Mission Statement Generator
http://www.unitedmedia.com/comics/dilbert/career/
bin/ms2.cgi

eToys
http://www.etoys.com/

foodlocker.com
http://www.foodlocker.com/

Greenlight
http://www.greenlight.com/default.jsp

GroceryWorks.com
http://www.groceryworks2.com/

Johnson Controls
http://www.jci.com/

MySoftware Company
http://www.mysoftware.com/

NetRatings
http://www.netratings.com/

PartsAmerica.com
http://www.partsamerica.com/

SBA – Business Plan
http://www.sbaonline.sba.gov/starting/
indexbusplans.html

SBANC - Small Business Advancement National Center at the University of Central Arkansas
http://www.sbaer.uca.edu/

SCORE – Service Corps of Retired Executives
http://www.score.org/

Toys 'R' Us
http://www.toysrus.com/

WeddingChannel.com
http://www.weddingchannel.com/

For Additional Review

Amor, Daniel. 2000. *The E-Business (R)evolution: Living and Working in an Interconnected World*. Upper Saddle River, NJ: Prentice Hall PTR.

Anders, George. 1999. "Della & James to Receive $45 Million From Investors," *The Wall Street Journal*, September 23. Available online at: http://weddings8.della.com/about_guest/pc_wsj_9-23-99.asp.

Bazdarich, Colleen. 1999. "An E-Commerce Company Is Born: Della & James Goes After the Bridal Registry Market," *CBS MarketWatch*, June 22. Available online at: http://weddings.della.com/about_guest/pc_cbs_6-22-99.asp.

Bounds, Wendy. 1999. "Several Major Retailers Say "I Do" to Wedding-Registry Web Site," *The Wall Street Journal*, June 9. Available online at: http://weddings8.della.com/about_guest/pc_wsj_6-9-99.asp.

Business Briefs. 2000. "Toys 'R' Us, Amazon.com Join Sites," *The Houston Chronicle*, August 11, 2C.

Business Plan Guidebook. 2000. Small Business Advancement National Center, University of Central Arkansas. Available online at: http://www.sbaer.uca.edu/.

"Click! Shop! GroceryWorks.com Delivers Fast, Efficient Online Grocery Shopping with Cisco-Eagle Order Picking System". 2000. Cisco-Eagle Case Studies. Available online at: http://www.cisco-eagle.com/CaseStudies/index.htm.

Cross, Kim. 2000. "The Ultimate Enablers: Business Partners," *Business 2.0*, February 1. Available online at:http://www.business2.com/content/magazine/indepth/2000/02/01/15658.

Deise, Martin V. et al. 2000. *Executive's Guide to E-Business: From Tactics to Strategy*. New York: John Wiley & Sons, Inc.

Golden, Bruce. 2000. "Forming a Board: What to Look for In Your Company's Directors," *Business 2.0*, March. Available online at: http://www.business2.com/content/magazine/indepth/2000/03/01/20718.

Haylock, C. F. and Muscarella, L. 1999. *NET Success: 24 Leaders in Web Commerce Show You How to Put the Internet to Work for Your Business*. Holbrook, MA: Adams Media.

It Starts With a Plan... A Marketing Plan. 1997. Online Women's Business Center. Available online at: http://www.onlinewbc.org/docs/market/index.html.

Krauss, Michael. 2000. "Make, Buy, or Partner – You Decide," *Marketing News*, 34(3), January 31, 8.

Lea, Wendy. 2000. "Dancing With a Partner," *Business 2.0*, March 1. Available online at: http://www.business2.com/content/magazine/indepth/2000/02/01/14428.

Lucas, James R. 1998. "Anatomy of a Vision Statement," *Management Review*, 87(2), February, 22(5).

Mazur, Laura. 2000. "Mission: Impossible," *Across the Board*, 37(4), April, 11.

Millard, Elizabeth. 1999. "Higher Earning: Business Plan Contests at Leading Graduate Schools Create an Embarrassment of Riches," *Business 2.0*, August 1. Available online at: http://www.business2.com/content/magazine/vision/1999/08/01/11414.

"Mission Statements: What are they for and how effective are they?" 2000. Total Research Strategic Marketing Services. Available on request at: http://www.totalres.com/.

Online Women's Business Center. 1997. "Purpose of the Marketing Plan." Available online at: http://www.onlinewbc.org/DOCS/market/mk_plan_why.html.

Paton, Jamie. 2000. "Update: Amazon Partners With Greenlight to Sell Cars," *TheStreet.com*, August 23. Available online at: http://www.thestreet.com/brknews/internet/1052049.html.

Ross, L. Manning. 2000. *businessplan.com: How to Write an eCommerce Business Plan*. Second Edition. Central Point, OR: The Oasis Press/PSI Research.

Stevenson, Howard W. et al. 1999. *New Business Ventures and the Entrepreneur*. New York, NY: Irwin/Mcgraw-Hill.

Turban, Efraim et al. 2000. *Electronic Commerce: A Managerial Perspective*. Upper Saddle River, NJ: Prentice-Hall, Inc.

Virtual Business Plan. 2000. BizPlanIt.com. Available online at: http://www.bizplanit.com/vplan.htm.

White, Jennifer. 2000. "Very Few Firms Ever Follow Those Mission Statements Hanging on the Wall," *The Business Journal*. 18(4), May 19, 29.

Williams, Edward E. et al. 1999. *Business Planning: 25 Keys to a Sound Business Plan*. The New York Times Pocket MBA Series. New York, NY: Lebhar-Friedman Books.

Willins, Michael. 2000. "E-commerce Union Created by Advance Auto Parts, CSK Auto," *Aftermarket Business*, February, 10.

CHAPTER **5**

Getting Your E-Business off the Ground

In this chapter, you will learn to:

Describe e-business startup financing issues

Discuss the role of angel investors in an e-business startup

Identify issues important to venture capital investors

Discuss the advantages and disadvantages of business incubators

Pitch your e-business idea to investors

In late August 1999 Stephen Knight, a licensed physician and MBA graduate from the Yale School of Organization and Management, knew he had a good e-business idea. Knight's idea was to make available to patients and pharmaceutical companies a Web-based database of clinical drug trials. Pharmaceutical companies run clinical trials on drugs being developed, but often the details of those trials are kept secret for competitive reasons. Many patients want to participate in the trials but do not know how to find them. Knight reasoned that if his company provided a comprehensive Web-based database of trial sites along with crucial medical information, both patients and pharmaceutical companies could win: patients could locate and enroll in the trials easily, and the pharmaceutical companies could fill their trials faster, get their drugs to market sooner, and save millions of dollars in the process.

Knight, who had just become president and chief operating officer of a medical supply firm, needed someone else to get the e-business off the ground, so he called on a friend from Yale Medical School, Robert Adelman. Adelman, a board-certified physician with a consulting and entrepreneurial background, agreed to become chief operating officer for the e-business venture, now named Veritas Medicine.

Knight was unsuccessfully searching for venture capital funding when he met Andrew Olmsted, head of development for Cambridge Incubator (CI). Olmsted suggested that Knight pitch the Veritas Medicine e-business idea to CI's chief executive officer, Timothy Rowe. As a last-ditch effort to secure financing, Knight agreed. By early September 1999 Knight had signed a deal with CI that changed Veritas Medicine from an entrepreneur's dream into an operating company with $834,000 in seed money, office space and furniture, phones, a T1 line, and a computer network. As part of the deal, CI would also provide Veritas Medicine with access to the professional services necessary for an e-business startup: Web developers, lawyers, public relations and marketing specialists, human resources specialists, and introductions to funding sources. Joining the Internet incubator jump-started Veritas Medicine and put it weeks or months ahead in the startup process. But was incubating Veritas Medicine worth it?

Startup Financing

As an entrepreneur starting a new e-business, you must be prepared to invest time, effort, and your own money to get your new e-business off the ground. When you work hard to build your e-business, you are creating value in the business called **sweat equity**. The cost of starting an e-business can include building a Web prototype, doing market research, paying rent, leasing office equipment, and other expenses. In addition to sweat equity, you must be prepared to invest cash in your e-business startup by drawing on personal savings, mortgaging personal assets, or borrowing from a bank with a personal loan. Some entrepreneurs even use cash advances on credit cards. Additionally, many entrepreneurs turn to their friends and family to help provide startup funds.

Friends and family

Friends and family investors are family members or friends who invest in a business. Many entrepreneurs successfully solicit startup money from their network of friends and family. A network of potential friends and family investors extends beyond immediate family members and friends to *their* families and friends and *their* families and friends. A solid round of friends and family investment can help cover a startup e-business's legal fees, Web site prototype development, and other out-of-pocket expenses until additional funding is secured. Also, friends and family who know and trust you are likely to stand by you during tough times in the startup process because they are largely investing in you rather than in your e-business idea. An example of successfully tapping into a friends and family network of potential investors to find startup financing is that of Lynn McPhee and Xuny, Inc. (pronounced "zoonie").

TIP

While some refer to friends and family investors as the three F's: friends, family, and fools—all friends and family investors would like to be like Mike and Jackie Bezos. A few years ago the Bezos gave son Jeff $300,000 for his e-business startup. Now their participation in Amazon.com could be worth millions.

With a B.A. from Stanford and experience in both investment banking and e-business, Lynn McPhee co-founded Xuny with two other entrepreneurs in July 1999. When looking for startup financing, McPhee let her parents, other family members, and friends know she was looking for investors. McPhee's friends and family network responded by providing more than $250,000 in startup financing for Xuny. Then, in November 1999, Xuny secured additional funding from angel investors in the retail and technology fields.

Xuny began as a B2C e-business selling cutting-edge urban, skate, and hip-hop streetwear fashions to an 18–30 year old target market from its Web site. Xuny's demographic target market is men and women born since 1961, often referred to as Generation X and Generation Y. Xuny continues to evolve with changes in the e-business economy. In addition to its original B2C e-business model, Xuny now is also following a B2B e-business model by developing turnkey Web sites for streetwear fashion designers and manufacturers. Additionally, in July 2000 Xuny announced the development of a network of Gen X/Gen Y content Web sites where Xuny (Figure 5-1) provides network participants with Web site content and merchandising technologies.

E-CASE   Cautionary Tale #1: Who You Know Still Counts

Colin Osburn graduated with a degree in entrepreneurship from Carnegie Mellon University and had experience as a Web content developer. His friend Staci Hester was formerly in corporate sales with Dell Computer Corp. Together the two friends, living in Austin, Texas, had what they thought was a great e-business idea: promoting unknown and emerging artists by selling their artwork over the Internet. Osburn and Hester recruited a third partner and then tapped family members for about $40,000 in startup funds. Osburn continued to work at another e-business while taking loans for an additional $25,000. Hester stayed home and worked on the details of starting the e-business. They spent much of the next six months and their startup funds working on their own to write and rewrite their business plan, rework their Web site, and conduct market research, all without any help from experienced e-business advisors. Osburn and Hester didn't know anyone who could give them good advice about starting their e-business, and, unfortunately, they didn't try to meet anyone.

(Continued on the next page)

Early in April 2000 Osburn and Hester traveled to Dallas, Texas, to once again pitch their e-business idea, this time to an economist with expertise in art. The economist's response was "wonderful idea, wrong time." He didn't think the art world was ready for the Internet. He also cautioned Osburn and Hester to wait because he thought that the e-business economy was ready for a fall. A few days after their trip to Dallas, Osburn pitched their e-business idea to Seedstage.com, an Austin-based Internet accelerator, who thought the e-business idea was good but agreed with the economist that the timing was wrong. In April and May 2000, e-business stocks suffered a reversal, and Osburn and Hester gave up on their e-business idea.

In hindsight, Osburn and Hester realized that not identifying experienced advisors in the beginning was the biggest mistake they made. They now understand that early in the development of their e-business idea they should have gotten to know some of the most influential people in the Austin, Texas, entrepreneurial community: the people with experience and money, or the people with connections to others with experience and money. Then they might have received the advice they needed about the viability of their e-business idea before getting money from family members or going into debt. Osburn admits that next time he starts an e-business he will make those important connections up front.

An advantage of soliciting financing from your friends and family network is that it may be the easiest money you will ever find. A disadvantage is the potential loss to people personally close to you. However, getting funds from friends and family members is a great way to acquire enough money to get started while you develop your business plan and e-business prototype. After tapping into your friends and family network, your next step will likely be getting touched by an angel.

Figure 5-1
Xuny

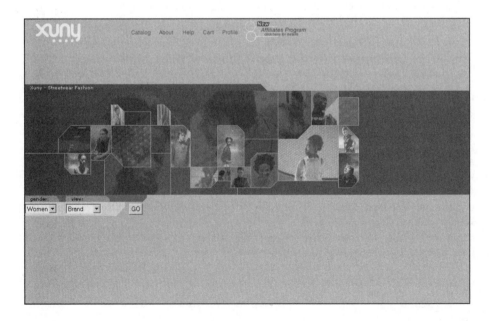

Angel investors

The term *angel investor* originally referred to wealthy investors in Broadway theatrical productions. Now the term **angel investor** commonly refers to an individual with money and time who enjoys the excitement of investing in the early stages of a new business and who is not averse to taking risks. Unlike friends and family investors, who typically invest in the entrepreneur, angel investors are primarily interested in the e-business idea. Angels provide small to medium investments of $10,000 up to around $300,000 as early-stage funding for startup businesses. Some less experienced angels may make a series of small investments in the $10,000–$25,000 range in a number of startups. Angels who invest $50,000–$250,000 in a single startup may also want to become involved in the business. Angels who invest $100,000 or more are usually experienced investors who are likely to invest only in a business in a field with which they are familiar. Angel investors generally take less of an equity position in a new e-business and are less involved in its management than some other funding sources. In fact, angels often assume an advisory role by sharing their business wisdom, experience, and contacts with a new e-business.

To find angel investors, first check with family, friends, professional acquaintances, and business advisors such as attorneys and accountants. Ask these connections to make a third-party introduction to angels they know. Another place to look for angel investors is on the Web. There are a number of e-businesses that help put potential investors together with entrepreneurs, such as Capitalyst and Garage.com (Figure 5-2).

Figure 5-2
Garage.com

Although most angels operate alone, many individual angels are now participating in investment clubs (also known as syndicates, associations, forums, or networks) where they can combine their investments with other angels and spread their investment risk. An investor who wishes to join an **angel investment club** must qualify as an **accredited investor** under federal securities law, which means he or she must have a minimum net worth of $1 million, or an individual income of at least $200,000 per year, or a household income of $300,000 per year. Additionally, each angel investment club has its own membership criteria, including a financial commitment to the club. Some angel investment clubs such as Silicon Valley's Band of Angels are loosely organized networks of investors who hold meetings where entrepreneurs pitch their business ideas. Then the individual members of the network do their own research, legal work, and negotiations, and make their own investment decisions.

Band of Angels, one of the first angel investment clubs, was founded in 1995 by Hans Severiens, a former venture capitalist, and his friend Jack Carsten, a former Intel vice president. Both realized the advantages of a network of angel investors: a pool of investors that could use its differing expertise to help individuals quickly decide if a new business was a good investment, and that could provide the capital a new business needed and spread the risk of that investment over a number of investors. As of June 2000 Band of Angels, which consists of approximately 140 high-tech executives, had invested $60.5 million in 109 companies. One Band of Angels high-profile investment was a $4 million investment in Sendmail, Inc.

Sendmail, Inc.'s software runs more than 75 percent of all e-mail servers on the Internet. Sendmail had its origins in software written in 1979 by Eric Allman, a graduate student at the University of California at Berkeley. The open-source software, which was free, made it possible to forward e-mail from one network to another. The Sendmail software became so popular that Allman began spending much of his time providing support to Sendmail users, including corporate users such as AOL and Earthlink. After years of giving free support to Sendmail users, Allman decided that there was a business opportunity in selling packaged warranted Sendmail software with full service and support. He teamed up with Greg Olsen, a Cornell graduate, in March 1998 to form Sendmail, Inc. In July 1998 Allman and Olsen raised $6 million from angel investors: $4 million from Band of Angels and another $2 million from individual angels. While the core Sendmail software is still free, Sendmail, Inc. sells customized versions of the software and provides software support to customers such as Coca-Cola Co., Pfizer Inc., Lucent Technologies, and Charles Schwab Corp. In 1999 Sendmail, Inc. sold three million copies of the Sendmail software, including 600,000 copies to commercial firms. By the end of 2000, Sendmail, Inc. expects to sell 900,000 commercial copies of the software. In April 2000 Sendmail, Inc. (Figure 5-3) acquired an additional $35 million in funding and had grown to more than 160 employees.

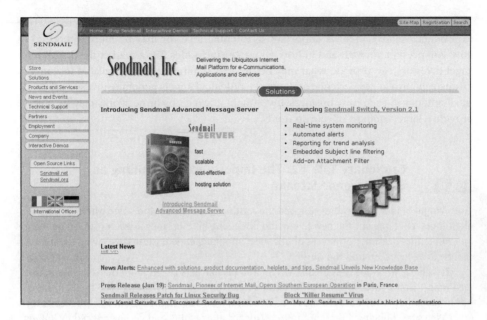

Figure 5-3
Sendmail, Inc.

Other angel investment groups such as WomenAngels.net, founded in January 2000, are more formally organized and invest in a business through the group's treasury. WomenAngels.net members range in age from 25 to 70 and have backgrounds in finance, entrepreneurship, or other professions. By April 2000 the WomenAngels.net (Figure 5-4) investment club had 85 members, each of whom contributed to a $6+ million investment pool for startup companies. It also had a waiting list of 50 potential members.

Figure 5-4
WomenAngels.
net

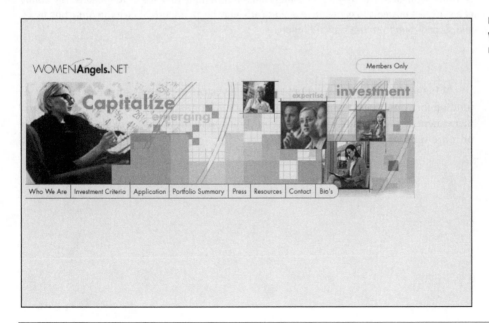

When presenting your e-business idea to an angel investor, remember that while experienced management is important, a potential investor may be more interested in how you present yourself and in how your e-business idea fills a market need. Although angel investors can provide what may be the first significant funds available to you and can usually provide experience and advice, individual angels generally invest a relatively small amount in each new business. Investment angel clubs or networks may be able to provide a larger amount of startup funds.

Cautionary Tale #2: The Importance of Planning an
E-CASE Early Revenue Stream

Jason Wright, a former stockbroker, loved the Austin, Texas, music club scene where he was often found checking out the new talent and discussing his e-business idea: a plan to use the Internet to broadcast performances of unknown and independent musicians, giving them a chance to be heard around the world.

To achieve his e-business dream, Wright and his partners raised more than $1 million in seed money and signed up 1,100 musicians and 22 club venues. By Spring 2000 the e-business was broadcasting 2,400 hours of music each month, repackaging the best of the broadcasts for other Web sites, building a presence in New Orleans and Seattle, and getting noticed by Rolling Stone magazine. Unfortunately, when the bottom fell out of tech stocks in April and May 2000, the venture capital the e-business needed to keep going dried up. Weeks later, when the seed money was gone, Wright and his partners had to close down their e-business.

While Wright's e-business may well have been a victim of market timing, he also believes that a big downfall for his e-business was its failure to generate sufficient revenues up front. He and his partners believed they had enough time to build their business revenue stream; their business model didn't start generating meaningful revenues until the e-business expanded into about eight different markets. The downturn in venture capitalists' interest in funding new e-businesses meant that Wright's e-business did not have time to reach eight markets. The ability of a startup e-business to generate revenues early may be an important factor in attracting first- and second-round venture capital funding.

Whether individually or as a group, angel investors are a great source of seed money funding for your e-business. After securing your startup seed money from your personal investment, family, friends, and angels, your first round of funding will likely be with one or more venture capital investors.

Venture capital investors

Venture capital (VC) firms such as Kleiner Perkins Caufield & Byers, Murphree Venture Partners, and Sigma Partners can raise hundreds of millions of dollars in funding for new businesses from sources such as endowments, insurance companies, and pension funds. VCs may invest $1 to $5 million in the early stages of a business and may invest several million over the life of a new business. To get VC funding, an e-business can expect to give up a large equity stake, perhaps 20 to 40 percent of the e-business. Additionally, VCs expect to have one or more seats on the e-business's board of directors and to be actively involved in the e-business. In addition to an exciting e-business idea backed by sound market research, VCs especially look for experienced people on a new e-business's management team.

VCs can provide a new e-business with more than just money. VCs generally have extensive knowledge about a specific industry and access to important contacts. Often VCs invest in a portfolio of complementary businesses, each of which adds value to the overall portfolio. VCs often limit their investments to businesses in a specific geographic area and/or industry. For example, Flatiron Partners (Figure 5-5), located in New York City, focuses its investments exclusively on e-businesses. Flatiron also focuses on first-round investments for e-businesses in the metropolitan New York City area that have a management team in place and have revenues. Additionally, professional business organizations and VCs in a particular geographic area sometimes host venture capital forums to bring together investors and entrepreneurs.

TIP

A study by the National Venture Capital Association (NVCA) and research partner Venture Economics indicates that VC investments totaled $24.5 billion in the second quarter of 2000, up 96 percent over the same quarter in 1999. According to the study, 81 percent of the second quarter of 2000 VC dollars went to Internet-related investments. According to a similar study by PricewaterhouseCoopers, the trend of VC Internet-related investments was moving away from B2C and even B2B e-business and toward telecommunications infrastructure and enabling software areas.

Figure 5-5
Flatiron
Partners

One example of a new e-business that successfully secured first-round funding from VC firms is the Atlanta, GA -based Global Food Exchange. When Mark C. Moore and Christopher Swann, two consultants at McKinsey & Co., compared notes they realized that they both were thinking about starting a B2B e-business. They also recognized in each other the entrepreneurial spirit necessary to start a new e-business. After brainstorming several ideas, they decided to create an online marketplace for the food trade. In May 1999 Moore and Swan left McKinsey & Co. and formed Global Food Exchange to connect buyers and sellers of large amounts of seafood, produce, meat, and poultry. Global Food Exchange's seed money came from Moore's and Swan's friends and family network.

By January 2000 Global Food Exchange had 42 employees, and the e-business's poultry and seafood sites were up and running. Additionally, Global Food Exchange had secured $810,000 in private funding seed money. In February 2000 Global Food Exchange secured $12.2 million first-round financing—in exchange for "a sizable chunk of equity" and two seats on the board—from three VCs with different but complementary approaches to investing. The three VCs are Charles River Ventures, which focuses on investing in B2B online marketplaces, Sigma Partners, which invests in early-stage technology companies, and Bain Capital, Inc., which focuses on businesses with exceptional management teams.

TIP

For a funny take on the VC world, check out The VC Web site.

In January 2000 Global Food Exchange had its most exotic transaction up to that point: arranging the sale of 31,000 pounds of perch between Uganda and Colombia. By April 2000 Global Food Exchange's international business had expanded, and it opened its first international office in Guadalajara, Mexico, with plans to open offices in Brazil, Chile, and Argentina by the end of 2000. Global Food Exchange also had 600+ registered customers such as ConAgra Poultry, Kroger Co., Food Lion, Sysco Co., Campbell's Soup, and Oscar Meyer, and projects more than $5 million in revenues in 2000. Based on strategy, execution, finances, and Web site design, Forbes named Global Food Exchange (Figure 5-6) one of the ten best B2B Web sites in the "Food and Beverage" category in its first annual "Best of the Web" survey in July 2000.

E-CASE IN PROGRESS foodlocker.com

foodlocker.com's principals worked many long hours and invested substantial sweat equity during the startup process. Additionally, they invested cash in the startup and tapped into their network of family and friends to help raise the seed money they needed. Another step in securing financing was to align foodlocker.com with a major regional food store chain, which became their most significant strategic partner. Then they had a private placement document prepared and began soliciting angel investors to complete their initial funding.

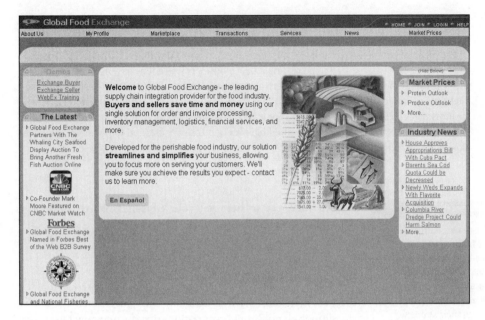

Figure 5-6
GlobalFood
Exchange.com

Venture capital firms can be great for supplying first- and second-round financing; however, expect to give up a substantial equity in your e-business, seats on your board of directors, and some management control. But if you can negotiate a good deal with one or more VCs, your e-business could be on its way to becoming profitable.

E-CASE Cautionary Tale #3: What Do VCs Really Want?

Ron Losefsky, who studied how to integrate and automate supply chains and other e-business topics at The University of Texas at Austin business school, felt that he understood how to use the Internet to revolutionize old-line industries. One day while waiting for his car to be repaired, he began thinking about ways to apply his e-business knowledge to the auto repair business. Losefsky came up with an e-business idea based on offering customers instant information about their cars, what was wrong with them, and where they could get them repaired. Then he wrote a business plan, built a demo, and hired a developer to create the Web site.

After several auto repair shops signed on, Losefsky pitched his e-business idea to an angel investor who offered $150,000. Losefsky combined the angel's investment with another $50,000 in seed money and $100,000 in personal credit card debt to get started. Soon, the e-business employed eight people and listed about 40 Austin, Texas, repair shops. The Web site was getting about 10,000 hits a month with about 700 people using the site on a regular basis. Losefsky thought the e-business was on its way to success.

Unfortunately, as quickly as it started, the e-business was effectively finished. The problem was the inability to attract additional funding from venture capitalists because of the lack of an experienced executive on the management team. Although Losefsky's board of advisors included people from major auto-related companies, he found it difficult to recruit a seasoned executive

(Continued on the next page)

for the e-business's management team because he didn't have VC funding at this early stage. Conversely, VCs were not interested in funding the new e-business because there were no veterans on the management team. Losefsky's e-business failed the same day that NASDAQ fell 355 points. In retrospect, Losefsky realized that creating a new e-business was not as easy as it first looked, and that even if an entrepreneur does everything he or she thinks is right, the e-business idea still might not be successful. Losefsky admitted that there is an intangible quality to creating a new business that can't be learned in school. For example, understanding that VCs look for experience on an e-business's management team could have alerted Losefsky early on to potential problems looking for VC funding.

TIP

One of the strategies for entrepreneurs starting a new e-business and for VCs investing in an e-business is to take that business public. Going public is not a simple matter; therefore e-businesses that are serious about going public should consult with their attorneys about the ramifications of a public stock offering, including how to file the appropriate documents. For a brief overview of the IPO process, check out the CCH Business Owners Toolkit – Initial Public Offering Web page.

Another way to help get your e-business off the ground is with a business incubator.

Business Incubators

Business incubators have traditionally been government- or university-supported nonprofit organizations that nurture new businesses. An example of a government-sponsored business incubator is the NASA Ames Technology Commercialization Center (ATCC). The ATCC helps businesses identify NASA technology with commercial potential, then provides access to a network of business and professional experts in law, finance, marketing, sales, management, and operations. The ATCC (Figure 5-7) also offers low-cost office space and other startup services.

University-sponsored business incubators include the Advanced Technology Development Center at Georgia Institute of Technology, the Austin Technology Incubator at the University of Texas, and the business incubator at Northwestern University Evanston Research Park. These and other nonprofit business incubators, such as the Houston Technology Center, can provide startup companies with management advice, office space, networking opportunities, and other critical startup services. For example, when Christopher Swan, co-founder of Global Food Exchange, decided it was impossible to run an international food exchange from his 600-square-foot apartment, he and his partner moved the e-business into 1,000 square feet at the Advanced Technology Development Center at a lease rate much below that charged by commercial landlords.

An interesting private nonprofit business incubator is the Women's Technology Cluster (WTC). Founded in February 1999 by Catherine Muther, a former vice president of corporate marketing for Cisco Systems, the WTC is based on two goals: to get more women-backed startups funded and to generate support for entrepreneurs to invest in their community. The WTC, funded with $2 million of Muther's own money and $750,000 from a variety of sources, including $250,000 from the City of San Francisco, only accepts companies with marketable ideas in the information technology area and an experienced management team. Each incubated company is required to place 2 percent of its equity into a fund that will provide money to be reinvested in the community.

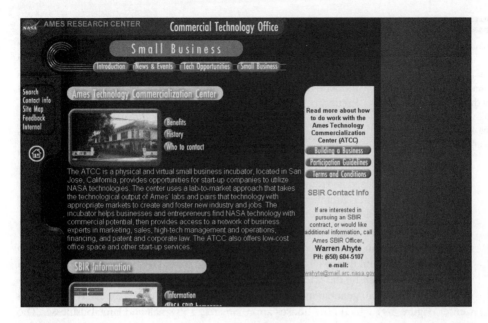

Figure 5-7

NASA Ames Technology Commercialization Center

Incubated businesses are encouraged to leave after two years, when they need more than 10 percent of the WTC physical space, or when they secure venture capital funding. WTC (Figure 5-8) incubated e-businesses include LevelEdge.com, a sports recruiting management portal, MsMoney.com, a financial services portal for women, and RosePlace.com, a senior care network.

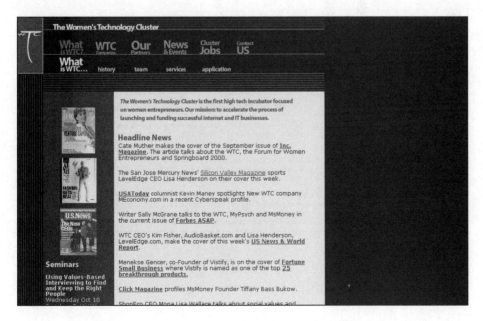

Figure 5-8

Women's Technology Cluster

The early success of many e-businesses encouraged other entrepreneurs to create their own e-business. Some entrepreneurs, recognizing that other entrepreneurs with a good e-business idea might need help getting started, are following a new business model—the commercial business incubator. **Commercial business incubators** offer startup e-businesses access to the same services offered by nonprofit incubators. One of the first commercial business incubators on record is the Batavia Industrial Center. In 1957 Masse Harris Ferguson, Ltd. occupied approximately one million square feet of building space and was Batavia, New York's, largest employer when it closed its Batavia factory, leaving thousands without work and an empty manufacturing complex. Members of a local family, Charles Mancuso and his oldest son, Benjamin, had actually worked as laborers in 1907 to help build the manufacturing complex. In 1959 the Mancuso family purchased the empty complex and charged another family member, Joseph, with putting people back to work. When Joseph Mancuso was unable to find an individual employer to fill the entire complex, he started renting space to startup and small businesses. Then he did whatever he could to help the startups and small businesses grow. Today the fourth generation of the Mancuso family still operates the Batavia Industrial Center (Figure 5-9), which now houses approximately 150 small and large businesses. The Batavia Industrial Center's tenants or graduate businesses employ approximately 5,000 people in the Batavia community.

Figure 5-9
Batavia Industrial Center

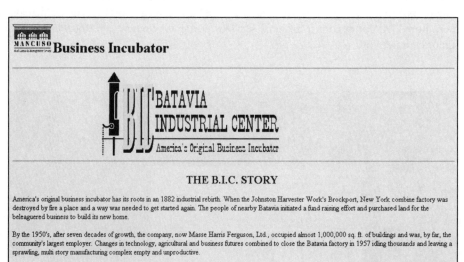

MANCUSO *Real Estate & Management Group* **Business Incubator**

BATAVIA INDUSTRIAL CENTER
America's Original Business Incubator

THE B.I.C. STORY

America's original business incubator has its roots in an 1882 industrial rebirth. When the Johnston Harvester Work's Brockport, New York combine factory was destroyed by fire a place and a way was needed to get started again. The people of nearby Batavia initiated a fund raising effort and purchased land for the beleaguered business to build its new home.

By the 1950's, after seven decades of growth, the company, now Masse Harris Ferguson, Ltd., occupied almost 1,000,000 sq. ft. of buildings and was, by far, the community's largest employer. Changes in technology, agricultural and business futures combined to close the Batavia factory in 1957 idling thousands and leaving a sprawling, multi story manufacturing complex empty and unproductive.

The Patriarch of an enterprising local family, Charles Mancuso and his oldest son, Benjamin, had actually worked as immigrant laborers to help build a portion of the complex in 1907. Now, in 1959, their family pooled its resources, purchased the vacant complex and placed Joseph L. Mancuso in charge of "putting people back to work." Unable to fill the huge building with a single employer he rented to start up and small enterprises and did whatever was necessary to help them grow. Sharing employees, equipment and expertise, the Batavia Industrial Center saw its tenants fortunes rise and local employment increase with their success.

Today, commercial business incubators are primarily interested in high-technology businesses that can become financially viable quickly and leave the incubator within six months to a year. A commercial business incubator's main goal is to accelerate the growth of the businesses in their portfolio of companies. In exchange for their services, a for-profit incubator may want 20 to 60 percent equity in a new business. Some commercial business incubators such as eCompanies, iHatch, and idealab! only incubate companies created from business ideas generated internally.

idealab!, one of the first high-tech commercial e-business incubators, was founded in March 1996 by Bill Gross and Marcia Goodstein. Gross, a graduate of the California Institute of Technology, was an entrepreneur with several business startups behind him when he co-founded idealab! with about $3 million from investors such as filmmaker Steven Spielberg, Compaq chairman Ben Rosen, and others. By September 2000 idealab! was incubating approximately 50 companies in various stages of development, by providing office space, network and technology infrastructure, graphic design, and marketing research. idealab! also provides legal, accounting, and business development services. Public companies incubated by idealab! include GoTo.com, eToys, Citysearch, NetZero, and Tickets.com. Because of its IPO (initial public offering) success with several incubated companies, idealab! (Figure 5-10) is often credited with causing the sudden growth of commercial business incubators in 1999 and early 2000, when more than 300 new commercial business incubators focusing on e-business were launched.

Figure 5-10
idealab!

Other commercial business incubators such as Cambridge Incubator, Redstone7.com, ThinkTank, Stonepath Group, and eHatchery accept outside applications to join the incubator. Jeff Levy, who previously co-founded a Web-based audience-tracking service that merged with a competitor in 1998, founded eHatchery in 1999. eHatchery provides new e-businesses with customized turnkey support and equity investment capital. eHatchery is backed by investors that include idealab!, Donaldson, Lufkin & Jenrette, Cox Enterprises, and United Parcel Service. As of September 2000, eHatchery (Figure 5-11) had three client e-businesses: eTour, a personal Web tour guide; VetExchange, a B2B exchange for veterinarians and their suppliers; and Tr@deUps, a Web site that allows consumers and businesses to "trade up" used merchandise for new products.

Internet accelerators and keiretsu providers

Some e-business incubators such as iStart Ventures and Katalyst (Figure 5-12) style themselves as Internet accelerators. An **Internet accelerator** is a commercial business incubator whose goal is to get a new e-business up and running quickly. Internet accelerators generally provide the same range of services as other commercial technology business incubators for an equity position in the e-business.

Keiretsu is a Japanese term that refers to a network of businesses that do business with each other as a means of mutual security. Commercial business incubators that use the **keiretsu model** offer startup e-businesses entry into a network of companies that do business with one another with the goal of serving the overall interest of the network. CMGI is an example of a commercial business incubator that follows the keiretsu model. CMGI began as the College Marketing Group (CMG), a direct marketing firm founded in 1968. By 1998 the company added the "I" to its name to become CMGI, and

Figure 5-11
eHatchery

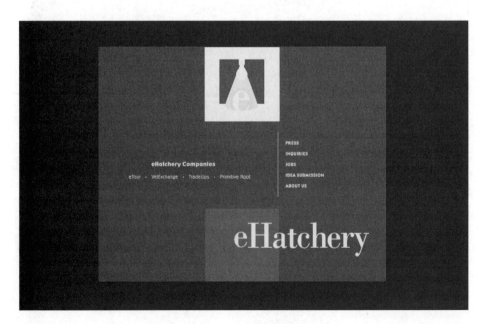

Figure 5-12
Katalyst

its focus was on creating a network of Internet-related businesses that could share technologies, talent, and experience. CMGI (Figure 5-13) owns and invests in a diverse network of B2B and B2C e-businesses such as AltaVista, a search tool; FindLaw, a law and government portal; yesmail.com, a permission e-mail marketing company; and MotherNature.com, an e-tailer of natural and healthy living products.

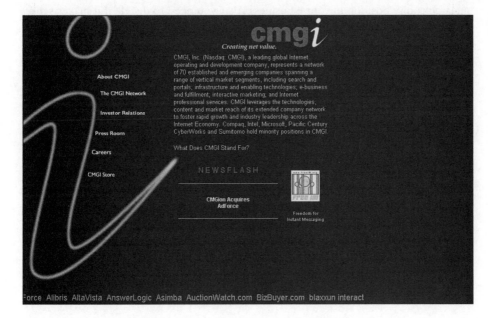

Figure 5-13
CMGI

Critics of commercial business incubators point out that commercial business incubators, as of this writing, do not yet have a successful track record for incubating profitable e-businesses. Even some of the public companies launched by idealab! have struggled since the fall of technology stocks in the spring of 2000. Some commercial business incubators have revised the original commercial business incubator model and are now charging their portfolio companies for rent, network access, and other professional services, in addition to their equity stake in the startup e-business, as a way to make the commercial business incubator model profitable.

Proponents of commercial business incubators counter that commercial business incubators are the least risky way to start an e-business if an entrepreneur requires some guidance and is willing to give up a large measure of both independence and equity for that guidance.

Before incubating your e-business startup, you should consider the following questions:

♦ Does the business incubator offer seed money or venture capital funds linked to the incubator? It may well be worth joining the incubator to get access to seed money or first-round venture capital financing.

♦ What specifically will the business incubator do to help your e-business? The startup services provided by business incubators vary. You should know up front which services are provided by the business incubator and which services you must provide yourself.

♦ What is the business incubator's track record with other e-business startups? Many commercial business incubators are new and have not yet graduated a profitable e-business. You should be comfortable about the level of experience that the incubator's principals and advisors bring to your e-business. You should also review the progress of the other e-businesses in the business incubator's portfolio.

♦ How much will it cost your e-business—in cash and equity—to be incubated? You need to get sufficient value from the incubator's services to warrant giving up a large equity position in your e-business. Also, you need to decide if it makes sense for your e-business to pay for the incubator's services, if that is part of the deal, as well as give up a large equity position.

♦ How long is the incubation period? Some business incubators allow just a few months for the incubation period. If that is the case with the incubator you are considering, you should decide if it is worth giving up a large equity position in your e-business to have it incubated for a short period of time.

♦ How do you feel about the business incubator's environment? Certainly you should visit the incubator's facilities and then trust your instincts about the physical environment and potential culture clashes. You should feel comfortable working in the environment and working with the incubator's principals and advisors.

After a careful review of the advantages and disadvantages of incubation for your e-business, you can decide if joining a business incubator is the right thing for your startup e-business.

Self-incubation

Some e-business startups like the idea of sharing office space with other entrepreneurs, exchanging ideas with entrepreneurs going through the startup process, and taking advantage of a mutual network of advisors. However, they don't want to give up a huge equity position to a commercial business incubator. One way some e-business startups enjoy the advantages of sharing resources without joining a commercial business incubator is to become part of an informal group of **self-incubators**.

For example, Lynn McPhee of Xuny looked at several commercial business incubators but decided instead to join with two other entrepreneurs to share resources: Kayla Bakshi of VenusSports, a Web site that provides content for female athletes, and Dave Park of Coolboard, which provides customized message boards for Webmasters.

The three e-business startups shared office space, office equipment, customer-service data, and information on potential business partners. Employees of the three startups held brainstorming sessions whenever they needed to exchange ideas. The self-incubation arrangement was useful for Xuny, VenusSports, and CoolBoard for several months as the e-businesses grew. But by May 2000 both VenusSports and Xuny had plans to move, allowing CoolBoard to take over the entire office space. Self-incubation may make sense during the early months of an e-business startup but will likely be outgrown by successful e-businesses within a year.

Whether you are soliciting funding from your friends and family network, angel investors, or venture capital firms, you need to be ready to pitch your e-business idea.

Pitching Your E-Business Idea

The first meeting with angel investors or VCs is a sales meeting. Your immediate objective in a first meeting is to get potential investors excited about your e-business idea and interested in pursuing more extensive discussions that might lead to financing. The first meeting is likely to be brief, about one hour in length. Instead of presenting your business plan, which you should be ready to provide upon request, you should present a brief **pitch document**, which is a short marketing document. A pitch document should be based on the Executive Summary portion of your business plan and be from one to three pages in length. It should briefly highlight the market need, how your e-business meets that need, potential profits, and how your e-business's management team can make it all happen.

Your verbal presentation should be based on your pitch document, and you can present it visually in an easy-to-understand slide show presentation, using Microsoft PowerPoint or a similar tool. Your presentation should clearly and briefly:

- Define your product or service
- Define who will buy your product or service and how much they will pay for it

- Define your key industry competitors
- Explain how much it will cost to provide the product or service
- Explain when the investors can expect your e-business to be profitable
- Illustrate the planned exit strategies both for investors and for your e-business principals
- Detail how much money you are looking for, and how it will be spent

Potential investors will ask questions during the presentation, to try to determine how well you understand your e-business, your target market, your competitors, and critical marketplace issues. If you do not have an answer to a question, don't try to pretend that you do. Simply acknowledge that you need to do further research on the issue and move on. Also be prepared to point out any potential risks or problems and how your e-business is positioned to handle those risks or problems.

During the presentation differentiate yourself and your management team from the competition by describing how your team's background and experiences position your e-business for success. Show a real commitment to your e-business, to create the feeling that your e-business idea is a viable, exciting opportunity for your investors. Learn as much as you can about your potential investors before meeting with them. Then try to establish a good fit between your e-business idea and their existing portfolio of businesses.

E-CASE Zipping Along and Getting It Right

It is critical to be prepared for each meeting with potential investors by having your handout materials organized and reviewed, by practicing your presentation, and by trying to anticipate all possible questions so that you are prepared to give decisive answers. Just ask Scott Kucirek of zipRealty.com about the importance of being prepared. zipRealty.com is an online real estate broker that does marketing, documentation, and matching of homebuyers and sellers. The e-business was founded in January 1999, although Kucirek and his partner Juan Mini had been working toward starting an e-business together all the way through business school at the University of California at Berkeley. By the time they graduated in May 1999, they had secured $1.7 million from Vanguard Venture Partners and a group of angels led by Barrington Partners. By August 1999 zipRealty.com was up and running. In December 1999 zipRealty.com closed a second round of financing in the amount of $16 million led by Benchmark Capital and participated in by the two original investors.

But Kucirek and Mini learned a few hard lessons along the way to getting that financing. Their first lesson was about being careful and being brief. The original business plan for zipRealty contained 40 pages of text with 20 pages of financial statements and projections. After scheduling presentations to investors they began revising pages in their handout material, and in their haste they mixed up the revised pages. The problem was discovered on the way to a presentation, and they had to stop, fix the errors, then recopy and rebind the handout material. Then during the presentation they noted inconsistencies in the name of their proposed e-business on several of their charts. Needless to say, no one expressed interest in investing in their e-business after that presentation. The good news is that Kucirek and Mini did get their business plan pared down to 20 pages and learned to carefully review their material before each presentation.

Another lesson Kucirek and Mini learned was the need to practice their presentation and try to anticipate potential investors' questions. The day before one investor meeting, Kucirek and Mini were suddenly asked to bring along two other members of their management team who had real estate experience, to discuss why buyers and sellers would use zipRealty's services. With only one day's notice, the zipRealty.com management team didn't have time to practice answers to anticipated questions. When the potential investors asked questions that the management team members were unable to answer, the presentation bombed. Kucirek and Mini and the others on the management team never made another unrehearsed presentation to potential investors. Luckily, the zipRealty.com (Figure 5-14) team learned from their mistakes, got it together, and began to receive serious interest from investors, leading to their first and second round of funding. It doesn't always work out so well.

Figure 5-14
zipRealty.com

If an investor is interested in your e-business, he or she may send you a term sheet. A **term sheet** is a list of the major points of the proposed financing being offered by the investor, and is used to start negotiations for the investment deal. The term sheet will include a valuation of your e-business. For example, if the investor proposes to invest $200,000 in your e-business for 10 percent equity, then the investor values your e-business at $2 million. Term sheets may also include demands for a certain class of stock, automatic buyouts in case of an acquisition, seats on your e-business's board of directors, and other contingencies. Because term sheets can be very complicated, you should have them reviewed by an attorney experienced in negotiations with investors before agreeing to any investment deal.

The deal between Veritas Medicine and CI was not a great one for Knight, Adelman, or future employees looking for equity in the new e-business. In exchange for $834,000 in seed money and access to facilities and professional services, CI got 51.22 percent ownership of Veritas Medicine. On the evening of February 9, 2000, Robert Adelman, Veritas's chief operating officer was beginning to appreciate the ramifications of the incubating deal over a Chinese take-out dinner with a potential angel investor. The angel investor was shocked to find out that Knight had handed over 51.22 percent of his company to the incubator, noting that, "You guys are on a very clear path to going public owning only your own shorts." Common equity ratios for a startup after its initial funding are approximately 30–40 percent held by investors, 40–50 percent held by the management team, and 20 percent equity set aside for employee incentives. With approximately 40 percent of Veritas's ownership pie held by the management team and 51.22 percent held by the incubator, Adelman realized that only a meager 8–9 percent pool was left for employee incentives, which might make it difficult for Veritas to attract key people.

Equity distribution wasn't the only problem. In addition to their equity position, CI billed Veritas Medicine up to $19,000 a month to cover Veritas's use of the CI infrastructure and professional services. Then Veritas found CI's early recruiting services to be less than helpful, and it ended up locating most of its staff itself through Monster.com. It also used an outside graphics house to design its Web pages. While CI had been useful in helping the Veritas management team understand what venture capitalists want to hear and providing venture capital contacts, as of July 1, 2000, none of the venture capital contacts had translated into financing. In fact, the venture capital contact that looked most promising was one Veritas had made on its own. And then there was the difference in business styles. Adelman acknowledges a clash in business styles between CI and Veritas because he felt that CI was looking for a Japanese style of business, a keiretsu, while his entrepreneurial business style was more "American cowboy."

On March 31, 2000, CI moved into new facilities. After a little more than five months in the incubator, Veritas did not move into the new facilities with CI. Instead, Veritas elected to remain in its old offices, paying rent not to CI but to the company that leased the old office space. Veritas added their own phone system and T1 line; however, Veritas still paid CI a fee to use its network and servers. Adelman admitted that taking this independent step was important, to create a level of autonomy for Veritas.

So, was the incubation of Veritas Medicine worth it? At the time of this writing, there is no clear answer to that question. Many experienced venture capitalists question the value of giving away so much equity to incubators. Knight himself acknowledges that he was naïve in negotiating the arrangement with CI. Although there is no doubt that access to CI's facilities and services put Veritas ahead initially, inexperience, different personalities, and different business styles seem to have made the Veritas Medicine–CI arrangement less than great. On September 11, 2000, Veritas Medicine (Figure 5-15) announced that $8 million in financing had been secured, and it launched its Web site. As of this writing, Veritas is still listed as part of the CI family of e-business startups after more than twelve months in the incubator.

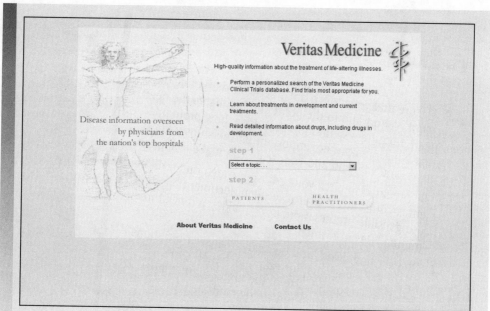

Figure 5-15
Veritas
Medicine

Summary

- Sweat equity is value added to a business by the entrepreneur's hard work and effort.
- In addition to sweat equity, an entrepreneur can expect to invest his or her own funds in the e-business by using savings, mortgaging personal assets, or taking out a personal loan.
- An entrepreneur's network of friends and family is an important source of seed money for a startup e-business.
- An angel investor is a wealthy individual who enjoys investing in startup businesses and is not afraid of taking some risk.
- Venture capital firms can raise hundreds of millions of dollars to fund new businesses and expect equity, a seat on the board of directors, and some management control over the businesses in which they invest.
- Business incubators can be nonprofit private, government, or university-sponsored organizations that nurture startup businesses.

- Commercial business incubators can provide a wide range of services for a startup company in exchange for a large equity stake, sometimes as high as 60 percent or more.
- Self-incubation allows a startup e-business to enjoy the advantages of sharing resources with other startups without giving up any equity.
- The first meeting with potential investors is really a sales meeting in which your objective is to excite interest in your e-business idea.
- A pitch document is a short marketing document that highlights the key facets of your e-business idea.
- Before you meet with potential investors, try to anticipate their questions and practice your presentation.
- Learn from each presentation and use the feedback you get to refine your pitch document and verbal presentation.

Pitching Your E-Business Idea

❑ Be sure that your "pitch document" or marketing document is short and to the point and highlights the need, your e-business solution to that need, how profitable the e-business can be, important partnerships, critical risks, and why you and your team can make it happen.

❑ Be prepared to pitch your e-business idea under different circumstances: with your pitch document handout, using a slide presentation, or by making a quick three-minute "elevator" pitch.

❑ Do your homework and be ready to answer questions about the e-business: for example, benefits to potential customers, size of the target market, how fast the target market is growing, and the basis for the e-business model.

❑ Try to anticipate all potential investor questions, but if you don't know the answer to a question posed by a potential investor, resist the urge to "wing it."

❑ When pitching your e-business idea, be passionate about it and demonstrate a commitment to it by relating any personal investment (time, effort, money) you have made in the e-business.

❑ Be able to describe how the e-business's management team's experience and track record position the e-business for success.

❑ Practice! Practice! Practice! Don't forget that pitching your e-business idea to others helps you refine it.

❑ Be prepared to pitch your e-business idea to potential investors many times. Listen to their feedback and use it to refine your presentation.

Key Terms

accredited investor
angel investment club
angel investor
business incubator
commercial business incubator

friends and family investors
Internet accelerator
keiretsu model
pitch document
private placement memorandum

self-incubators
sweat equity
term sheet
venture capital (VC) firms

Review Questions

1. Sweat equity is usually value added to a startup business by:
   a. Friends and family investors.
   b. Commercial business incubators.
   c. Entrepreneurs.
   d. Angel investors.

2. Internet accelerators are:
   a. Nonprofit business incubators.
   b. Angel investors.
   c. Commercial business incubators.
   d. Part of a friends and family network of potential investors.

3. A private placement memorandum is:
   a. A negotiating document sent to an entrepreneur by a potential investor.
   b. Used to secure a personal bank loan.
   c.. Part of the business plan for a new e-business startup.
   d. An offering document prepared by a qualified attorney that discloses the benefits and risks of an investment.

4. Which of the following is a disadvantage to an entrepreneur when getting startup funding from a member of his or her friends and family network?
   a. It may be the easiest money an entrepreneur can find.
   b. Friends and family members are investing in the entrepreneur rather than in the e-business idea.
   c. The risk of losing Uncle George's retirement funds may upset family relationships.
   d. Each friend and family investor must have a minimum net worth of $1 million.

5. The company generally credited with developing the first e-business commercial business incubator model is:
   a. Cambridge Incubator.
   b. idealab!
   c. The Women's Technology Cluster.
   d. Amazon.com.

6. A pitch document is a list of the major points of a proposed investment deal. **True or False?**

7. An angel investor is typically an individual with money and time who enjoys the excitement of investing in startup businesses and is not averse to taking risks. **True or False?**

8. An accredited investor must have a minimum net worth of $2 million. **True or False?**

9. Venture capital firms can be a good source of first- and second-round funding for a new startup e-business. **True or False?**

10. Commercial business incubators that follow a keiretsu model create a portfolio of businesses that do business with one another to the mutual benefit of the entire portfolio. **True or False?**

Exercises

1. Using Internet search tools or other relevant sources, locate information on an e-business that was launched with startup money solicited from a friends and family network. Write a one- or two-page paper describing the e-business and noting how its principals used the friends and family network of potential investors to fund the e-business.

2. Using Internet search tools and the keywords "term sheet" and "private placement document" or other relevant sources, find and print or copy (with permission) a sample term sheet and a sample private placement document.

3. Using the Garage.com Web site, select three venture capital firms and then write a one- or two-page paper describing each firm and giving examples of the types of e-businesses in which it has invested. Then compare and contrast the investment focus of the three firms.

4. Review the Web sites for the WomenAngels.net and The Washington Dinner Club angel investment clubs (or two similar angel investment clubs). Write a one- or two-page paper describing each investment club, its membership requirements, and its investment focus.

5. Using Internet search tools, business periodicals, or other relevant sources, locate information about an entrepreneur who solicited funding for a new e-business startup from angel investors and/or venture capitalists. Write a one- or two-page paper describing the entrepreneur, the e-business, and the entrepreneur's experiences in seeking funding for his or her startup e-business. Indicate whether or not the entrepreneur was successful and the reasons for his or her success or lack of success.

CASE PROJECTS

◆ 1 ◆

You have just graduated from college and have a great idea for your own e-business. Your aunt introduces you to the president of a new commercial business incubator who wants you to consider bringing your e-business idea to his firm, which can provide office space and equipment, access to a computer network, professional services, and introductions to sources of funding. After reviewing the terms of the incubation deal, you realize you will have to give up 60 percent equity in your e-business if you join the business incubator. Using the e-business model of your choice, write a one- or two-page paper describing your e-business and the advantages and disadvantages of joining this incubator. Make a decision for or against joining the business incubator and state your reasons.

◆ 2 ◆

You and your two best friends are considering starting a new e-business. Using the e-business model of your choice, write a one- or two-page description of your e-business idea. Then create an outline of the steps you and your friends will take to develop equity in the startup, get seed money, and solicit first-round funding.

◆ 3 ◆

You and three associates have developed a new wireless technology that will make it easy for C2C e-businesses to transact business over wireless devices, and the three of you want to start a new e-business to market the technology. When you discuss your e-business idea at a family reunion, several family members offer to lend you money to get started, including your father, who offers to cash out his retirement plan. Consider the advantages and disadvantages of accepting your family members' offers, and then make a decision about whether to accept their offers. Write a one-page explanation of your decision.

TEAM PROJECT

You and three classmates have just developed new software that allows B2B exchanges to process buy and sell transactions much more quickly than with other tools currently being used. You and your classmates want to create a new e-business to sell that process to B2B exchanges, but you need seed money to get started. A professional associate has secured a third-party introduction to a group of angel investors, and you are meeting with them next week to pitch your e-business idea. Now you and your team must prepare for the meeting. Create a one-page pitch document and a five- to ten-slide presentation highlighting your e-business idea, using Microsoft PowerPoint or another tool. Try to anticipate the kinds of questions you may be asked, and be prepared to answer many questions about your new process, the B2B exchange marketplace, and your team's ability to successfully manage the new e-business.

Using your pitch document and your slide show, you and your team must pitch your e-business idea to a group of angel investors consisting of classmates selected by your instructor. Your team will have between 45 minutes and an hour to make your presentation and answer any questions about your e-business idea posed by the group.

Useful Links

ACE-*Net* – The Angel Capital Electronic Network
https://ace-net.sr.unh.edu/pub/

Allstocks.com – Venture Capital Links
http://www.allstocks.com/links/html/
    venture_capital.html

American Entrepreneurs for Economic Growth (AEEG)
http://www.aeeg.org/

Brainspark – European Internet Incubator
http://www.brainspark.com/

Business Incubator Links
http://www.unc.edu/depts/dcrpweb/courses/261/
    grinnell/olinks.htm

BusinessFinance.com – Business Funding Directory
http://www.businessfinance.com/

campsix – Internet Incubator
http://www.campsix.com/

Capital Growth Interactive
http://www.capitalgrowth.com/

Direct Stock Market
http://www.dsm.com/

EarlyBirdCapital.com
http://earlybirdcapital.com/

eCorporation –Internet Incubator
http://www.ecorporation.com/

e-cradle – U.K. Internet Incubator
http://www.e-cradle.co.uk/

EntreWorld.org — Starting Your Business – Finances
http://www.entreworld.org/Channel/
    SYB.cfm?Topic-Finc

ExperTelligence, Inc. – B2B Hub Internet Incubator
http://www.expertelligence.com/

FinanceHub
http://www.financehub.com/

Grassroots Venture Capital – How Venture Capital Investing Works
http://www.grassrootsvc.com/GrassRootsWeb/
   HowVCIWorksFrame.htm

Harvard Business School – Baker Library Industry Guide: Venture Capital & Private Equity
http://www.library.hbs.edu/industry/venture.htm

HOTventures – Internet Incubator
http://hotventures.com/home_body.html

inc.com – Downloadable Sample Term Sheet
http://www.inc.com/freetools/details/
   1,7532,AGD2_CNT61_SUB14_TOL20085,00.html

Investopedia
http://www.investopedia.com/

IPO.com
http://www.ipo.com/

MagPortal.com — Magazine Articles on Small Businesses
http://magportal.com/c/bus/small/

National Business Incubation Association (NBIA)
http://www.nbia.org/

OffRoad Capital
http://offroadcapital.com/home/index.html

Oingo – Capital Access and Financing for Entrepreneurs
http://www.oingo.com/topic/8/8588.html

Policy.com
http://www.policy.com/

Private Equity Network
http://www.nvst.com/pnvHome.asp

Private Equity Network – Business Valuation Forum
http://forums.nvst.com/pfrmHome.asp

PrivateInvestor.com
http://www.privateinvestor.com/welcome/

Rare Medium – Internet Incubator
http://www.raremedium.com/

Redherring.com's A-Z Guide to Incubators
http://www.redherring.com/insider/2000/0119/
   resources/vc-fea-incubator-az.html

Safeguard Scientifics, Inc.
http://www.safeguard.com/index.shtml

Sample Long Form Term Sheet
http://www.orrickemerging.com/areas/venture/
   document/4termsht/2longfrm.htm

SBA Office of Advocacy — ACE-*Net* (Angel Capital Electronic Network)
http://www.sbaonline.sba.gov/ADVO/acenet.html

Seedstage.com
http://www.seedstage.com/

StartupZone, Inc.
http://www.startupzone.com/

Stybel, Peabody, & Associates – Questions to Ask About Angel Financing
http://www.stybelpeabody.com/quesang.htm

The Capital Connection
http://capital-connection.com/

The Dinner Club
http://www.thedinnerclub.com/

The eMedia Club
http://www.emediaclub.com/

The Financial Opportunity Exchange
http://www.vfinance.com/

The Institute of Business Appraisers
http://www.instbusapp.org/

The Washington Dinner Club
http://www.washingtondinnerclub.com/

TheAngelPeople.com
http://www.theangelpeople.com/index.html

ThinkTank – Internet Incubator
http://www.thinktank.com/

vcapital.com
http://www.vcapital.com/Welcome+to+Venture+
   Capital+Online%21.htm

Venture Capital Web Links
http://pacific.commerce.ubc.ca/evc/vc_title.html

VentureLine – Business Valuation Tools
http://www.ventureline.com/default.htm

Links to Web Sites Noted in This Chapter

**Advanced Technology Development Center –
Georgia Institute of Technology**
http://www.atdc.gatech.edu/home.html

AltaVista
http://www.altavista.com

Amazon.com
http://www.amazon.com

Ames Technology Commercialization Center (ATCC)
http://ctoserver.arc.nasa.gov/incubator.html

Austin Technology Incubator
http://www.ic2-ati.org/

Bain Capital
http://www.baincap.com/

Barrington Partners
http://www.barringtonpartners.com/

Batavia Industrial Center
http://www.iinc.com/mancusogroup/INCUBATR.html

Battery Ventures
http://www.battery.com/

Benchmark Capital
http://www.benchmark.com/home.html

Business Owner's Toolkit – Initial Public Offerings
http://www.toolkit.cch.com/text/p10_2500.asp

Cambridge Incubator
http://www.cambridgeincubator.com/index.htm

Capitalyst – Startup Funding
http://www.capitalyst.com/

Charles River Ventures
http://www.crv.com/

Citysearch
http://www.citysearch.com/

CMGI, Inc.
http://www.cmgi.com/

CoolBoard.com
http://www.coolboard.com/

eCompanies
http://www.ecompanies.com/

eHatchery
http://www.ehatchery.com/hatchware/web/main/
front.html

eTour, Inc.
http://www.etour.com/

eToys
http://www.etoys.com/html/
welcome.shtml?etys=welcome

FindLaw
http://www.findlaw.com

Garage.com
http://www.garage.com/

Global Food Exchange
http://www.globalfoodexchange.com/index.shtml

GoTo.com, Inc.
http://www.goto.com/

Houston Technology Center
http://www.houstontechcenter.org/

idealab!
http://www.idealab.com/

iHatch.com
http://www.ihatch.com/

iStart Ventures
http://www.istartventures.com/index.html

Katalyst
http://www.katalyst.com/

Kleiner Perkins Caufield & Byers
http://www.kpcb.com/

LevelEdge
http://www.leveledge.com/

MotherNature.com
http://www.mothernature.com

MsMoney.com
http://www.msmoney.com/index.asp

Murphree Venture Partners
http://www.murphco.com/

National Venture Capital Association (NVCA)
http://www.nvca.org/

NetZero
http://www.netzero.net/

Northwestern University Evanston Research Park
http://www.researchpark.com/

Redstone7.com
http://www.redstone7.com/

RosePlace.com, Inc.
http://www.roseplace.com/

Sendmail, Inc.
http://www.sendmail.com/

Sigma Partners
http://www.sigmapartners.com/

Stonepath Group
http://www.nvholdings.com/

The VC
http://www.thevc.com/

The Women's Technology Cluster
http://www.womenstechcluster.org/

ThinkTank
http://www.thinktank.com/

Tickets.com
http://www.tickets.com/

Tr@deUps
http://www.tradeups.com/

TRUSTe
http://www.truste.org/

Vanguard Venture Partners
http://www.vanguardventures.com/

VenusSports.com
http://www.venussports.com/

Veritas Medicine
http://www.veritasmedicine.com/

VetExchange
http://www.vetmedcenter.com

Venture Economics
http://www.ventureeconomics.com/

WomenAngels.net
http://womenangels.net/

yesmail.com
http://www.yesmail.com

Xuny.com
http://www.xuny.com/index.jhtml

zipRealty.com
http://www.ziprealty.com

ZoneTrader
http://www.zonetrader.com/

For Additional Review

Amor, Daniel. 2000. *The E-Business (R)evolution: Living and Working in an Interconnected World.* Upper Saddle River, NJ: Prentice Hall PTR.

Ashbrook, Tom. 2000. *The Leap: A Memoir of Love and Madness in the Internet Goldrush.* Boston, MA: Houghton Mifflin Company.

Bahls, Jane Easter. 1999. "Cyber Cash," *Entrepreneur, 27(3),* March, 108.

Brady, Monica. 2000. "Business Incubators," *National Public Radio, Morning Edition,* January 19. Audio available online at: http://npr.org/.

Brooker, Katrina. 2000. "First: Survival of the Fittest," *Fortune,* 142(3), July 24, 34.

Business Week. 1999. "An Angel Who Doesn't Flit Away," *December 21.* Available online at: http://www.businessweek.com/smallbiz/9912/f991221.htm.

Business Week. 2000. "Angels Who Hope to Make a Killing," June 12, 198.

Business Week. 2000. "How A VC Does It: Venture Capitalist Robert E. Davoli," July 24, 96.

Carbonara, Peter and Overfelt, Maggie. 2000. "The Dot-Com Factories: Internet Incubators Claim to Be the New Way to Create Great Companies. But Beware, or Risk Getting Caught in the Gears," *Fortune Small Business,* 10(5), July – August, 47.

Case, Thomas L. et al. 2000. "Attracting High-Technology Firms to Georgia," *Department of Information Systems, Georgia State University, prepared for the Intellectual Capital Partnership Program of the University System of Georgia,* September. http://www.icapp.org.

Copeland, Lee. 2000. "Equity Exchanges Cement Start-up Relationships: Trades Can Cut Partnership Costs and Seal a Deal, But They May Also Pose Risks," *Computerworld,* March 20, 38(1).

Dagres, Todd. 2000. "Going For Round Two," *Business 2.0*, March 1. Available online at: http://www.business2.com/content/magazine/indepth/2000/03/01/11126.

Darrow, Barbara. 2000. "Deep-Pocket Executives: An Answer to the Prayers of Industry Start-Ups," *Computer Reseller News*, April 17, 156.

Dery, Mark. 2000. "King of Capital: We Spent Morning Till Night With Silicon Alley Venture Capitalist JERRY COLONNA," *TIME Digital Magazine: Day in the Life*, 5(3), July. Available online at: http://www.time.com/time/digital/magazine/articles/0,4753,50339-1,00.html.

Dintersmith, Ted. 1999. "Getting to Know the Term Sheet," *Mass High Tech*, January 25–31. Available online at: http://www.crv.com/crvhtml_bak/03h_99-01-25.htm.

Eisenberg, Daniel. 2000. "Testing Time for the VCs," *TIME Magazine Online*, 155(25), June 19. Available at: http://www.time.com/time/magazine/articles/0,3266,47142-1,00.html.

Evanson, David R. and Beroff, Art. 1999. "Burnt Offerings? Should You IPO?" *Entrepreneur*, 27(7), July, 56.

Fletcher, Scotty. 2000. "GlobalFoodExchange.com Is Filling Its Plate," *dBusiness.com*, January 26.

Greenberg, Paul A. 1999. "E-Incubators Paving the Way for Startups," *E-Commerce Times*. December 16. Available online at: http://www.ecommercetimes.com/news/articles/991216-1.shtml.

Harrington, Cynthia. 2000. "Hold 'Em…or Not: Take a Look at Two Entrepreneurs Balancing the Eternal Choice Between Selling Equity and Borrowing," *Entrepreneur*, 28(7), July, 78.

Haylock, C. F. and Muscarella, L. 1999. *NET Success: 24 Leaders in Web Commerce Show You How to Put the Internet to Work for Your Business*. Holbrook, MA: Adams Media.

Jordan, Peter. 2000. "Show Me the Money: The Path from Idea to IPO Is Not Always an Easy One," *VARBusiness*, 16(13), June 26.

Jorgensen, Barbara. 2000. "Dangerous Liaisons?" *Electronic Business*, May. Available online at: http://www.findarticles.com/cf_0/m0GSY/5_26/62140104/p1/article.jhtml.

Kagle, Bob. 2000. "The VC Meeting: What To Do When You Finally Get Through the Door," *Business 2.0*, March 1. Available online at: http://www.business2.com/content/magazine/indepth/2000/03/01/11112.

Kawasaki, Guy. 1999. "Let the Hard Times Roll! Why Too Much Capital Can Kill You," *Business 2.0*, July 1. Available online at: http://www.business2.com/content/magazine/indepth/1999/07/01/12978.

Klein, Karen E. 2000. "Be Prepared: The Art of Pitching to VCs," *BusinessWeek Online*, March 30. Available online at: http://www.businessweek.com/smallbiz/0003/sa000330.htm.

Klein, Karen E. 2000. "To Angels, There's Nothing Heavenly about an Unprepared Entrepreneur," *BusinessWeek Online: Smart Answers*, February 3. Available online at: http://www.businessweek.com/smallbiz/0002/sa000203.htm.

Kucirek, Scott. 1999. "Raising Money Is Like Theater: You Better Have Your Act Together: ZipRealty's Team Tripped, Goofed, and Rambled Before They Got It Right," *BusinessWeek Online: Net Journal*, October 8. Available online at: http://www.businessweek.com/smallbiz/news/coladvice/diarynj/nj991008.htm.

Kvinta, Paul. 2000. "Frogskins, Shekkels, Bucks, Moolah, Cash, Simoleons, Dough, Dinero — Everybody Wants It. Your Business Needs It. Here's How to Get It." *Ziff Davis Smart Business for the New Economy*, August 1, 74.

Levine, Daniel S. 2000. "Email Company Gets Message: You've Got Cash," *San Francisco Business Times*, 14(36), April 7, 3.

Libsohn, Ralph J. and Mandelbaum, Judah. 2000. "Incubation Alternatives for Internet Startups," *netcommerce Magazine*, July. Available online at: http://www.netcommercemag.com/july00/july_finance.html.

McDonald, Marci. 2000. "A Start-Up of Her Own," *U.S. News & World Report: Cover Story*, May 15. Available online at: http://www.usnews.com/usnews/issue/000515/women.htm.

Monaco, Monta. 2000. "Angels on a Mission," *Office.com*, March 20. Available online at: http://www.office.com/global/0,2724,507-16704-16772,FF.html.

Nance-Nash, Sheryl. 2000. "Women Angels to the Rescue," *FSB/finance*, September 4. Available online at: http://www.fsb.com/fortunesb/articles/0,2227,744,00.html.

Osborne, D. M. 2000. "A Network of Her Own," *Inc. Magazine*, September 1. Available online at: http://www.inc.com/incmagazine/articles/details/print/0,7570,ART20125,00.html.

Park, Andrew. 2000. "Former Austin, Texas–Area Internet Entrepreneurs Give Tips on Starting Firms," *Austin American-Statesman, Knight-Ridder/Tribune Business News*, July 5.

Park, Lark. 2000. "The Trouble With Incubators," *The Standard.com*, July 24. Available online at: http://www.thestandard.com/article/display/0,1151,16959,00.html.

Pham, Alex. 2000. "Internet E-Mail Company Expands to Meet Customer Demand," *The Boston Globe*, April 26.

Press Release. 2000. "GlobalFoodExchange Named in Forbes' 'Best of the Web' Survey," Atlanta, GA, July 5. Available online at: http://www.globalfoodexchange.com/gfeJuly5_2000.shtml.

Press Release. 2000. "GlobalFoodExchange.com Secures First Round Financing from Top Venture Firms," Atlanta, GA, February 14. Available online at: http://www.globalfoodexchange.com/gfeFeb14.shtml.

Redherring.com. 2000. "How a Farming Town Hatched a New Way of Doing Business," January 19. Available online at: http://www.redherring.com/insider/2000/0119/resources/vc-fea-incubator-hstry.html.

Redherring.com. 2000. "The Dean of Silicon Valley Incubation," January 19. Available online at: http://www.redherring.com/insider/2000/0119/resources/vc-fea-incubator-rob.html.

Redherring.com. 2000. "The New Girls' Club," January 19. Available online at: http://www.redherring.com/insider/2000/0119/resources/vc-fea-incubator-wom.html.

Redherring.com. 2000. "Where Ideas Turn to Gold," January 19. Available online at: http://www.redherring.com/insider/2000/0119/resources/vc-fea-incubator-idea.html.

Reynolds, Kate. 2000. "Popular Incubators Help Hatch Fledgling Firms," *The Business Journal*, 18(10), 32.

Rodriguez, Karen. 2000. "Corporate and Venture Investors Deliver $35M to Sendmail," *The Business Journal*, 17(53), April 14, 12.

Roha, Ronaleen R. 1999. "Touched by an Angel," *Kiplinger's Personal Finance Magazine*, 53(9), September, 125.

Roussel, Tara. 1999. "Smart Startup: zipRealty Has Graduated Into Major Venture Funding," *San Francisco Business Times*, 14(18), December 10, 32.

Salvage, Bryan. 2000. "Dot.Com Food Marketplace Gets Funding from Venture Firms," *MeatingPlace.com*, February 16. Available online at: http://www.globalfoodexchange.com/gfeFeb16_1.shtml.

San Diego Business Journal. 2000. "What You Need to Know About the New Breed of Incubators: Business Incubators and Accelerators Versus Venture Capital Companies or Angels," 21 (27), July 3, 24.

Schaff, William. 2000. "Divine Isn't What It Seems To Be—Incubators Are the Poor Person's Way to Play the Private Equity Markets—But Not Necessarily the Smart Way," *InformationWeek*, August 14, 158.

Schneider, Gary P. and Perry, James T. 2000. *Electronic Commerce*. Cambridge, MA: Course Technology.

Schuch, Beverly. 2000. "Angel Investing," *CNNfn Business Unusual Transcripts*, July 18. Available online at: http://library.northernlight.com/EE20000718040000077.html?cb=0&sc=0 doc.

Semas, Judith Harkham. 1999. "Hatching Success," *TechWeek, November 29*. Available online at: http://www.techweek.com/articles/11-29-99/success.htm.

Sherrid, Pamela. 1997. "Angels of Capitalism," *U.S. News & World Report*, 123(14), October, 43–45.

Singer, Thea. 2000. "Inside an Internet Incubator," *Inc.*, 22 (10), 92–100. Available online at: http://www.gte.com/ldnewsletter/articles/gen_management/InsideIncubator/iii1-6.html.

Sobieski, Ian Patrick. 2000. "Get Off the Ground With an Angel," *Business 2.0*, March 1. Available online at: http://www.business2.com/content/magazine/indepth/2000/03/01/11111.

Stefanova, Kristina. 2000. "In Small Steps, Women Break Into 'Old Boys Club' of Venture Capital Funds," *The Washington Times*, 14-15(2) D, August 7. Available online at: http://www.library.northernlight.com/UU20000808050072701.html?cb=0&sc=0 doc.

Stein, Tom. 2000. "Business Services: Three Young Startups Invent Their Own Incubator Model," *Red Herring Magazine*, May. Available online at: http://www.redherring.com/mag/issue78/mag-business-78.html.

Stevenson, Howard W. et al. 1999. *New Business Ventures and the Entrepreneur*. New York, NY: Irwin/McGraw-Hill.

The Economist (US). 2000. "Hatching a New Plan: Stocks of Four Largest Publicly-Listed Internet Business Incubators— CMGI, Internet Capital Group, Safeguard Scientific, and Softbank—Have Plummeted," 345(8183), August 12, 53.

The Economist. 1999. "Brands Hatch," December 25. Available online at: http://www.economist.com/displayStory.cfm?Story_ID=271000.

Turban, Efraim, et al. 2000. *Electronic Commerce: A Managerial Perspective*. Upper Saddle River, NJ: Prentice-Hall, Inc.

Van Osnabrugge, Mark and Robinson, Robert J. 2000. *Angel Investing: Matching Start-Up Funds with Start-Up Companies*. San Francisco, CA: Jossey-Bass.

Walsh, Mark. "Web Incubators May Lay Eggs; Net Market Climate, Devaluations Leaving Start-Up Investors Cold," *Crain's New York Business*, 16, June 19, 3.

Weintraut, J. Neil. 2000. "Told Any Good Stories Lately? Ditch Your Business Plan. Tell the Moneymen Why They Should Care," *Business 2.0*, March 1. Available online at: http://www.business2.com/content/magazine/indepth/2000/03/01/13738.

Wolk, Martin. 2000. "Execs Get Roughed Up By Angels," *MSNBC News: PC Technology 2000*, June 29. http://www.msnbc.com/news/427226.asp#BODY.

Building Your E-Business

By 1995 Jeff Dachis had acquired a B.A. in dance and dramatic literature from SUNY Purchase, received a master's in education from New York University, and done stints as a disk jockey, actor, model, and marketing professional. His high-school pal, Craig Kanarick, took a different path. Kanarick, with a B.A. in philosophy and a B.A.S. in computer science from the University of Pennsylvania and an M.S. in visual studies from MIT, pursued a career in IT (information technology) by designing network protocols, simulation software, and digital media. Dachis and Kanarick ran into each other on a New York street and renewed their friendship. After Kanarick introduced Dachis to the Mosaic Web browser and the Web, it wasn't long before they proved to be the perfect match to take part in the coming Internet revolution.

Intrigued by the possibilities, Dachis and Kanarick sat down in Dachis's Manhattan kitchen and decided that they wanted to create an Internet services company that could offer clients a mixture of technology, business modeling, and marketing, including branding and visual identity development. When the time came to name the new Internet services company, Dachis and Kanarick picked a list of 100 names from a dictionary and then, using other brainstorming techniques, narrowed the list to 10 possibilities. Dachis took the list of 10 names to the bank and selected one of the names to open the new company's business account. As it turned out, Dachis's selection set the tone for everything that followed.

Understanding Legal Issues

One of the first things an e-business entrepreneur should do is establish a relationship with an attorney experienced in business startups and, if possible, with e-business startups. In fact, an experienced attorney is critical for a startup e-business in the development of its business plan, in establishing company valuation, in incorporating the new e-business, and in setting up its equity structure. Also, an established attorney can provide leads to angels and other investors and should participate in any negotiations with investors.

But entrepreneurs have a need for other legal assistance related to both startup activities and ongoing business activities. An e-business is still a business, subject to laws and regulations, the same as brick-and-mortar businesses. Legal issues for an e-business can be complicated because business on the Web moves across traditional jurisdictional boundaries, and because the growth in business on the Web is leading to changes in laws affecting e-business.

A startup e-business should seek legal advice when dealing with such issues as copyrights, trademarks, employee benefits and compensation, retirement plans, and personnel policies. Understanding and solving marketplace issues such as taxation, content liability, and information privacy may also require the advice of an experienced attorney.

TIP

Another sound reason for an e-business to establish a relationship with an experienced attorney is the rapid rate of change in the laws directed at e-businesses. For example, on October 1, 2000, the Electronic Signatures Act (ESA) went into effect. This law gives records containing digital or electronic signatures sent over the Web the same legal validation as signed paper documents. This is seen as both a boon to the consumer and a major opportunity for digital signature solution providers. Also, it is likely that many brick-and-mortar businesses such as title companies and car dealerships, as well as government agencies, will now put systems in place to handle digital or electronic signatures.

Copyrights and trademarks

A **copyright** is a form of protection for the author of published or unpublished original work, including writing, drama, music, art, and other intellectual property. A copyright takes effect the moment the work is created. Virtually everything on the Web is copyrighted, whether the copyright statement is visible or not; however, as a matter of form and best practice, an e-business's Web site should contain a copyright notice. **Fair use** allows the limited use of copyrighted material under certain circumstances. However, use of text, pictures, sound, or other copyrighted material on an e-business's Web site without the copyright holder's permission is a violation of copyright laws. Both The U.S. Copyright Office Web site and The Copyright Website (Figure 6-1) provide good information on copyright issues.

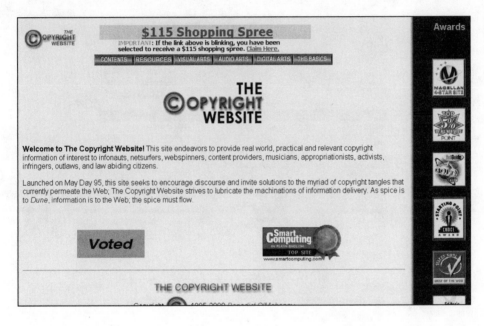

Figure 6-1
The Copyright Website

In 1996 Michael Robertson discovered that one of the most popular search terms on the Web was MP3. He immediately bought the domain name MP3.com for $1,000, and in 1998 the now well-known e-business was born. MP3.com's business model was simple: offer a system for accessing music files online, attract a huge number of viewers, and sell advertising to generate revenue. At first, MP3.com was very successful, with up to 500,000 visitors a day. There was only one problem with that kind of success . . . it attracted the attention of the music industry and the courts, who have ruled that MP3.com is liable for copyright violations! As of this writing, MP3.com (Figure 6-2) is in the process of negotiating settlements with record companies and recording artists and may reposition itself from being the preeminent download e-business to a music industry portal.

Figure 6-2
MP3.com

A **trademark** is a distinctive symbol, word, or phrase used to identify a business's products and distinguish them from other businesses' products. Services are distinguished by distinctive words, phrases, or symbols called **service marks**. Trademarks and service marks make it easier for a consumer to identify the source of a product. For example, the word "Nike" and the Nike "swoosh" symbol identify shoes manufactured by Nike. An e-business must take steps to register its own trademark or service mark with the U.S. Patent and Trademark Office and must take care that its Web site content does not infringe on the trademarks or service marks of others. The U.S. Patent and Trademark Office and the International Trademark Association (Figure 6-3) Web sites are a good source of information about trademark issues.

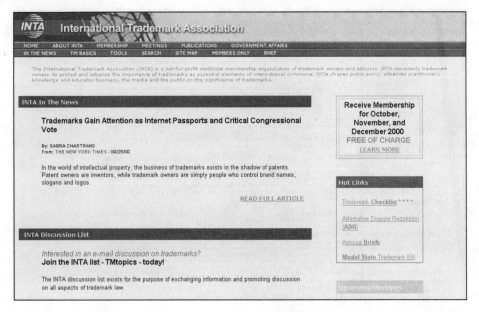

Figure 6-3
International Trademark Association

Content liability

An e-business may be exposed to potential liabilities for defamation, libel, copyright infringement, obscenity, indecency, and other liability, on the basis of the content and hyperlinks at its Web site. To protect against potential lawsuits, an e-business should have its attorney draft an appropriately worded disclaimer of liability to be posted on its Web site where it is easily accessible by viewers. Figure 6-4 illustrates the foodlocker.com Web site's terms of use and liability disclaimer.

Information privacy

One of the challenges for an e-business is the collection and analysis of customer information. Many potential customers may be hesitant to purchase products or services online or to provide valid information, for fear that the information will be misused or sold to others. An e-business should have a clear policy on how it handles and secures the information it collects at its Web site. Additionally, an e-business should take steps to make customers aware of its information privacy policies before it collects the information, including how that information will be secured from possible unauthorized or fraudulent use. One way an e-business can build customer confidence about information privacy is to become a TRUSTe licensee.

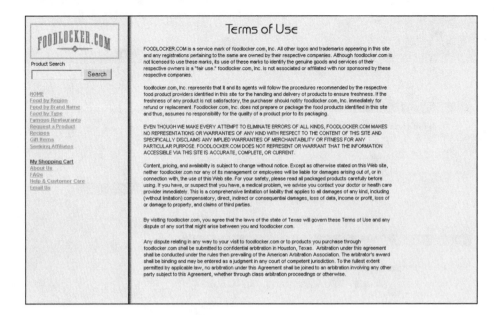

Figure 6-4
foodlocker.
com Terms
of Use

TRUSTe (Figure 6-5) is a nonprofit organization that advocates information privacy policy disclosure, informed user consent, and consumer education about information privacy issues. E-businesses that are licensed to display TRUSTe's online seal, or "trustmark," agree to make customers aware of what personal information is being gathered, with whom it will be shared, and how it will be used, and to provide options to control the way the information is disseminated. TRUSTe licensees include e-businesses such as AOL.com, E*TRADE, and eBay.

Figure 6-5
TRUSTe

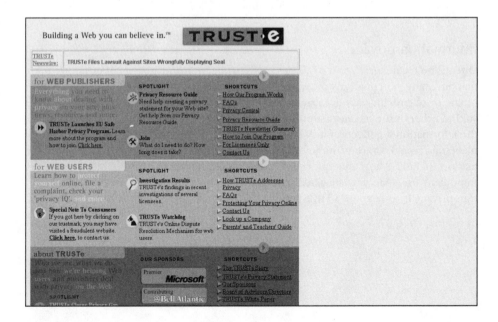

Another concern of many consumers is the automatic collection of data at Web sites they visit when surfing the Web. This concern is building demand for software that allows Web surfers to remain anonymous. E-businesses such as Anonymizer.com and Zero-Knowledge Systems (Figure 6-6) offer software to protect the privacy of individuals browsing the Web, as well as software solutions for e-businesses and government agencies that need to protect the privacy of their customers and clients.

Figure 6-6

Zero-Knowledge Systems

DoubleTrouble

Two recent examples seem to confirm consumers' information privacy concerns. Early in the year 2000, DoubleClick, the e-business advertiser, announced that it was creating a huge database of consumer profiles, using data collected from individuals browsing the Web, together with demographic data maintained by Abacus Direct, a data collection firm it owns. Privacy advocates fought so vigorously against creation of the database that DoubleClick's stock lost 60 percent of its value between January and March 2000, and DoubleClick (Figure 6-7) had to suspend creation of the database pending adoption of online privacy standards.

In July 2000 the Federal Trade Commission sought a court order to bar bankrupt toysmart.com from selling the names, addresses, and purchasing behavior of its former customers, in violation of its own information privacy policy. (The FTC has since settled the case.)

(Continued on the next page)

Figure 6-7
DoubleClick

Taxation

The taxation of business conducted on the Web is one of the toughest issues facing government. Sales taxes are the largest single source of revenue for many state and local governments, which fear that increased sales on the Web will cause a serious decline in sales tax revenues that will have to be made up by increased taxes on property and income. As of this writing, e-businesses are not required to collect sales taxes from out-of-state customers unless the e-business has a physical presence in the customer's state. Instead, buyers are responsible for paying their state's applicable sales taxes, although few if any do so. However, there are several ongoing initiatives, supported by state and local government officials, to find an efficient method of collecting sales taxes on Web-based purchases.

Although the Internet Tax Freedom Act, originally passed by Congress in 1998 and recently extended to 2006, places a moratorium on new Internet taxes, e-businesses are still subject to the same property and corporate income taxes as brick-and-mortar businesses. It is important that an e-business's management stay informed about changes in tax law. A good source of information on current e-business tax issues is EcommerceTax.com (Figure 6-8).

> **TIP**
>
> In July 2000 the General Accounting Office (GAO) reported that state and local governments would lose between $300 million and $3.8 billion in tax revenues in 2000 because of purchases over the Internet. According to the GAO's estimates, Texas with $26 to $342 million and California with $23 to $533 million in lost taxes are the biggest losers.

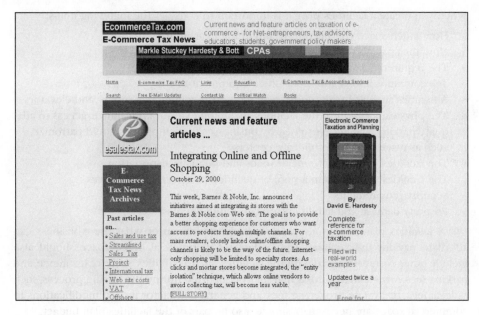

EcommerceTax.com
E-Commerce Tax News
Current news and feature articles on taxation of e-commerce - for Net-entrepreneurs, tax advisors, educators, students, government policy makers

Markle Stuckey Hardesty & Bott | CPAs

Home   E-commerce Tax FAQ   Links   Education   E-Commerce Tax & Accounting Services

Search   Free E-Mail Updates   Contact Us   Political Watch   Books

esalestax.com

E-Commerce Tax News Archives

Past articles on...
- Sales and use tax
- Streamlined Sales Tax Project
- International tax
- Web site costs
- VAT
- Offshore

Current news and feature articles ...

Integrating Online and Offline Shopping
October 29, 2000

This week, Barnes & Noble, Inc. announced initiatives aimed at integrating its stores with the Barnes & Noble.com Web site. The goal is to provide a better shopping experience for customers who want access to products through multiple channels. For mass retailers, closely linked online/offline shopping channels is likely to be the way of the future. Internet-only shopping will be limited to specialty stores. As clicks and mortar stores become integrated, the "entity isolation" technique, which allows online vendors to avoid collecting tax, will become less viable.
[FULL STORY]

Electronic Commerce Taxation and Planning

By David E. Hardesty

Complete reference for e-commerce taxation

Filled with real-world examples

Updated twice a year

Free for

Figure 6-8
Ecommerce
Tax.com

An e-business should certainly seek the advice of an experienced tax accountant when dealing with tax issues. It also bears repeating that an e-business should establish a relationship with an experienced attorney to help sort through the legal and tax issues it faces.

Sooner or later a startup e-business will outgrow the back bedroom. When that happens, it's time to move into suitable office space.

Leasing Commercial Office Space

Many startup e-businesses begin at an entrepreneur's kitchen table and quickly expand to temporary office space, often a spare room converted into an office. Signs that an e-business has outgrown its temporary home office space include the need to add employees, a need for more storage, insufficient electrical outlets, insufficient phone lines, a need to add telecom services, and the need for more office equipment for which there is no space. Also, if an e-business needs conference room facilities, or if meeting with clients and business associates in a home office undermines an e-business's credibility, then it's time to look for commercial office space.

If an e-business has major facility needs, such as a call center or product distribution center, it is likely that management has already considered those needs and outlined them in its business plan. However, for many e-business startups, the details about commercial office space needs may not have been considered in the business planning process. If that is the case, it is important to develop a detailed facilities plan before the search for commercial office space begins. Failure to plan adequately for a move to commercial office space might lead to a bad experience with a poor business location, inadequate space for future growth, or poor access to facilities for clients, business associates, and employees.

When you create a facilities plan, it is important to consider the following items:

- How much space you need now and in the near future
- How the space is to be divided into offices or work areas
- The requirements for the electrical wiring infrastructure
- Telecommunications and other office equipment needs
- Amenities such as break rooms, reception areas, conference rooms, and elevators
- The physical location of the facilities, including exterior appearance, access to adequate parking, and access to nearby businesses that employees could patronize, such as restaurants and childcare centers
- The facilities' proximity to major roads and public transportation
- The facilities' proximity to a pool of qualified prospective employees
- Security requirements
- Property damage and liability insurance requirements

A facilities plan should also include a determination of what the e-business can afford to pay for the facilities. The budget for commercial office space should also include other costs associated with the facilities, such as utilities, security, insurance, and parking fees. Any estimated costs that may be incurred during the leasing process, such as legal fees, real estate brokerage fees, and general contractor fees (if modifications to commercial space are necessary), should also be part of the facilities plan budget.

After developing a facilities plan, it is a good idea for the management of an e-business to look for opportunities to save money on office space. One way to do that is to look for office space with a nonprofit business incubator. Another way is to share office space with other businesses, so that the costs of the common areas (reception areas, conference rooms, and break rooms) can be split among the businesses. Subletting space from another business is another good way to save money. Some e-businesses find ways to trade services with strategic business partners in exchange for office space.

Many communities offer special tax abatements and other financial incentives to businesses locating in their communities. Older, transitional communities may have established economic development zones with special incentives for businesses locating there. And, finally, an e-business should consider telecommuting options for some of its employees. Perhaps allowing employees to telecommute could allow the e-business to downsize its office space requirements.

It is a good idea for an e-business to work with an experienced and reliable commercial real estate broker who can help locate commercial office space that meets its needs, and help negotiate the lease on the selected space. It is also a good idea for an e-business's attorney to review all lease agreements.

One of the major reasons it is necessary to move from a home office to commercial space is to accommodate additional employees.

Hiring Employees

The key to success for any business—and an e-business is no exception—is its people. Finding key people in a tight labor market, as has been the case in the U.S. for the past several years, can be difficult. Additionally, knowing whom to hire first may be confusing for a startup e-business. Usually, each additional round of financing allows an e-business to increase its staff. While every e-business startup is different, an e-business might consider its first key hire to be the chief information or technology officer. Even if much of the technology is to be outsourced, it is critical to have a key person with superior technology skills to help design and monitor the e-business systems development and operations. The next hires might be the internal technology staff, if needed, such as software engineers, Web developers, product managers, and network administrators.

Because maintaining an accurate picture of an e-business startup's financial position and **burn rate** (the rate at which the e-business is using its cash reserves) is critical, the next hire should likely be a chief financial officer. A chief financial officer can also play a key role in developing additional financing. Along with a chief financial officer, an e-business should consider hiring an experienced marketing professional to begin working on business development.

After securing the next round of financing, it may be time for an e-business to concentrate on hiring employees for the marketing and customer support teams. Also, with a growing number of employees, it will likely be necessary to hire someone to handle the e-business's human resources functions.

Leads on finding qualified and talented employees can come from many sources. Investors, boards of advisors or directors, family, friends, strategic business partners, and current employees are all great sources for new employee leads. When hiring senior managers or technology workers, an e-business might use a professional recruiting firm. Professional recruiters, also known as headhunters, have access to experienced, talented people who are not actively looking for employment but who might consider moving to a new e-business to take advantage of a perceived opportunity. Usually an entrepreneur finds that he or she must hire an experienced chief executive officer or other senior management to add credibility to a startup e-business and satisfy investors' requirements. In addition to leads from advisors, an entrepreneur can turn to one of several well-known global professional recruiters, such as Heidrick & Struggles International, Korn/Ferry International, and Spencer Stuart (Figure 6-9), that focus on filling senior management positions.

TIP

As of September 1, 2000, the overall U.S. unemployment rate was 4.1 percent, largely unchanged since October 1999. Contributing to this low unemployment rate is the change in demographics between the larger number of "baby boomers" nearing retirement and the smaller number of "Generation X" employees available to take their place, as well as the explosion of technology-based businesses.

Information technology (IT) jobs continue to be the hottest, most in-demand jobs in the new economy. The demand for skilled IT workers in the U.S. continues to outstrip supply. For example, the Information Technology Association of America estimated that in 2000 approximately one-half of the new IT positions went unfilled and that one in every twelve IT jobs remained vacant. In this market an e-business may be required to use professional recruiters to fill its technology positions, such as Technical Search Corporation or TransQuest Ventures (Figure 6-10), which focus on recruiting skilled technology workers.

Figure 6-9

Spencer Stuart

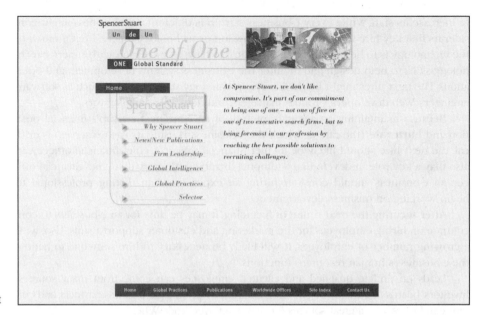

Figure 6-10

TransQuest
Ventures

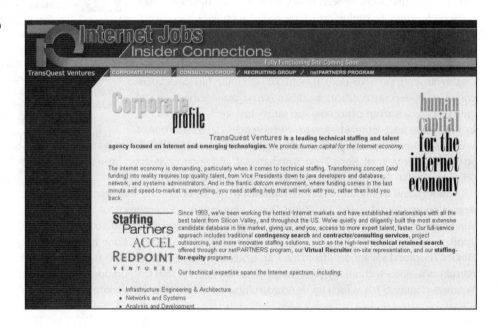

In a tight labor market, a startup e-business must find ways to compete with other businesses to hire and keep talented employees. It is important for an e-business's management to know what the salary and benefit expectations are for prospective employees in their region. The Web is a good source of information on salary and benefits standards for various regions of the country. E-businesses such as SalariesReview.com (Figure 6-11) sell salary and bonus information by region. Government or nonprofit organizations or business magazines often post free salary survey information at their Web sites.

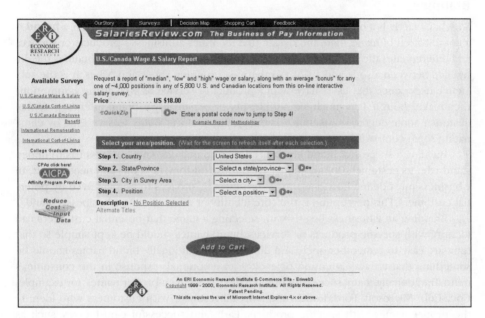

Figure 6-11
Salaries
Review.com

Often a startup e-business must offer prospective key employees not only a competitive salary but also some equity in the e-business in the form of stock options. However, despite offering competitive salaries and stock options as compensation, many e-businesses find that compensation alone is insufficient to attract—and keep—talented employees. An attractive corporate culture, or workplace environment, is often more of a drawing card than salary. A corporate culture that promotes and rewards creativity and personal growth may be important in attracting talented and in-demand senior managers and technical staff.

A **perquisite**, or "**perk**," is an incidental benefit extended to employees. To enhance the workplace environment, many e-businesses offer their employees perks not generally offered by traditional companies, such as subsidized cafeterias, onsite dry-cleaning facilities, onsite gyms, onsite childcare, casual attire every workday, massages, the option to bring a pet to work, telecommuting, and other unusual incentives. The management of a startup e-business must be aware of what combination of compensation and perks is going to be necessary to attract talented employees in their geographical area.

After hiring a marketing professional and marketing staff, an e-business should begin building its brand.

Building Your Brand

Building a brand is more than just a URL and a logo. A brand involves all aspects of a customer's experience. In traditional media, the effectiveness of branding is measured in awareness. Online, an effective brand leads to more than just awareness; it leads to action taken by a customer.

Branding

So what exactly is a brand? According to the American Marketing Association, a **brand** is a combination of name, logo, and design that identifies a business's products or services and differentiates the products and services from those of competitors. A brand is the difference between a soda and a Coke, or between running shoes and Nikes. It is the subjective experience that users have with a product or service, as well as the assumptions they make about it. It is much like an identity or personality. A brand offers a potential customer some degree of assurance about quality and the ability to save time by eliminating some comparison-shopping.

Building a successful brand involves a lot of homework. An e-business has to understand who its customers are and understand those customers' wants and needs. Next, an e-business must define how it wants its products or services to be perceived by potential customers. This perception is the core identity or essence of an e-business's brand.

Branding an e-business begins with selecting a name that potential customers can identify with specific products or services. Brand names should be kept simple so that they are easy to remember, spell, and understand. Additionally, brand names should be something that attracts attention or evokes an emotional response in the consumer's mind. Traditionally, many successful brand names have been proper names; for example, Coca-Cola, Microsoft, Ford, Kodak, and Sony all resonate with customers who identify the proper names with specific products. Early and successful e-businesses, such as Yahoo!, Amazon.com, eBay, and priceline are also branded with proper names. Internet brand names with a little "snap" or those that evoke an emotional response can be memorable. For example, consider Hotmail as a name for an e-mail provider, or EarthLink as a name for an ISP. Both names have "snap" and make an immediate mental link to the products or services the e-businesses offer.

Another way to make an e-business's brand name memorable is to personalize it by naming the e-business after an individual. While not many people are likely to remember PC Limited and its products and services, most people will immediately recognize the e-business by the name it uses today, Dell Computer Corporation. Switching the brand name from PC Limited to the proper name Dell (based on the name of the founder, Michael Dell) enhanced the publicity potential of the e-business. All the attention and publicity Michael Dell receives directly benefits the Dell Computer brand.

Ask Jeeves (Figure 6-12), an information portal, is personified by Jeeves the butler. The e-business decided on a brand that would keep it from being just another search engine. The brand is focused on personal service, and Jeeves the butler, embodies personal service and gives the brand personality. Jeeves is the distinctive digital hook that resonates with users and sticks in their minds.

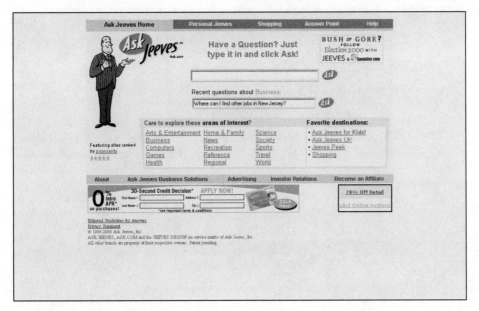

Figure 6-12
Ask Jeeves

Generic names based on common words are often used by new e-businesses. The Web abounds with e-businesses with names like cars.com, buy.com, and Office.com. Unfortunately, generic names like these often lack the interest or "snap" necessary to enable potential customers to remember them and to make the mental link between brand name and product. Some marketers question whether or not generically named e-businesses can build a successful brand over the long term.

Some marketers are also critical of the common practice of naming an e-business to include a reference to the Web or the Internet, such as Cyberchefs or TimeTech, thinking that such names are limiting in the long term. With this in mind, an e-business should consider the following issues when settling on a name:

◆ Do prefixes and suffixes such as "cyber," "net," "tech," and ".com" shorten the usefulness of the name?

◆ Does a descriptive name such as "perfumesonline" too closely identify the e-business with one product?

Many companies look to professionals such as The Namestormers, NameLab, and Catchword to help them create a memorable brand name for their business. Catchword (Figure 6-13), a name development and brand strategy company, has successfully worked with a number of companies in a variety of industries, including the e-businesses e-Stamp (Web-based postage), Chemdex (B2B exchange for life science companies, researchers, and suppliers), and RightWorks (B2B exchange software).

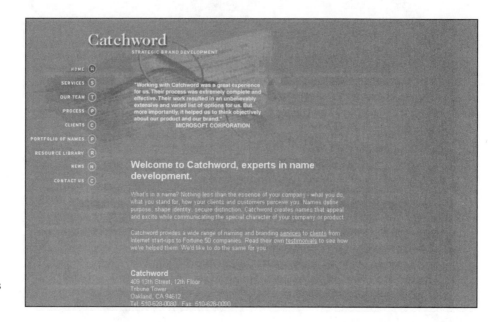

Figure 6-13
Catchword

Part of branding an e-business is the selection of a memorable and easy-to-enter domain name.

Domain names

A **domain name** identifies a Web site on the Internet and allows a Web browser to locate that site. For example, the domain name "www.foodlocker.com" identifies the foodlocker.com Web site located on a particular Web server. The "foodlocker" portion of the domain name identifies the organization or entity associated with the domain name. The "www" identifies the server that handles Internet requests. The "com" portion is called the top-level domain (TLD) and identifies the general category in which the domain name is registered. For example, the top-level domain "com" indicates the commercial domain.

There are seven original TLDs, as shown in Table 6-1.

TABLE 6-1
Original TLDs

Type of Organization	TLD Name
Commercial	com
Educational	edu
Government	gov
International treaty organization	int
Network provider	net
Nonprofit	org
Military	mil

In July 2000 the Internet Corporation for Assigned Names and Numbers (ICANN), the organization chosen by the U.S. government to take over Internet naming duties, approved a resolution to create several new TLD names. In November 2000, ICANN approved seven new domain names (Table 6-2).

TABLE 6-2
Approved global TLD names

Type of Organization	TLD Name
Unrestricted commercial	biz
Information service	info
People's names	name
Professionals	pro
Museums	museum
Airline industry	aero
Cooperative	coop

Organizations can search for and register domain names at one of several domain registration services, such as Network Solutions or register.com. However, the availability of short, meaningful domain names is limited. In fact, most English words have already been registered as domain names. For this reason, some companies must either buy a good domain name from someone who has already registered it, or must invent a new word for a domain name. Often, the name of an e-business is the same as its domain name, such as Amazon.com, foodlocker.com, or register.com. Sometimes, an e-business's name and domain name are not the same, either to make it easier to key in the domain name in a Web browser, such as bn.com for Barnes & Noble and ey.com for Ernst & Young, or because a similar e-business has already registered the domain name. For example, the registered domain name for *The New York Times* is times.com, while the registered domain name for the *Los Angeles Times* is latimes.com. When a domain name is the same as or closely associated with an e-business, the domain name is another way to brand the e-business.

Because a domain name can be closely associated with an organization, selling domain names has become big business. There are two types of companies or individuals that register multiple domain names for resale: legitimate domain name brokers and cybersquatters. **Domain name brokers** register generic domain names that they think will be easy to resell. **Cybersquatters** register domain names, often a company name or acronym, with the intent of selling the domain name to its rightful owner. Cybersquatting is now illegal. The U.S. Anticybersquatting Consumer Protection Act became law in November 1999 and prohibits the bad-faith registration of, trafficking in, or use of a domain name that is a registered trademark or that is identical or confusingly similar to a famous mark, whether registered or not.

TIP

A domain name is not a trademark. In order to function as a trademark, the domain name must serve as a product identifier, and not merely as a Web address. If the domain name acts separately as a product identifier, the domain name must be registered with the U.S. Patent and Trademark Office as a trademark or service mark. If registered trademark owners want to use their trademarks as their domain names, they must file for domain name registration with a registrar accredited by ICANN.

Brand names, domain names, and trademarks are becoming progressively more interchangeable in consumers' minds. Since domain names have economic value, many e-businesses are registering their domain name as a trademark. Several domain name registrars such as register.com and NameProtect.com (Figure 6-14) also search for registered trademarks when you use their search tool to search for registered domain names.

An e-business can elect to register its domain name itself or to have its technology service provider register it. Various technology service providers can also help an e-business select the software, hardware, and other technologies, such as Web site hosting, that are unique to an e-business.

Selecting the Technology

An e-business entrepreneur must make many technology decisions for an e-business. Software must be selected to handle all of the aspects of the e-business, including such important functions as sales transactions, inventory monitoring, accounting, and internal recordkeeping. The e-business's principals must decide whether to hire and maintain the in-house technology expertise needed to develop the e-business's internal computer systems and Web site, or to outsource these functions. Table 6-3 summarizes several technology software options.

Figure 6-14
NameProtect.
com

Name	Description	Sample Vendors
ERP – enterprise resource planning	System that integrates all aspects of a business	J.D. Edwards PeopleSoft Oracle SAP
CRM – customer relationship management	System that manages the customer base	Same as ERP vendors
ISV – independent software vendors	Develop and sell applications for various operating systems	Oracle PeopleSoft SAP BroadVision, Inc.
VAR – value-added resellers	Modify, repackage, or sell existing software	Microage ASG (Viasoft) Merisel, Inc.

Table 6-3
Technology software options

Defining the enterprise

An enterprise is a business organization. When discussing e-business, the term **enterprise** is often used to identify any organization that uses computers and a network to interact with employees, suppliers, and customers. The amount of information that an enterprise must keep track of is tremendous. Also, this information must be processed quickly. For example, it may be necessary to rapidly update product and pricing changes on a Web site. There may also be a need for the customer to verify that the product he or she placed in a shopping cart is actually in inventory. Vendors need to quickly route orders for fulfillment. Several systems have been developed to help e-businesses store and process their information while using the power of the network to make the information quickly available to employees, strategic partners, and customers.

Enterprise resource planning systems

Enterprise resource planning (ERP) refers to a system that integrates all aspects of a business, including planning, manufacturing, human resources, accounting, finance, sales, and marketing. Traditional ERP systems had their origins in the 1960s in manufacturing systems, which began to incorporate raw material planning, purchasing, shop floor management, and distribution systems. In the early 1990s other enterprise activities, such as engineering, project management, accounting, finance, and human resources were being incorporated into ERP systems. Many e-businesses use an ERP system, which is tied into its Web site, to handle these tasks. An ERP system can encompass, but is not limited to, the following functions:

◆ *Accounting*: Part of running a successful business is accurate bookkeeping. Accounting systems can help automate the processes of tracking sales, accounts receivable, accounts payable, cash flow, liabilities, and so forth.

- *Inventory management:* The ability to deliver on what is promised is important in any business. Inventory management systems can be used to guarantee availability of goods and maintain accurate inventory records.
- *Resource planning:* One of the biggest constraints many businesses face is a limit on their resources. They may only be able to make a given number of parts on their assembly line each day. They may only have a few delivery trucks to deliver the orders. There may only be a dozen technicians to provide installation and service. It is necessary to be able to effectively allocate these resources. This is where resource planning is very helpful.

ERP systems are usually built on top of databases that store all of the data needed by the different systems. These databases must be capable of handling large numbers of records and many requests for data at the same time. In addition, the databases must protect the data from being changed while another system is using the data, and from accidental erasure.

Today, vendors, suppliers, customers, and outside sales representatives use the Internet to access an enterprise's ERP system to get needed information. Complete ERP systems are generally found in very large enterprises such as Fortune 2000 companies. For small and midsized enterprises, it may not be necessary or cost-effective to fully integrate all aspects of the business. For these enterprises, an ERP solution might consist of integrating two or three critical applications using the Web. There are many companies that provide ERP systems, or components of ERP systems. The price of such systems can run from thousands of dollars to millions of dollars. Leading ERP providers for both brick-and-mortar enterprises and e-businesses are J.D. Edwards, PeopleSoft, Oracle, and SAP (Figure 6-15).

Figure 6-15
SAP

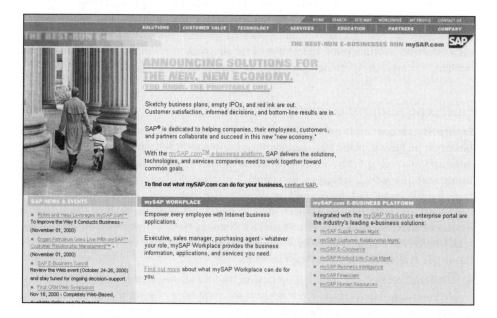

Customer relationship management systems

In today's e-business economy, getting and keeping customers is a top priority. **Customer relationship management (CRM)** systems, sometimes called e-CRM systems, use software and the Internet to help an enterprise manage its customer base. For example, an enterprise might maintain a customer database that contains data describing customer relationships in enough detail so that the e-business can match customer needs with product plans and offerings, remind customers of service requirements, know what other products a customer has purchased, and access other useful information. CRM applications should be able to connect to different data sources such as sales figures, call center activities, Web transactions, and mobile access transactions, in order to gather relevant information about customer interactions. ERP systems providers such as SAP and Oracle also provide CRM systems.

An important facet of CRM systems is the ability to manipulate the stored customer and transaction data to find patterns of behavior.

TIP

There are other systems that can assist with the operation of an e-business. Software that provides content management can help the e-business quickly update information on its Web site, without requiring the technical skills of a Web designer to make each change. Selling to customers can become more complex than just placing items on the electronic store shelves. A small business can easily handle a small number of customer inquiries and requests, but as the business grows, it may need to use enterprise systems software that assists in the management of customer contact and customer data.

Data mining

A tremendous amount of valuable information is hidden away in the large customer and sales databases of most enterprises. Because of the large volume of data, this data is not easy to examine in its raw form. Using data mining or Web mining principles, marketing researchers can determine many interesting statistics and correlations that can help a business better attract and sell to its customers.

Data mining doesn't just present data in new ways. True **data mining** actually discovers previously unknown relationships among the data. Data mining looks for hidden patterns in groups of data. For example, data mining can help e-tailers identify customers with common interests. A **data warehouse** is a database that contains huge amounts of data, such as customer and sales data. Data mining can be used to extract new patterns of data from a data warehouse, which can then be used by management to aid in decision making. **Web mining** applies traditional data mining principles to data from Web sources: Web server data, Web server logs, and Web transactions. Because most businesses have both corporate data warehouses and Web server data, data mining can integrate data from both sources to help discover hidden and relevant patterns and relationships in the data.

Many companies, such as American Express and Wal-Mart, have been using data mining since before the Web existed. However, the Web presents an ideal opportunity to use data mining, because every action performed by a customer at a Web site generates an electronic record. In fact one potential problem with collecting customer data from a Web site is that too much data can be collected. Imagine the volume of recorded transactions if Amazon.com kept a record of every click every customer made at its Web site!

E-businesses can use data mining software to determine how one event leads to another event. For example, an e-business can use data mining to show how a customer's purchase of a product leads to the customer's later purchase of another product. Data mining software can also be used to identify associations between events, such as what percentage of new car buyers shopped online for their new car. By discovering these hidden patterns, e-businesses can attempt to forecast future results. Oracle's Darwin and Information Discovery's Data Mining Suite (Figure 6-16) are data mining software applications.

Figure 6-16
Information Discovery

E-CASE · Reel 'Em In!

Reel.com, an online movie superstore, first attempted to attract customers by allowing deep discounts on purchases. Despite this, Reel.com could not determine which customers were likely to come back to the Web site for more shopping. In an attempt to identify return customers, Reel.com began using a consumer-behavior-tracking software package named LifeTime from the Boston startup Verbind.

After analyzing two years' worth of Reel.com transactions, management discovered that 33 percent of all customers bought from Reel.com at least one more time after their initial purchase. Also, 71 percent of customers who made a second purchase made this purchase within 30 days of the first purchase. Using this data, Reel.com targeted new customers with offers for comparable products within the 30-day window when they were more responsive. Reel.com's management also discovered that 55 percent of their customers who bought four or more movies preferred older releases. Reel.com forecasts an additional $1 million in revenues by retaining those customers.

Finally, Reel.com uses the customer preference data to send customers targeted and well-timed special offers, in an attempt to influence their customers' buying behaviors. For example, when Reel.com determines that a customer is a discount shopper who likes James Bond movies, they send him or her a coupon offering a discount on the 007 movies. Reel.com (Figure 6-17) estimates that by using the right offer they can keep customers for an additional 100 days, and can add $5 million to their annual revenues.

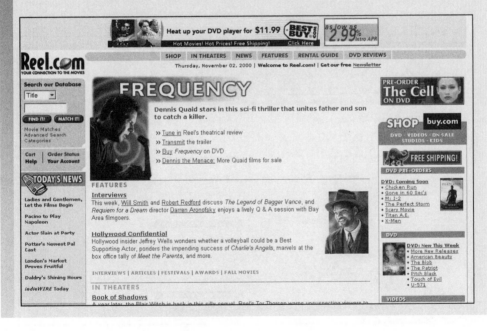

Figure 6-17
Reel.com

In addition to the major ERP and CRM systems application vendors, there are many other software vendors from which to choose.

Independent software vendors and value-added resellers

Independent software vendors (ISVs) are companies that develop and sell software applications for a variety of operating system platforms such as Microsoft, Apple, IBM, and Hewlett-Packard. Some ISVs concentrate their efforts on developing software for specific operating systems, while other ISVs focus on a particular application area such as engineering, accounting, or Web security. Leading ISVs include Oracle, PeopleSoft, SAP, and BroadVision, Inc. (Figure 6-18).

A **value-added reseller (VAR)** is a company that modifies an existing software product by adding its own "value," such as a special computer application, and then resells the software as a new product or software package. VARs include MicroAge, ASG (Viasoft) (Figure 6-19), and Merisel, Inc.

Many e-businesses, choosing to concentrate internal resources on their core business, elect to outsource all of their internal computer systems development, including ERP and CRM systems. Additionally, an e-business must also decide whether to host its Web site internally or outsource its Web hosting to an ISP or Web hosting company.

Figure 6-18
BroadVision

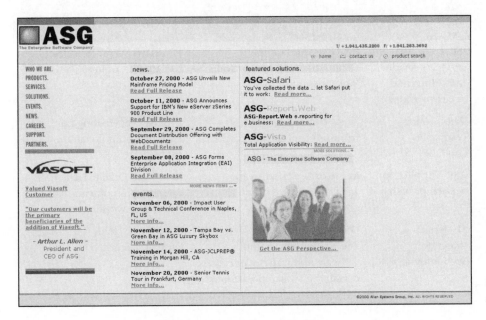

Figure 6-19
ASG (Viasoft)

Identifying Technology Service Providers

There are many different kinds of technology service providers available to a startup e-business. These different service providers offer a range of overlapping services and are known by an array of confusing acronyms. Additionally, as the e-business economy continues to evolve, companies originally positioned to provide one set of services are rapidly changing their business model and providing new services to better take advantage of e-business evolution. The wide array of technology services available includes access to the Internet, Web site hosting, complete software and hardware outsourcing, Web development, Web strategy consulting, and all conceivable combinations of these and other technology services. Table 6-4 shows some technology service providers.

Internet service providers

Many e-businesses prefer to outsource their Web site operations because they do not have the in-house staff to monitor and maintain their Web sites or guarantee 100 percent reliability of their servers. Many e-businesses also find that it is more cost-effective to outsource their Web site operations. E-businesses can outsource their Web site operations to an ISP. **Internet service providers (ISPs)**, sometimes called **Internet access providers (IAPs)**, are the companies that provide connections to the Internet. ISPs often offer Web hosting services, because they already have the hardware, software, and Internet connection necessary to operate Web sites. An e-business that outsources its Web site operation to an ISP rents space on the ISP's servers located at the ISP's facilities. Another way to outsource Web site operations is to use a Web hosting company.

Table 6-4
Technology
service
providers

Name	Description	Sample Vendors
ISP – Internet service provider IAP – Internet accessprovider	Provides connection to the Internet and may provide Web hosting services	EarthLink, AT&T WorldNet, XO, Hawk Communications
Web hosting companies	Hosts commercial Web sites from Internet data centers	Affinity, WebHosting.com, Exodus Communications, Interliant
ASP – application service provider	Rents or leases software applications and other computer services	Microsoft, Oracle, Dell, Qwest Communications, USInternetworking
SI – systems integrator, Web integrator	Oversees modifications to and implementation of software applications; integrates existing software applications with Web applications	iXL, Viant, Scient, Sapient, Razorfish

Web hosting companies

Web hosting was a $1.8 billion industry in 2000 and is growing dramatically. For **Web hosting companies**, such as Affinity and WebHosting.com, managing and maintaining outsourced commercial Web site operations at Internet data centers is their primary focus. One of the largest Web hosting companies is Exodus Communications. Founded in 1994, Exodus Communications (Figure 6-20) was a pioneer in the Internet data center market and by 1999 was ranked number 11 in *Business Week's* "Info Tech 100" profile. In January 2000, Exodus was running 17 host centers around the world with 39,000 servers, and was expecting to have double the number of hosting centers by 2001. In June 2000, Exodus was hosting Web sites for more than 2,800 customers, including Yahoo!, priceline, Hotmail, and British Airways. In September 2000, Exodus Communications announced the acquisition of GlobalCenter, the Web hosting subsidiary of Global Crossing Ltd., and a new 10-year network services agreement with Global Crossing Ltd.

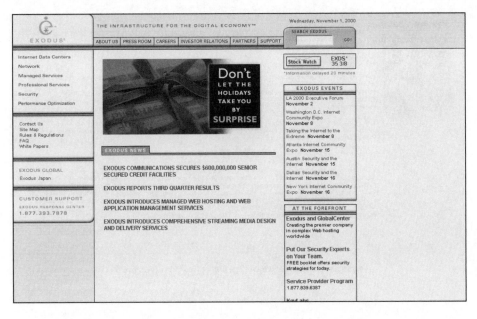

Figure 6-20
Exodus
Communi-
cations

Using a Web hosting service provides an e-business with several advantages. An ISP or Web hosting service has a staff available 24 hours a day with the technical knowledge to keep the servers and network connections running. Because the servers are located at the ISP's or Web hosting service's facility, connectivity to the Web site is not limited by the **bandwidth** (data transmission rate) between the customer and the host. An ISP or Web hosting company can provide a level of redundancy that an e-business may not be able to provide for itself, including redundant hard drives on the servers, backup power from generators, and redundant connections to the Internet.

Also, the ISP's or Web hosting company's costs for the hardware, software, and operation personnel are spread over many e-businesses. This often makes outsourcing Web site operations a more cost-effective choice for an individual e-business. It is important, however, to be sure that an ISP or Web hosting company can provide scalable servers that can grow to accommodate an increasing volume of Web transactions. It is also important to know that the ISP or Web hosting company has enough fast connections and communication lines to handle customer transactions and provide sufficient bandwidth.

There are several Web sites that are useful for locating ISPs or Web hosting companies, such as ISPs.com, ISPcheck, the Hosting Repository, HostSearch, and the Web Host Guild. The Web Host Guild, founded in 1998, is an organization dedicated to raising the standards of the Web hosting industry. Web hosting companies must pass a stringent Web Host Guild certification evaluation to be able to display the Guild Seal on their Web site and to be listed in the searchable directory of Web Host Guild members (Figure 6-21).

Creativedata.net

Creative Data Concepts Limited, Inc., was founded in 1993 and has been devoted to Web development, Web hosting and Electronic Commerce services since 1996. We have extensive database integration experience as well as network configuration and management. We offer a wide range of hosting services including server co-location services, managed server co-locations, Windows NT Web hosting, support for ColdFusion, Active Server Pages, iCat, Intershop, Simple Commerce and a variety of other commerce-enabling technologies.

Creative Data Concepts, Ltd., works with each client to help them develop a strategic plan to meet the specific goals of their business. Our service is unmatched in its ability to ensure that each business achieves their goals and realizes their objectives. We help businesses achieve the full potential of their Web presence.

Our Vision

Our vision is for our reputation to precede us. We are consistently striving to be known as the company that provides our clients with the resources and services needed to achieve the results they deserve from their Web site. We want it to be known that we can reduce disappointing results that may be associated with other Internet Service Providers and Web Development Companies. We can help business cut time and expenses while reducing the valuable time they spend consulting with their clients. We want our clients to be confident that they can achieve success by partnering with us and using our services. We need your assistance to make our vision a reality. Business owners you know may need an introduction to know that services exist which, can solve their business promotion issues and maximize their ability to increase name recognition and profitability.

Back to WHG Members

WHG News | Who We Are | What We Do | WHG Members | Why Certification? | Apply For Certification | News Flash

Figure 6-21
Web Host
Guild

One disadvantage of outsourcing Web site operations is lack of control. For example, an ISP or Web hosting company may restrict the kinds of software programs that can be installed on its servers. One solution to this lack of control is for an e-business to provide its own servers and software and simply co-locate its servers at the ISP or Web hosting company facilities. E-businesses that **co-locate** their servers rent space and Internet connectivity for their own servers from an ISP or Web hosting company. Co-locating servers allows an e-business to maintain control over its servers while taking advantage of the benefits of the high-speed transmissions available from an ISP or Web hosting company. Web hosting companies or other technology companies that provide Web hosting services, such as Interliant (Figure 6-22) often also provide co-location services.

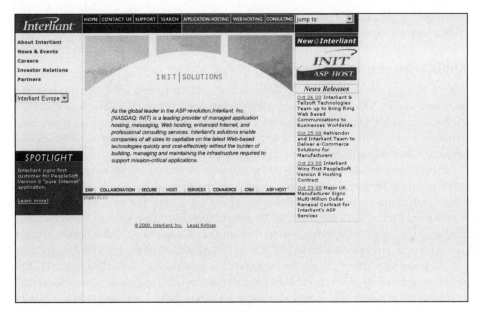

Figure 6-22
Interliant

E-CASE IN PROGRESS foodlocker.com

In setting up its e-business, foodlocker.com had many issues to deal with, from legal issues to office space to technology. For legal issues, foodlocker.com retained the services of a law firm that one of the founders had used for previous business ventures. Because of the founder's prior experience with the law firm, foodlocker.com knew the legal firm's experience and capabilities, including familiarity with startup companies attempting to secure private-placement (angel) financing.

Initially, foodlocker.com had no office space — all work was done out of the founders' homes. In order to hire employees and begin operations, foodlocker.com had to rent office space. foodlocker.com did not want to sign a long-term lease, so there were limited options available. The choice was either one of several executive suites or a new technology-oriented executive suite designed for dot-com companies, which included Internet connections, use of high-quality printers, and IT support. Before choosing one of these two options, foodlocker.com created a strategic partnership with a local specialty food store chain that then provided foodlocker.com with office space at the chain's main campus. The office space included Internet connectivity, office furniture, and use of the phone system. foodlocker.com was able to pay a greatly reduced rent because of its relationship with its strategic partner and because the strategic partner's office space and equipment were not currently being used.

Because the office space included Internet connectivity, it was no longer necessary to choose an ISP. The only service needed was Web hosting. foodlocker.com selected the Yahoo! Store for their online store system on the basis of a list of criteria and a cost-benefit analysis. Secondary Web hosting was also set up with another company with which the founders were familiar. The secondary Web hosting was used to provide both functionality not offered by the Yahoo! Store and a platform for development work.

Another alternative is to rent or lease application software or to purchase other technology services from an application service provider.

Application service providers

Many e-business startups lack the in-house technology expertise to get their e-business idea launched, so they outsource their technology needs. Another reason many e-businesses outsource their technology needs is to reduce the costs associated with developing and maintaining in-house technology facilities and expertise. In either case, many e-businesses now outsource their technology needs to application service providers.

Application service providers (ASPs) deliver and manage software applications and other computer services for multiple customers from remote data centers, using private networks or the Internet. Many large software and hardware companies are jumping on the ASP bandwagon. Microsoft, Oracle, SAP, Dell Computer, IBM, AT&T, and Qwest Communications among others are now providing ASP services. In August 2000, Gartner, the market research group, predicted that the worldwide ASP market will grow to more than $25.3 billion by 2004.

Different ASPs offer a diverse menu of services. However, ASPs typically provide IT infrastructure, software licenses, application support, database administration, system administration, software upgrades, and technical-problem resolution for their clients. Client companies, who access the software applications remotely using leased communication lines or the Internet, may pay a monthly fee for these services or may pay a per-transaction fee. One leading ASP is USInternetworking (Figure 6-23), which partners with major companies such as PeopleSoft (human resources and financial applications), Microsoft (e-business, exchange hosting, and messaging applications), and Sieble Systems (customer information applications) to provide a variety of packaged application software. USInternetworking also provides and manages strategically located enterprise data centers for hosting client operations.

Figure 6-23
USInternet-
working

There are many advantages for a startup e-business that outsources all or part of its technology needs to an ASP. An e-business startup should consider using an ASP in order to:

- Quickly get the e-business launched. ASPs already have the equipment, software applications, and Internet technology expertise to help launch a startup e-business more quickly.
- Focus efforts on the principal e-business activities. Outsourcing computer systems and Web management can allow a startup e-business's management team to concentrate on core business functions.
- Take advantage of an ASP's superior level of IT experience and services.
- Reduce hardware, software, and personnel costs.

Although there are many advantages to a startup e-business in using an ASP, there are some disadvantages as well. How well an e-business's applications perform depends on how well the ASP performs. Additionally, it may be very expensive to customize application features.

Before contracting with an ASP, an e-business should learn about the ASP's business practices: how big its data centers are, how many locations it maintains, what its management policies are, how it monitors capacity and performance problems, how its staff handles hardware and/or software changes, and whether or not it provides user training. A service agreement with an ASP should define the services to be provided, including response time for online transactions and service calls, delivery time for paper documents, and procedures to be followed in case of special or unusual circumstances such as natural disasters or Internet hacking.

E-CASE **A Technology Solution to a Paperwork Nightmare**

In 1998, when Jon Fisher was running the press office for a candidate for the office of California Secretary of State, he interacted with many union members and construction companies. Both union members and contractors expressed frustration with the time and cost involved in requesting, completing, submitting, and archiving municipal building permits. Fisher quickly realized that a technology solution rather than a political solution was needed. He partnered with a technology analyst and incorporated NetClerk, Inc. in January 1999.

NetClerk's forms management service is set up to provide homeowners and contractors with a way to save time and money by filling out and submitting building permits online. The completed permits are then delivered to the appropriate municipality or county office via fax, e-mail, or courier, depending on municipality or county requirements. As of this writing, NetClerk charges homeowners $50 for three permits and allows contractors to subscribe and pay $100 to $2,000 per month, depending on the volume of permits processed. As of September 2000, NetClerk's services were available in 900 cities across the country.

(Continued on the next page)

From the beginning, NetClerk was committed to using the best available technology and turned to one of the leading technology service providers, Oracle, to get up and running quickly. NetClerk engaged Oracle Consulting to help it develop and implement its new service in only six months. NetClerk used Oracle Consulting to help it select its hardware (Sun Microsystems Enterprise servers and Cisco firewalls) and software (Oracle8 Database Server, Oracle Application Server, Oracle Alert, Oracle Financials, and Oracle Purchasing application software). NetClerk (Figure 6-24) also outsourced other aspects of its building permit services, such as its faxing services, and co-located its servers with one of Oracle's strategic partners, Exodus Communications.

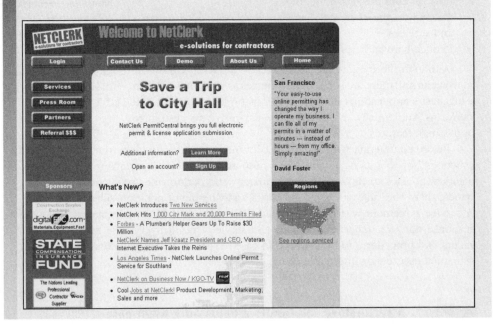

Figure 6-24
NetClerk

When a client business requires customized software application services, some ASPs partner with system integrators to provide the needed customization.

System integrators and Web integrators

System integrators (SIs) oversee the implementation of software applications. They also work with client businesses to configure specific application functions and enhancements, as well as providing consulting and technical support. **Web integrators** typically consult on Web strategies, Web application development, and integration of Web applications with older computer systems and ERP applications. Web integrators include iXL, Scient, Viant, and Sapient. For example, when deregulation of the energy industry allowed the trading of energy as a commodity, the principals at HoustonStreet Exchange saw an opportunity to develop an online energy-trading portal. HoustonStreet Exchange looked to a Web integrator, Sapient (Figure 6-25), to develop the necessary proprietary software and handle the project management needed in the development of the portal.

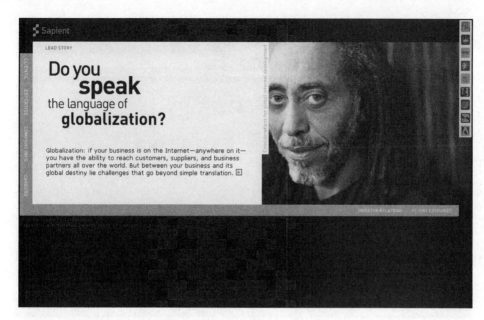

Figure 6-25
Sapient

At the bank, Dachis looked at the list of 10 names and chose the only one that stood out, Razorfish. Today, Razorfish is a publicly traded company (RAZF) with a home office in New York and branch offices in England, Germany, Italy, Norway, Sweden, and Japan. As of this writing, Razorfish has 1,800 employees worldwide and clients that range from Charles Schwab and eBay to the U.S. Navy. Originally known for its Web design services, Razorfish acquired a number of back-end systems providers and consultants, such as I-Cube, and expanded its expertise to provide a complete menu of Web strategy, design, and technology services.

Along the way, Razorfish used its rather unusual name to create a brand synonymous with creativity and a unique corporate culture. Razorfish's culture, based on speed of actions and creativity of solutions, is found in everything the company does: from the clients and projects it chooses, to the people it hires, and even to the size of its business cards. Razorfish has, in the past, turned down clients and projects that weren't a good "cultural fit" and turned away as many as 80 percent of job applicants also considered not a good "cultural fit," even in a tight labor market. Razorfish eschews standard 3.5 × 2 inch business cards in favor of 1.5 × 3.5 cards. For employees, the company culture rewards creativity with good pay and interesting perks: for example, bringing pets to work, three-day parties in Las Vegas, whitewater rafting in Oregon, and seminars in Sweden.

In September 2000, Razorfish was named the second-fastest-growing technology company in Deloitte & Touche's "Fast 50." Using a memorable name and unique corporate culture to brand the company, Razorfish (Figure 6-26) has successfully differentiated itself from other Web integrators.

(Continued on the next page)

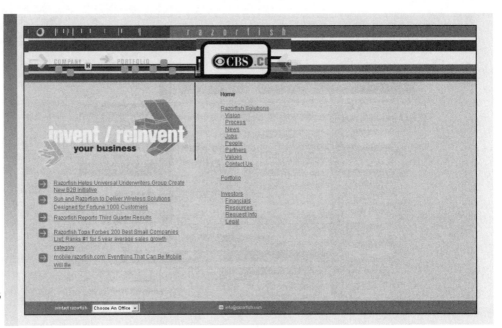

Figure 6-26

Razorfish

Summary

- One of the first things an entrepreneur should do is establish a relationship with an experienced attorney and accountant.
- A copyright is a legal protection for the author of an original work, including writing, drama, music, art, and other intellectual property.
- A trademark or service mark is a distinctive symbol, word, or phrase that identifies a business and its products and services.
- An e-business should post a liability disclaimer on its Web site to protect it from lawsuits.
- An e-business should have a clear policy about how it handles and secures the information it collects at its Web site.
- The laws affecting e-business taxation continue to change and should be monitored.

- An e-business should create a facilities need plan before looking for commercial office space and should use the services of a commercial real estate broker to locate and negotiate for such space.
- Startup e-businesses often hire needed employees as funds become available to cover salaries; first hires should be senior management, including a chief technology officer and a marketing professional.
- A brand is a combination of name, logo, and design that differentiates an e-business's products and services from those provided by other e-businesses.
- E-businesses may elect to outsource their technology needs with ISPs, Web hosting companies, ERP and CRM system providers, ASPs, or Web or system integrators.

❑ Do your homework — research several ISPs, Web hosting companies, ASPs, and Web integrators, and then choose the approach that best meets your e-business needs.

❑ When evaluating technology service providers, look for those that have a good reputation and those that will provide you with a list of current customers you can interview.

❑ Determine whether or not ISPs and Web hosting companies have sufficient servers, communication lines, and bandwidth to quickly respond to a growing volume of transactions at your e-business's Web site and their other customers' Web sites.

❑ Understand the level of service you expect to receive, including domain name registration, content management, backup and data recovery operations, server maintenance, and Internet connection management.

❑ Determine how easy it will be to update your Web site, manage files, and perform other housekeeping tasks with different service providers.

❑ Be sure the service provider offers technical support 24 × 7 × 365, and try to evaluate whether or not the service provider has adequate staff to support all its customers.

❑ Consider how service providers will handle scheduled and unscheduled downtime and how quickly their technical support staffs can respond to problems.

❑ Address issues of confidentiality and security, and determine each service provider's internal security features.

❑ Identify all of the costs, including setup fees, monthly fees, and costs for additional bandwidth. Consider agreeing to a shorter period of time for any service agreement with an automatic renewal feature rather than a long-term agreement that locks you into a relationship that is not working for your e-business.

❑ Specify how your e-business can get out of a service agreement if the relationship with the service provider does not work, including how your e-business can transition its technology needs to another service provider.

Key Terms

application service providers
(ASPs)
bandwidth
brand
burn rate
co-located server
copyright
customer relationship
management (CRM)
cybersquatters

data mining
data warehouse
domain name
domain name brokers
enterprise
enterprise resource planning (ERP)
fair use
independent software vendors
(ISVs)
Internet access providers (IAPs)

Internet service providers (ISPs)
perquisite (perk)
service mark
system integrators
trademark
value-added reseller (VAR)
Web hosting companies
Web integrators
Web mining

Review Questions

1. One of the first things an entrepreneur should do is to:
   a. Hire an ASP.
   b. Create an ERP system.
   c. File a state sales tax return.
   d. Establish a relationship with an experienced accountant.

2. An ASP:
   a. Focuses on Web hosting services.
   b. Is a nonprofit organization focusing on information privacy.
   c. Delivers and manages software applications for multiple customers from remote locations.
   d. Is part of a friends and family network of potential investors.

3. Web integrators typically provide:
   a. Web application development and integration with older computer systems.
   b. Accredited domain name registration.
   c. Space in which an e-business can co-locate its servers.
   d. Assistance in locating commercial office space.

4. Advice on solving business problems related to taxation, content liability, employee benefits and compensation, and equity distribution is best solicited from a(n):
   a. ISV.
   b. Attorney.
   c. VAR.
   d. IAP.

5. ERP is:
   a. A system that connects an e-business's Web site with other functions such as accounting, inventory management, and resource allocation.
   b. Used to manage an e-business's customer information.
   c. A company that develops and sells software applications.
   d. A company that oversees the implementation of software applications.

6. At the time of this writing, e-businesses are required to collect sales taxes in every state in which their customers reside. **True or False**?

7. A data warehouse is a huge facility in which e-businesses co-locate their servers. **True or False**?

8. The key to success for any e-business is in its people. **True or False**?

9. Information privacy is not a major issue with consumers who buy products or services on the Web. **True or False**?

10. In a tight labor market, an e-business may need to use a combination of compensation, equity, and perks to hire and keep talented employees. **True or False**?

Exercises

1. Using Internet search tools or other relevant sources, locate three e-businesses following different e-business models, and then review and print the terms of use or other liability disclaimers available at their Web sites.

2. Review the Web sites for iXL, Scient, Viant, and Razorfish (or four Web integrators of your choice). Using your Web site analysis, business periodicals, and other relevant sources, write a one- or two-page paper describing two of the Web integrators, including their services, clients, strategic partners, and management team.

3. Using Internet search tools and other relevant resources, locate and review the current salaries for your region for the following positions: chief information officer (CIO), accountant, marketing director, Web designer, Webmaster, database programmer, and executive assistant. Write a one-page paper listing each position, the salary range for each position, and your data sources.

4. Using Internet search tools and other relevant resources, locate three professional recruiting firms in your area and one global executive recruiting firm. Then write a brief paragraph about each firm, describing its recruiting business scope.

5. Using the EcommerceTax.com Web site, Internet search tools, and other relevant resources, locate two or three articles on the current status of e-business sales taxes. Write a one-page paper summarizing the articles.

CASE PROJECTS

◆ 1 ◆

You and a fellow business school graduate have started a new e-business following the B2B exchange model, and you have been working in a temporary office in a spare bedroom. Although you have outsourced the technology development for your e-business to a Web integrator, you now find that you need to hire a receptionist and a director of marketing, and there is no room for two additional employees in the home office. Additionally, you are ready to set up meetings with prospective clients and need a more professional atmosphere in which to conduct those meetings. Create a detailed facilities plan for the commercial office space you need, including the area in which you want to locate your office. Then research the available commercial office space in that area. Write a two- or three-page detailed facilities plan, including a budget, a description of the commercial office space you want to lease, and the reasons for your choice.

◆ 2 ◆

You are creating a startup e-business following the B2C model and need to decide whether to use an ISP or a Web hosting company to host your Web site. Use Web sites such as ISPs.com, the Hosting Repository, and HostSearch to locate and review two ISPs and two Web hosting companies. Then write a one- or two-page paper defining your B2C e-business idea and analyzing the selected ISPs and Web hosting companies. Make a decision on which company you want to host your Web site, and give the reasons for your decision.

◆ 3 ◆

You are the sales director for a B2C e-business. Next week, the management team is having a three-day brainstorming session on how to increase the number of repeat customers and how to target specific customers on the basis of their previous buying habits. The CEO asks you to prepare a short presentation on tools available to help the e-business gather and analyze customer information. Using Internet search tools and other relevant sources, research data mining and Web mining concepts and tools. Then prepare a one - page outline you can use to guide your presentation.

TEAM PROJECT

You and three classmates are starting an e-business following the B2B e-business model. It's now time to decide on a name for the e-business. You all agree that the business name and the related domain name should be part of the branding process the e-business uses to differentiate its products or services. Meet with your associates and (1) define the e-business, (2) use whatever brainstorming techniques you want to come up with an appropriate e-business name and domain name, and (3) verify that the domain name you select is available.

Using Microsoft PowerPoint or another presentation tool, create a 5- to ten-slide presentation and then present the e-business name and proposed domain name to a group of classmates selected by your instructor, who will evaluate your choice, using the following criteria:

◆ What personalities does the brand name convey?
◆ What attributes does the name suggest about the products or services offered?
◆ Who are the customers the e-business is trying to reach?

After your classmates' analyses, reveal the e-business idea, including products and services to be offered. Have your classmates determine how successfully the e-business name and domain name link to the e-business idea.

Useful Links

100 Top Web Hosting Sites – The Complete Web Hosting Portal
http://www.100topwebhostingsites.com/

Absolute Domain Names
http://www.absolutedomainnames.com/

All About ASP
http://aspindustry.com/

American Marketing Association
http://www.ama.org/

An Atlas of Cyberspaces
http://www.geog.ucl.ac.uk/casa/martin/atlas/atlas.html

ASP Industry Glossary
http://www.aspindustry.org/glossary.cfm

ASP Sites Resource Guide
http://www.aspsites.com/

ASPnews.com
http://www.aspnews.com/

Berkman Center for Internet and Society at Harvard Law School
http://cyber.law.harvard.edu/

Boardwatch Directory of Internet Service Providers (ISPs)
http://thelist.internet.com

ClickZ Network
http://www.clickz.com/

Copyright & Fair Use – Stanford University
http://fairuse.stanford.edu/

CRM-Forum
http://www.crm-forum.com/

Data Warehousing Knowledge Center
http://www.datawarehousing.org/

Domains Auction
http://www.domainsauction.com/

Electronic Commerce & Internet Taxation
http://www.caltax.org/ecommerce.htm

Electronic Commerce Knowledge Center
http://www.commerce.org/

Electronic Privacy Information Center
http://epic.org/

e-newsletters
http://e-newsletters.internet.com/

Enterprise Resource Planning Resource Center
http://www.cio.com/forums/erp/

Enterprise Systems Journal
http://www.esj.com/

ERP Fan Club and User Forum
http://www.erpfans.com/

ERP-People.com
http://www.erp-people.com/

FedStats – One-Stop Shop for Government Statistics
http://www.fedstats.gov/

findAspace.com – Checklist of important space attributes
http://www.findaspace.com/Resources/space_
    checklist.htm

Forrester Research, Inc.
http://www.forrester.com/Home/0,3257,1,FF.html

Franklin Pierce Law Center — Intellectual Property Mall
http://www.ipmall.fplc.edu/

GreatDomains.com
http://www.greatdomains.com/

Intellectual Property Center
http://www.ipcenter.com/

ISP Buyer's Guide – CNET
http://www.cnet.com/internet/0-3762-7-2518426.html?
    tag=st.int.3762.prl.3762-7-2518426

ISPcheck – ISP and Web Hosting Search
http://www.ispcheck.com/

ITtoolbox Portal for CRM
http://www.crmassist.com/

ITtoolbox Portal for ERP
http://www.erpassist.com/

John Marshall Law School – Center for Information Technology and Privacy Law
http://www.jmls.edu/InfoTech/index.html

Kdnuggets – Data Mining, Web Mining, Knowledge Discovery, and eCRM
http://www.kdnuggets.com/index.html

law.com – Intellectual Property Practice Center
http://www.law.com/professionals/iplaw.html

NewRegistrars.com
http://www.newregistrars.com/

RadioWallStreet
http://www.radiowallstreet.com/

searchCRM.com
http://www.searchcrm.com/

TechRepublic ERP Focus
http://www.erpsupersite.com/

The Intellectual Property Law Server
http://www.intelproplaw.com/

The Internet Marketing Center
http://www.marketingtips.com/

The List – The Definitive ISP Buyer's Guide
http://thelist.internet.com/

The Privacy Page
http://www.privacy.org/

TopHosts.com – Web Hosting Resource
http://www.tophosts.com/

U.S. Department of Labor – elaws Advisors
http://www.dol.gov/elaws/

Vertex Inc. – eTax Central
http://www.vertexinc.com/etax_central/etax_central_
    home.asp

WebHarbor.com – ASP Industry Portal
http://www.webharbor.com/

Web Hosting Guide – CNET
http://webisplist.internetlist.com/

Web Marketing Info Center
http://www.wilsonweb.com/webmarket/

Webopedia
http://www.webopedia.com/

whatis?com
http://www.whatis.com/

Links to Web Sites Noted in This Chapter

Absolute Domain Names
http://www.absolutedomainnames.com/

ACT!
http://www.act.com/home/default.php3

Activespace
http://www.activespace.com/

Affinity
http://www.affinity.com/

Amazon.com
http://www.amazon.com/

American Marketing Association
http://www.ama.org/

Anonymizer.com
http://www.anonymizer.com/

ASG (Viasoft)
http://www.asg.com/

Ask Jeeves
http://www.ask.com/

AT&T Solutions
http://www.att.com/solutions/

Barnes & Noble
http://www.bn.com/

BigForms.com
http://www.bigforms.com/

Bitlocker
http://www.bitlocker.com/site?c=1

Blockbuster Inc.
http://www.blockbuster.com/

British Airways
http://www.britishairways.com/regional/usa/

BroadVision
http://www.broadvision.com/OneToOne/SessionMgr/
    home_page.jsp

buy.com
http://www.buy.com/selectcountry.asp

cars.com
http://www.cars.com/carsapp/national/?srv=parser&act=
    display&tf=/index-default.tmpl

Catchword
http://www.catch-word.com/

Chemdex
http://www.chemdex.com/

Critical Path
http://www.cp.net/

Cyberchefs
http://www.cyberchefs.com/

Dell Computer Corporation
http://www.dell.com/us/en/gen/default.htm

DoubleClick
http://www.doubleclick.net/us/

EarthLink
http://www.earthlink.com/

eBay
http://www.ebay.com/

EcommerceTax.com
http://www.ecommercetax.com/

Ernst & Young
http://www.ey.com/global/gcr.nsf/US/US_Home

e-Stamp
http://www.e-stamp.com/cgi-bin/e-stamp.dll/index.jsp

Exodus Communications
http://www.exodus.com/

foodlocker.com
http://www.foodlocker.com/

FormSite.com
http://www.formsite.com/

Gartner
http://gartner5.gartnerweb.com/public/static/home/
    home.html

Global Crossing Ltd.
http://www.globalcrossing.com/

Hawk Communications
http://hawkcommunications.com/

Heidrick & Struggles International, Inc.
http://www.heidrick.com/

Hosting Repository
http://www.hostingrepository.com/

HostSearch
http://www.hostsearch.com/

Hotmail
http://lc4.law5.hotmail.passport.com/cgi-bin/login

HoustonStreet.com
http://www.houstonstreet.com/

IBM
http://www-3.ibm.com/e-business/overview/28210.html

ICANN – List of Accredited Registrars
http://www.icann.org/registrars/accredited-list.html

Information Discovery, Inc.
http://www.datamining.com/

interliant
http://webhosting.interliant.com/

International Trademark Association
http://www.inta.org/

ISPcheck
http://www.ispcheck.com/

ISPs.com
http://www.isps.com/

iXL
http://www.ixl.com/

J.D. Edwards
http://www.jdedwards.com/

Korn/Ferry International
http://www.kornferry.com/

Los Angeles Times
http://www.latimes.com/

Merisel, Inc.
http://www.merisel.com/

MicroAge
http://www.microage.com/

Microsoft Corporation
http://www.microsoft.com/

MP3.com
http://www.mp3.com/

NameLab
http://www.namelab.com/

NameProtect.com
http://www.nameprotect.com/

NetClerk
http://www.netclerk.com/htdocs/index.html

Network Solutions
http://www.networksolutions.com/

New York Times on the Web
http://www.times.com/

Office.com
http://www.office.com/

Oracle
http://www.oracle.com/

Oracle – Data Mining Products
http://www.oracle.com/datawarehouse/products/datamining/

PeopleSoft
http://www.peoplesoft.com/

priceline.com
http://www.priceline.com/

Qwest Communications International, Inc.
http://www.qwest.com/

Razorfish
http://www.razorfish.com/

Reel.com
http://www.reel.com/

register.com
http://www.register.com/

RightWorks
http://www.rightworks.com/home/index_flash.html

SalariesReview.com
http://www.salariesreview.com/surveys/national_pay.cfm

SAP
http://www.sap.com/

Sapient
http://www.sapient.com/home/home.asp

Scient
http://www.scient.com/flash/index.html

Siebel Systems
http://www.siebel.com/

Spencer Stuart
http://www.spencerstuart.com/

Technical Search Corporation
http://www.techsearch.com/

The Copyright Website
http://www.benedict.com/

The Namestormers
http://www.namestormers.com/

The Web Host Guild
http://www.whg.org/index.htm

TimeTech Investors
http://www.time-tech.com/

TransQuest Ventures
http://www.transquest.net/

TRUSTe
http://www.truste.com/

U.S. Copyright Office – Library of Congress
http://lcweb.loc.gov/copyright/circs/circ1.html

U.S. Patent and Trademark Office
http://www.uspto.gov/web/menu/tm.html

USInternetworking, Inc.
http://www.usi.net/index2.html

Verbind
http://www.verbind.com/

Viant
http://www.viant.com/

Viasoft, Inc.
http://www.asg.com/

WebHosting.com
http://www.webhosting.com/

WorldNet
http://www.world-net.net

Xchange – e-CRM
http://www.xchange.com/default.asp

xDSL.com – Analysis of DSL Technologies
http://www.xdsl.com/default.asp

XO
http://www.world-net.net/

Yahoo!
http://www.yahoo.com/

Zero-Knowledge Systems
http://www.zeroknowledge.com/default.asp

For Additional Review

Aaker, David A. et al. 2000. "Brand News: Want To Give Your Stock a Boost? Then Win Over the Public With a Stellar Brand Strategy," *Business 2.0*, September 15. Available online at: http://www.business2.com/content/channels/ebusiness/2000/09/15/18546.

Akdeniz, Yaman. 2000. "New Privacy Concerns: ISPs, Crime Prevention, and Consumer's Rights," *International Review of Law, Computers, & Technology*, 14(1), March 1, 55. Available online at: http://www.bileta.ac.uK/99papers/aKdeniz.htm.

Apicella, Mario. 2000. "Selecting an ERP Service Provider," *InfoWorld*, 22(26), June 26, 50.

Berst, Jesse. 2000. "R. I. P. – The Death of Online Privacy," *ZDNet AnchorDesk*, October 16. Available online at: http://www.zdnet.com/filters/printerfriendly/0,6061,2641042-10,00.html.

Birritteri, Anthony. 2000. "Internet Eases ERP Roll Outs," *New Jersey Business*, 46(7), July 1, 66.

Booker, Ellis. 2000. "ABCs of ASPs: Renting Marketing Software Across the Net Often Means Saving Money and – More Importantly – Time," *B to B*, 85, August 14, 29.

Bott, Ed. 2000. "We Know Where You Live Work Shop Bank...And So Does Everyone Else!" *PC/Computing*, March, 80.

Bucholtz, Chris. 2000. "Signed Sealed and Delivered: Show Clients How to Sign on the Cyber-Dotted Line," *VARBusiness*, 16(18), September 4, 67.

Business Week. 2000. "Give Me That Old Time Economy," 3678, April 24, 99.

Business Wire. 2000. "Web Hosting Discussion on RadioWallstreet.com: Bob Weissman is Telecom," September 6. Audio available online at: http://www.radiowallstreet.com/NASApp/RWS/EventPage?ID=38130.

Caggiano, Christopher. 2000. "Five Ways to Save Money on Office Space," *Inc. Magazine*, April 24. Available online at: http://www.inc.com/articles/details/0,3532,AGD8_ART18579_CNT56_GDE38,00.html.

Carpenter, Phil. 2000. *eBrands: Building an Internet Business at Breakneck Speed*. Boston, MA: Harvard Business School Press.

Caton, Michael. 2000. "Let Others Handle Your Mail," *eWeek*, June 19, 61.

Caton, Michael. 2000. "Some ASPs Taking New Tacks," *eWeek*, August 28, 47.

Caulfield, Brian. 1999. "Making a Business of E-mail Hosting," *Internet World*, 5(26), August 15, 59.

CCH Business Owner's Toolkit. 2000. "Business Facilities Outside the Home." Available online at: http://www.lycos.com/business/cch/guidebook.html?lpv=1&docNumber=P04_1200.

Cherry Tree & Co. 1999. "Application Service Providers (ASP) Spotlight Report," October. http://www.webharbor.com/download/ct_whppr.pdf.

Cirillo, Rich. 2000. "Who Are You Really? A Web Integrator? An ASP? A VAR?" *VARBusiness*, 16(13), June 26, 40.

Cope, James. 2000. "Web Hosting," *ComputerWorld*, 34(5), January 31, 61.

Corbitt, Terry. 2000. "ISPs: Making the Right Choice," *Credit Management*, August 1, 32 – 33. Available online at: http://library.northernlight.com/AA20000814030007298.html?cb=0&sc=0#doc.

Crane, Elizabeth. 2000. "Double Trouble – DoubleClick Learned the Hard Way, So You Don't Have To," *Ziff Davis Smart Business for the New Economy*, October 1, 42.

Cross, Kim. 2000. "Need Options? Go Configure," *Business 2.0*, February, 121.

Daniel, Lucas. 2000. "Swimming With Razorfish," *Web Techniques*, 5(17), July, 40.

Donahue, Sean. 1999. "New Jobs for the New Economy: From Email Channel Specialists to Chief Community Strategists, the Internet is Creating New Workplace Roles," *Business 2.0*, July 1. Available online at: http://www.business2.com/content/magazine/indepth/1999/07/01/19788.

Donahue, Sean. 2000. "Supply Traffic Control," *Business 2.0*, February, 130.

Doyle, T. C. 2000. "New Economy, New Culture," *VARBusiness*, 16(14), July 10, 27.

Dugan, Sean M. 2000. "NET PROPHET: Caught Between a Rock and a Hard Place: How Do You Make Online Privacy Policies Stick?" *InfoWorld*, 22(39), September 25, 104.

Durlacher Research. 1999. "Application Service Providers," July. http://www.webharbor.com/researchreportsf.shtml.

Farmer, Melanie Austria. 2000. "ASPs Growing by Leaps and Bounds, Analysts Say," *CNET News.com*, August 9. Available online at: http://news.cnet.com/news/0-1007-200-2480640.html?tag=st.ne.1007.

Feuerstein, Adam. 1999. "Bay Area Firm Aims to Manage World's E-Mail," *Sacramento Business Journal*, 16(39), December 10, 31.

findaspace.com, inc. 2000. "When to Leave the Spare Bedroom." Available online at: http://www.findaspace.com/Resources/moving_out.htm.

Flanagan, E. B. 2000. "The Race Inside the ASP Market," *VARBusiness*, 16(18), September 4, 22.

Follett, Jennifer Hagendorf and Torode, Christina. 2000. "ASPs: Evolution of the Species – Application Hosting Providers Cultivate More Complex Service Offerings," *Computer Reseller News*, September 4, 71.

Fortune. 2000. "The Trauma of Rebirth: Faced With Changing Internet Currents, Companies Are Redefining Themselves. Luminate Found That It's Hard To Do," 142(5), September 4, 367ff.

Georgia, Bonny L. 2000. "E-Mail the Way It Should Be," *P/C Computing*, April, 117.

Geraghty, Jim. 2000. "The Fight Over Internet Taxation," *Policy.com*, March 22. Available online at: http://www.policy.com/news/dbrief/dbriefarc576.asp.

Gilman, Hank and Ioannou, Lori. 2000. "How to Hire Smart: Tired of the Same Old 'C' Players Dragging Down Your Company? How Do You Recruit Only the Cream of the Crop? We Asked the Expert," *Fortune Small Business*, 10(5), July-August, 58.

Gonzales, Dan and Johnson, Robb. 1999. "Too Big for the Townhouse Basement, But Too Small for a Commercial Landlord," *Washington Business Forward*, August. Available online at: http://www.staubach.com/staubach/home.nsf/main/knowledge-townhouse.

Gonzales, Dan. 2000. "The Science of Securing Office Space," *Business 2.0*, March, 153.

Green, Jeff. 2000. "Employees By the Round: Who's the Key Hire? Who Should You Put Off Until Round Three?" *Business 2.0*, March, 168.

Harvin, Robert. 2000. "In Internet Branding, the Off-Lines Have It," *Brandweek*, 41(4), January 24, 30.

Hawley, Noah. 2000. "Brand Defined," *Business 2.0*, June 13. Available online at: http://www.business2.com/content/channels/marketing/2000/06/13/12401.

Heselbarth, Rob. 2000. "NetClerk Reports Getting Permits Online is a Success," *Contractor*, 47(3), March, 3.

Hilton, Lisette. 2000. "Smart Tech Companies Know It's About the People," *South Florida Business Journal*, 20(38), May 5.

Information Technology Association of America. 2000. "Executive Summary – Bridging the Gap: Information Technology Skills for a New Millennium," July. Available online at: http://www.itaa.org/workforce/events/wfwc1.ppt.

International Data Corporation. 1999. "The ASPs' Impact on the IT Industry: An IDC-Wide Opinion," September. Available online at: http://www.idc.com/Store/Free/PDFs/20323.pdf.

Jastrow, David. 2000. "Getting the ASP Model Right – VirtualSellers.com Offers Soup-to-Nuts Solution for Handling Online Transactions," *Computer Reseller News*, August 21, 119.

Karpinski, Richard. 2000. "E-jibberish Comes to Hub Branding Efforts, " *B to B*, 85, June 19, 11.

Kotler, Philip. 1997. *Marketing Management: Analysis, Planning, Implementation, and Control*. Upper Saddle River, New Jersey: Prentice Hall.

Leon, Mark. 2000. "Trading Spaces," *Business 2.0*, February, 127.

Lukas, Aaron. 1999. "Tax Bytes: A Primer on the Taxation of Electronic Commerce," CATO Institute, December 17. Available online at: http://www.freetrade.org/pubs/pas/tpa-009es.html.

Maselli, Jennifer. 2000. "Growing Pains – ASPs and Their Customers Still Face Challenges With Pricing, Performance, and Training," *Information Week*, July 24, 42.

Maselli, Jennifer. 2000. "Which ASP is Right For You?" *Information Week*, August 28, 146.

Mazurkiewicz, Greg. 1999. "Getting Permits in Line? Soon You'll Get Them Online," *Air Conditioning, Heating, and Refrigeration News*, 208(15), December 13, 10.

Meeks, Brock N. 2000. "New Domain Name Suffixes Selected: ICANN Selection Process Fraught with Contention and Rancor," *MSNBC*, November 16. Available online at: http://www.msnbc.com/news/491013.asp.

Miller, Jeff. 2000. "Web Hosting Fastest Growing Field for Tech Entrepreneurs," Business First of Columbus, Inc., 16(32), March, 44.

Murray, Lori. 2000. "Finding the Perfect Host," Business First of Columbus, Inc., 17(1), September 1, 31.

National Consumers League. 2000. "Online Americans More Concerned about Privacy Than Health Care, Crime, and Taxes, New Survey Reveals," Press Release, Washington, D.C. October 4. Available online at: http://www.natlconsumersleague.org/pressessentials.htm.

Neuman, Alica. 2000. "A Better Mousetrap Catalog," *Business 2.0*, 117, February.

Nucifora, Alf. 2000. "Immutable Internet Branding Laws Can Ease Chaos," *Puget Sound Business Journal*, 21(14), August 11, 21.

Ojala, Marydee. 2000. "The Dollar Sign: The Business of Domain Names," *Online*, 24(3), May 1, 78. Available online at: http://www.onlineinc.com/onlinemag/OLtocs/OLtocmay00.html.

Oracle Business White Paper. 2000. "Internet Relationship Management and Personalization Powered by Data Mining Insights," July. Available online at: http://www.oracle.com.

Perez, Ernest. 2000. "Free Web Database Hosting from Soup to Nuts," *EContent*, 23(3), June, 86.

Pickering, Carol. 2000. "They're Watching You," *Business 2.0*, February, 135.

Ponnuru, Ramesh. 2000. "The Tax Man Cometh – How Will He Handle the Internet?" *National Review*, 52(11), June 19.

Press Release. 2000. "Razorfish Named as the Second Fastest-Growing Technology Company in Deloitte & Touche's 'Fast 50,'" September 21. Available online at: http://www.razorfish.com/news/search/pressreleases/pr_1_258a.htm.

Rector, Susan D. 2000. "Battle for Domain Names Switches to Online Arbitration," Business First of Columbus, Inc., 16(48), July 2, 32. Available online at: http://library.northernlight.com/UU20000804170007835.html?cb=0&sc=0#doc.

Ries, Al and Ries, Laura. *The 11 Immutable Laws of Internet Branding*. New York, NY: HarperCollins Publishers.

Rizzo, Katherine. 2000. "Big Losses for Government: Internet Costs State, Local Governments a Lot in Taxes," *The Associated Press*, Washington, July 25.

Roberts-Witt, Sarah L. 2000. "Loudcloud," *Internet World*, 6(13), July 1, 54.

Robinson, Edward. 2000. "Click and Cover: As the Privacy Debate Rages On, a Handful of Entrepreneurs Is Gambling on Business Models Built Around Your Online Identity," *Business 2.0*, August 22. Available online at: http://www.business2.com/content/magazine/indepth/2000/08/22/17926.

Rothfeder, Jeffrey. 2000. "MP3.com Ignored One Tiny Detail: It's Illegal," *SmartBusinessMag.com*, November, 42.

Rosencrance, Linda. 2000. "More.com Denies It Violated Privacy Policy," *Computerworld*, September 25, 28.

Russell, Keith. 2000. "Student Reaps Benefits of Selling Domain Names," *The Tennessean*, 14(153), August 2, Business Section.

Schibsted, Evantheia. 2000. "A Clean, Well-Lighted (and Ethernet-Ready, Nicely Ventilated, Espresso Bar-Equipped) Place: Stealing a Page From Starbucks' Playbook, TechSpace Hatches a Global Incubator Franchise," *eCompany*, 1(1), June, 267.

Schneider, Gary P. and Perry, James T. 2000. *Electronic Commerce*. Cambridge, MA: Course Technology.

Schneider, Jason. 2000. "The Hiring Crisis: How to Find, Keep, and Motivate Employees in the New Economy," *Smart Business*, 13(7), July, 84.

Selinger, Marc. 2000. "How Well Do Free ISPs Connect?" *The Washington Post: Fast Forward, FINAL E01*, August 11. Available online at: http://library.northernlight.com/UU0000811030001069.html?cb=0&sc=0#doc.

Seltzer, Larry. 2000. "Web Hosting Guide," *Internet World*, 6(9), May 1, 76.

Senia, Al. 2000. "Why I Still Play the Dot-Com Game: Internet Start-ups Remain a Hot Market for Solution Providers," *VARBusiness*, 16(16), August 7, 24.

Smith, Eric J. 2000. "ICANN Expanding Web Domain Names," *American Metal Market*, 108(139), July 2, 8. Available online at: http://library.northernlight.com/UU20000727210007977.html?cb=0&sc=0#doc.

Sound Consulting. 2000. "Understanding the ASP Market," Sponsored by Software & Information Industry Association (SIIA). Available online at: http://www.siia.net/divisions/enterprise/guide_to_asps.pdf.

Stafford, Jan. 2000. "Solve Your Customer Puzzle – e-CRM Completes the Missing Piece of the CRM Puzzle," *VARBusiness*, May 15, 77.

Stein, Nicholas. 2000. "Winning the War to Keep Top Talent: Yes! You Can Make Your Workplace Invincible!" *Fortune*, 141(11), May 29, 132.

Strugnell, Anne-Christine. 2000. "There's More to Outsourcing than ERP," *Profit Magazine*, February. Available online at: www.oracle.com/oramag/profit/00-Feb/p10erp.html.

Symanovich, Steve. 2000. "Douglas Hickey," *San Francisco Business Times*, 14(37), April 14, 14.

Timon, Clay. 2000. "10 Tips for Naming," *Business 2.0*, March, 151.

Torode, Christina and Follett, Jennifer Hagendorf. 2000. "Empowering SMBs With Hosted Services – Selling ASP Services Presents New Opportunity for SMB Solution Providers," *Computer Reseller News*, August 21.

Torode, Christina. 1999. "No. 4: The Channel Mavericks – Web Integrators Robert Howe of Scient and Jeff Dachis of Razorfish are Running Two of the Channel's Most Watched Startups as They Create the Model for Deploying E-commerce Solutions", *Computer Reseller News*, November 15, 135.

Treese, G. Winfield and Stewart, Lawrence C. 1998. *Designing Systems for Internet Commerce*. Reading, Massachusetts: Addison Wesley Longman, Inc.

Turban, Efraim, et al. 2000. *Electronic Commerce: A Managerial Perspective*. Upper Saddle River, NJ: Prentice-Hall, Inc.

U S. Copyright Office. 1999. "Copyright Basics (Circular 1)," June. Available online at: http://lcweb.loc.gov/copyright/circs/circ1.html.

U.S. Department of Labor. "Futurework—Trends and Challenges for Work in the 21st Century." Available online at: http://www.dol.gov/dol/asp/public/futurework/report.htm.

United States General Accounting Office Report to Congressional Requesters. 2000. "Sales Tax: Electronic Commerce Growth Presents Challenges; Revenue Losses Are Uncertain." Available online at: http://www.gao.gov/.

Vaas, Linda. 2000. "Customer Privacy Lockdown," *eWeek*, October 15. Available online at: http://www.zdnet.com/eweek.

Vaas, Linda. 2000. "Laws Pull Shades on User Data," *eWeek*, October 16. Available online at: http://www.zdnet.com/zdner.

Van Winkle, William. 2000. "What's in a Domain Name," *Home Office Computing*, 18(8), August, 81. Available online at: http://library.northernlight.com/UU20000829180014243.html?cb=0&sc=0#doc.

VARBusiness. 2000. "The Young New-Age Workforce," 16(13), June 26, 133.

VARBusiness. 2000. "What's Happening to Our Market: Welcome to the Age of the Great Divide," 16(13), June 26.

Virzi, Anna Maria. 2000. "Doug Hickey: Critical Path's CEO Is on Acquisition Binge Hoping to Make His Brand Well Known," *Internet World*, 6(8), April 15, 63.

Vizard, Michael. 2000. "From the Editor in Chief: The Future Is Bright for ASPs if They Adopt the Electrical-Utility Model," *InfoWorld*, 22(36), September 4, 87.

Walsh, Mark. 2000. "Jeffrey Dachis, 33, President and CEO, Razorfish, Inc." *Crain's New York Business*, 16, January 31, 20.

Wazeka, Robert. 2000. "Internet Privacy," *Success*, 47(4), September, 64.

Wayner, Peter. 2000. "Protecting Your Property in Cyberspace: The Web's Rise as a Platform for Commerce and Communications Is Driving Many Organizations to Protect Their Images, Sound, and Text as Well as Trade Secrets and Application Code," *Computerworld*, January 10, 68(1).

Wilder, Clinton. 1999. "Internet Branding – Starbucks, Williams-Sonoma Plan to Leverage Their Brand Names on the Web," *Information Week*, June 28, 77.

Woolley, Scott. 2000. "Gold Mine or Glut?" *Forbes*, June 12, 23.

CHAPTER 7

Designing an E-Business Web Site

In this chapter, you will learn to:

Set Web site goals and objectives

Analyze Web site structure

Recognize effective Web design techniques

Identify Web development tools

Discuss Web site testing and maintenance issues

Michael Bates and Stuart Wagner began discussing the Web site design for their e-business, food-locker.com, late in 1999. Bates's own experience with difficulty finding regional specialties, such as Texas barbecue sauce and New Orleans coffee and beignets, coupled with a highly mobile society, reinforced Bates's and Wagner's enthusiasm about the viability of their e-business idea.

After identifying foodlocker.com's potential customers as both experienced and inexperienced Internet users across the U.S. who were looking for a taste of home, Bates and Wagner decided that the primary goal of the Web site was to make it easy for anyone to find the food items at the Web site and to purchase them online. A second goal was to choose an intermediate technology solution that would allow them to quickly create an inexpensive and simple demo Web site they could use when meeting with potential suppliers and investors. A third goal was to avoid purchasing any hardware or software until they were certain which technology would work best for foodlocker.com in the long term. To meet these Web site goals, Bates and Wagner took it one step at a time.

Setting Web Site Goals and Objectives

The first thing that an e-business entrepreneur should do, before any Web pages are created, is determine the goals, objectives, and overall purpose of the e-business's Web site. Without clear, established, and measurable goals and objectives, the Web site may not have the focus it needs to be successful. To help determine the goals and objectives, you should ask and answer questions such as the following. Will the Web site:

- Allow customers to order products and services online?
- Provide technical support for products and services?
- Advertise products and services?
- Build the e-business's image and brand?
- Collect information about current and potential customers?
- Provide links to related Web pages?
- Provide general or industry information?
- Recruit employees?

A quick look at this list of sample questions clearly indicates that most e-business Web sites will likely have multiple goals and objectives. The answers to these and similar questions are used to determine the Web site's overall purpose.

Although many national law firms have long had a Web presence, there is an industry-wide trend toward revamping those Web sites. Additionally, many smaller local law firms are launching new Web sites. Several law firms in Birmingham, AL, such as Burr & Forman LLP; Maynard, Cooper & Gale, P.C.; and Lightfoot, Franklin & White, LLC (Figure 7-1), participated in this trend by setting new marketing, recruiting, and client contact goals for their Web sites.

Unlike most e-businesses, many law firms have not used their Web sites primarily to attract new business. Traditionally, new law firm clients come from referrals. However, some law firms are now setting marketing-related goals for their Web sites. For example, some local Birmingham law firms have determined that using their Web sites to describe their practice and areas of expertise is useful in attracting out-of-area clients who need the services of a local firm. Maynard Cooper, for example, recognizes that more and more of their potential corporate clients around the country are using the Internet. Therefore, one goal of Maynard Cooper's Web site is simply to establish an effective Web presence to attract those clients.

Another important goal for law firm Web sites is attracting new attorneys. Because today's law students are technologically savvy, law firms are increasingly recognizing the importance of a Web presence in the recruitment process. For example, recruiting was the primary goal behind Lightfoot, Franklin & White's Web redesign. Another firm, Burr & Forman, plans to include online chats with the previous year's new associates at their Web site. Finally, more and more law firms are adding additional content to their sites with the goal of enabling existing clients to keep up to date with industry news and client seminars. For example, Burr & Forman's Web site redesign includes a client resource center containing industry links.

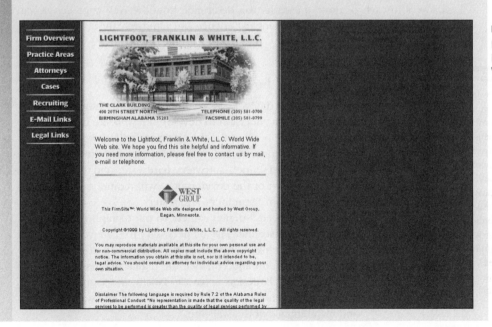

Figure 7-1
Lightfoot, Franklin & White LLC

Considering the Web site's target audience

After you establish the Web site's goals, objectives, and overall purpose, the next step is to consider the Web site's target audience. In too many instances, e-businesses design Web sites around the needs of the e-business rather than the needs of the target audience. It is critical to consider both the information the target audience wants and needs and the tools the target audience uses to access the Web. Again, a series of questions should be asked and answered:

- Is the audience composed of experienced Internet users, novice users, or a mix of both? Experienced Internet users will likely be able to handle a more sophisticated level of Web site complexity than novice users.
- What type of browser will the audience be using—Netscape, Internet Explorer, WebTV, or other specialized browsers? Some design techniques that are supported by later Web browser versions, such as animation and frames, may not be supported by earlier versions of the same browser or by other browsers.
- At what speed does the audience connect to the Internet—at 28.8 Kbps over a modem or over a high-speed dedicated connection? A Web site designed to be viewed successfully over a high-speed dedicated connection may be problematical for viewers using a slow modem connection.
- At what screen resolution does the target audience view Web pages—640 × 480, 800 × 600, or higher? Web sites designed to be viewed at 800 × 600 resolution without horizontal scrolling will greatly annoy those who view the site using a monitor with 640 × 480 resolution.

Answers to each of these questions will determine how the Web page design can enhance or detract from a visitor's viewing experience and, ultimately, the ability or failure of the Web site to meet its goals.

TIP

It is dangerous for an e-business to assume that all members of its target audience are using the latest Web browser versions, fast Internet connections, and large monitors, and then to design its Web site to include as many of the latest and greatest "bells and whistles" as possible. It is better to design a Web site to satisfy the most common technological constraints. To get a free, quick analysis of a Web site's loading time at different communication speeds, and of browser compatibility issues, check out the Web Site Garage site.

Planning the budget

When planning the Web site, another important issue to consider is the budget. The decision to develop the Web site in-house or to outsource some or all of the development will have an impact on the budget. In-house Web development may require the addition of technical personnel, software, hardware, and office space. Outsourcing part of the development will require contracts with Web designers, programmers, and testers.

You may need a digital camera to photograph products in a digital format. A scanner may be necessary to scan existing photos into a digital format. Software such as Adobe Photoshop may be needed to clean up and enhance the digital photos. The process may require Web design software. Finally, the e-business's staff must expend time and effort to participate in the design, testing, and maintenance of the Web site.

Once you have established the Web site's purpose, audience, goals, objectives, and budget, it is time to look at Web site structure and design.

Analyzing the Web Site's Structure

An e-business must carefully organize the information used as the Web site's content. A Web site must show visitors what information is available at the site, how to quickly find the information they want and need, and how to get additional information, if necessary. That information may include, but is not limited to:

◆ E-business's name
◆ Slogan, logo, or trademark
◆ Statement of mission or purpose
◆ Information on products or services
◆ Press releases and testimonials
◆ Employment information
◆ Contact information
◆ Maps to physical locations
◆ Web site map
◆ Customer support information
◆ Purchase or customer information forms

TIP

The business of developing Web sites is a $10 billion industry worldwide, and growing. In November 1999 ActivMedia published the results of a survey of 1,000 e-businesses, which suggests that the average cost of a Web site for a serious e-business is around $37,000. Media and portal e-businesses' average Web site development costs were $78,000, B2C site costs averaged $68,000, with "bricks-and-clicks" Web site costs averaging less than $30,000. While 34 percent of the e-businesses surveyed reported Web site development costs in the $1,000 to $5,000 range, e-businesses with larger budgets, 9 percent of which planned to spend more than $100,000 to develop a Web site, skewed the overall average.

Designing the Web site to be both attractive and well organized is the best way to give customers what they want at the site. Before considering the Web site content, you must design the site structure. Some Web site designs consist of a single level of separate and unrelated pages to which viewers link directly from a home page. Although easy to use, this flat structure can be somewhat boring. Some Web sites have multiple layers of linked pages creating a complicated structure that requires viewers to click through several pages to find the information they need. Viewing such unnecessarily complex Web sites can frustrate potential customers who want to find information quickly. One way to achieve balance in the structure of a Web site is to limit the number of linked pages and include as much important information as possible in the first three levels of linked pages.

When designing the structure of a Web site, it is a good idea to create a flow chart that diagrams the Web site's navigational structure. A good way to do this is to use a technique called storyboarding. A **storyboard** is a blueprint for a design, originally used to show copy, dialogue, and actions for a movie or television production. It consists of a board or panel containing a series of small drawings or sketches that roughly depict the sequence of actions. The storyboarding process can be used to good effect when designing the structure of a Web site.

To create a storyboard, first outline the contents of each Web page on individual sticky notes or index cards—one note or card for each page, including the home page. Then stick the notes or cards onto a whiteboard or bulletin board and arrange them in a hierarchical order of importance. The top note or card should represent the home page. The next level of notes or cards should be the other major pages at the Web site. The following levels should be the notes or cards for pages that provide details of each major area of the Web site.

Then draw connections from page to page. This presents a good picture of how a viewer would click through the site from page to page. It also helps identify potential navigation problems a viewer might experience. For example, if a Web site structure is 10 levels deep, a viewer might have to click 10 times through those 10 levels of pages to find useful information. Most users want to find information within three to five mouse clicks. The remedy for a Web site with ten levels is to organize the site into a flatter structure with fewer levels, and to include a search tool that allows viewers quick access to pages below the third to fifth levels. Figure 7-2 illustrates a sample storyboarding process. After completing the storyboarding process, you should make a flow chart (Figure 7-3) of the Web site's final structure.

Figure 7-2
Storyboarding process

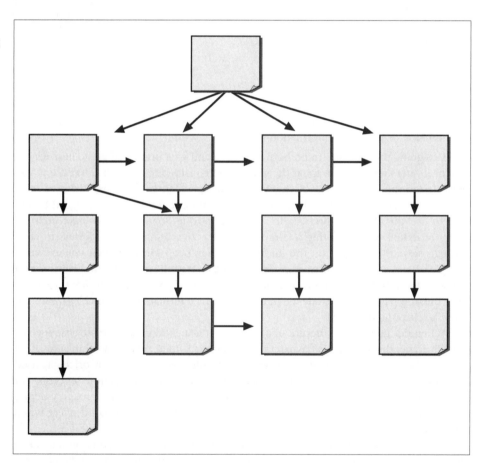

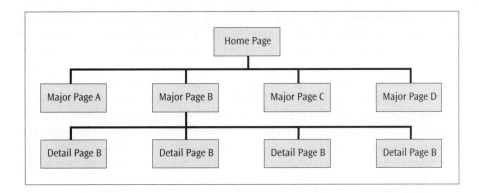

Figure 7-3
Sample Web site structure flow chart with three levels

After you have determined the Web site structure, the next step is to plan the overall design of the pages at the site.

Recognizing Effective Web Design

When doing business on the Web, the customer is truly in charge and can easily switch to an e-business's competitor with a click of the mouse. A viewer forms an impression of a Web site within the first few seconds of a visit. Good design can enhance the probability of a favorable impression as the viewer responds to visual cues. A viewer with a favorable impression of a Web site is more likely to become a customer.

Good Web site design should support an e-business's Web site message without distracting from that message. For example, if viewers are thinking or talking about the design of an e-business's Web site, instead of about the Web site content, the design is detracting from the Web site's message. Web site design techniques include the use of text, color, graphics, sound, video, and Web technology such as animation to convey the Web site's message.

Maintaining consistency

It is important that the Web design emphasize consistency in its presentation. For example, there are several elements that should appear on every page at a Web site:

- E-business name
- Contact information
- Logo or trademark
- Update date
- Copyright information
- Navigation elements

TIP

E-businesses should recognize that new services and interfaces that operate through PDAs, cell phones, pagers, instant messenger clients, and other computing services will increasingly require flexible information sources. Information design, rather than site design, may take on increasing importance in the near future, because you can't put a Web page designed to fit a browser window on a cell phone's tiny screen. Also, the kind of information a user wants sent to his or her cell phone—used for short, perhaps urgent communication—is very different from the kinds of information the same user may want to view in his or her Web browser. Some Web designers suggest that the long-term success of an e-business may depend its ability to structure its information in standardized formats and make it available on scalable hardware and software platforms that can handle huge amounts of information exchange.

The e-business name and contact information are important because customers may print a hard copy of individual pages (not necessarily the home page) from a Web site, and will want the e-business name and contact information available on the printout. The e-business logo or trademark is used in the branding process. Because much of the information at an e-business Web site is dynamic, such as product availability, product pricing, and shipping policies, each page of a Web site should include the most recent update date. While an e-business's Web site has copyright protection whether a notice is posted or not, it is best practice to post a copyright notice to remind viewers of that protection. Typeface and colors should also be consistent throughout the Web site. Finally, a viewer who cannot find an efficient way to move from page to page at a Web site may become confused and will ultimately have a less than enjoyable visit.

Using navigation elements

Navigation elements are important because a viewer may not always enter a Web site from its home page. He or she may enter the Web site from any page at the site when using a search engine or a hyperlink from a different Web site. When this happens, the viewer must have a way to get to the e-business's home page or other major pages at the e-business's site. Navigation elements are also important to assist the viewer in finding information on other pages at the Web site.

An **internal hyperlink** is a connection between two pages at the same Web site. A well-designed Web site should have a logical navigation scheme, based on internal hyperlinks, that is easy for viewers to understand and to use. That navigation scheme should include an easy way for viewers to connect to all the major pages at the Web site from each page, including the home page. Because viewers do not necessarily begin viewing a Web site from the home page, all pages should contain an internal hyperlink back to the home page.

A **navigation bar** is a series of icon or text internal hyperlinks to major pages at a Web site. Navigation bars using icons are often positioned at the top of a page, while navigation bars using text are often positioned on the left side or at the bottom of a Web page. The position of navigational bars should be consistent for all pages at the Web site. Figures 7-4 and 7-5 illustrate icon and text navigation bars on the eBay home page.

Figure 7-4
Icon navigation bar

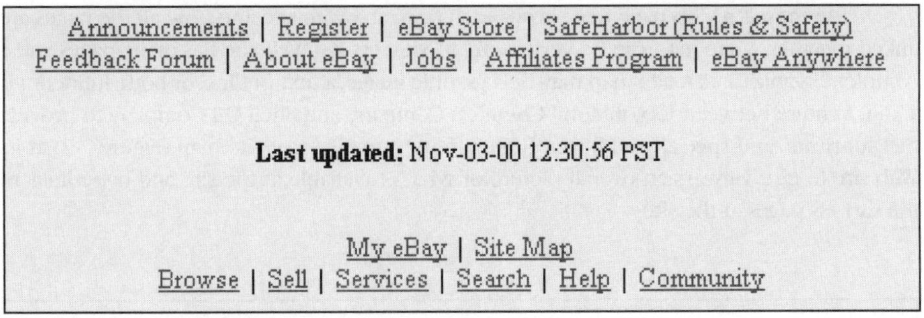

Announcements | Register | eBay Store | SafeHarbor (Rules & Safety)
Feedback Forum | About eBay | Jobs | Affiliates Program | eBay Anywhere
Last updated: Nov-03-00 12:30:56 PST
My eBay | Site Map
Browse | Sell | Services | Search | Help | Community

Figure 7-5
Text navigation bar

Another very effective way to help viewers with navigation is to use a hierarchical **navigational outline** showing all the levels of links between the home page or another major page and the page currently being viewed. Viewers using this outline can quickly move up or down in the hierarchy and easily understand the relationship of the page being currently viewed to the page on which they started. Hierarchical navigational outlines are great visual cues to the linking relationships among pages and should be used in addition to other navigational hyperlinks such as navigation bars. Figure 7-6 illustrates the navigational outline a viewer might see when shopping for a toy car for a 3-4 year old child at Amazon.com.

amazon.com.
VIEW CART | WISH LIST | YOUR ACCOUNT | HELP
TODAY'S FEATURED STORES
WELCOME
DIRECTORY
TOYS & GAMES
BOOKS
ELECTRONICS
COMPUTER & VIDEO GAMES
HEALTH & BEAUTY
NEW CARS
MUSIC
TOYSRUS.COM
teamed with Amazon.com
SEARCH TOYS
AGE RANGES
BRANDS
CATEGORIES
TOP SELLERS
BACK TO BASICS
GIFT IDEAS

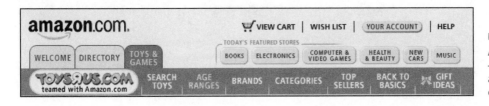

Figure 7-6
Amazon.com — toy cars for a 3–4 year old child

A **site map** is a Web page that shows each page at a Web site and how all the pages are linked together. A site map can be very useful to viewers if a Web site has many pages and a complex organization. A site map may be a graphic image, a text outline, or both. Infineum is a joint venture between ExxonMobil Chemical Company and Shell Oil Company to provide fuel, lubricant, and specialty additives. Infineum uses an effective site map (Figure 7-7) at its Web site to give viewers an overall picture of what is available at the site and hyperlinks to the various pages at the site.

Figure 7-7
Infineum site map

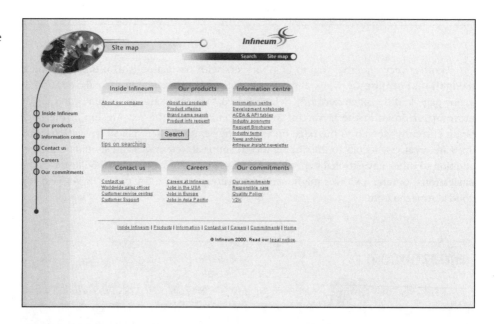

TIP

For a fun (but not so funny) look at poor Web design methods, including confusing navigation techniques, take the Sucky Pages Tour at Vincent Flander's Web Pages That Suck Web site.

Another important navigational technique is the inclusion of top-of-page hyperlinks at the bottom (or wherever necessary) of each Web page that a viewer must scroll vertically. A **top-of-page hyperlink** is text or a graphic image that is linked to a position at the top of the current page. Top-of-page hyperlinks enable a viewer to quickly return to the top of the same page after scrolling.

Other Web design considerations include whether to use text or graphics or a combination of both to convey the Web site message; what typeface, color combinations, background colors or images, and page layouts to use; and whether or not the site should use frames, animation, and multimedia. A startup e-business may want to get advice from a professional Web designer when resolving these design issues.

Using splash pages

The first page at a Web site is very important and should do several things. First, it should make the identity of the e-business clear. No viewer should ever enter a Web site and not be able to identify the e-business. The entrance into a Web site is just the first of many opportunities to build the e-business's brand. Second, a first page should quickly communicate a basic description of the purpose of the Web site.

Some Web designers encourage the use of a splash page as a Web site's first page. A **splash page**, sometimes called an **entry page**, is a Web page that usually contains big, flashy, sometimes animated graphic images (and occasionally sounds) and is used to create a showy entrance to a Web site. After the animated graphics and sounds finish playing, the Web site home page automatically loads in the Web browser. Most splash pages also contain a text hyperlink to the home page that the viewer can click to bypass the animation and get right to the Web site's home page. Gossamer Threads, a Web technology firm in Vancouver, B.C., (Figure 7-8) uses an attractive and understated splash page as a doorway into its Web site. Web technology firms, graphic design firms, and related e-businesses often use a splash page to provide viewers with a preview of the quality of the technological or design services they offer.

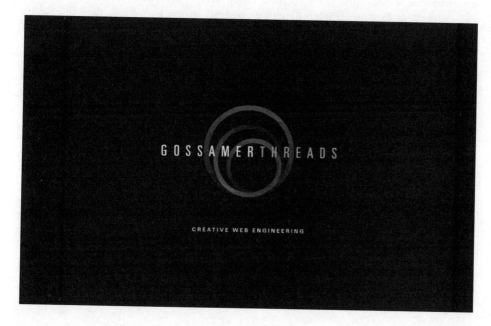

Figure 7-8
Gossamer Threads

Unfortunately, some splash pages take too long to load in a Web browser because of the size of the graphic images or because of the animation. Also, many viewers who are looking for quick access to good information may be annoyed by the additional download time taken by a splash page. Because first impressions are so critical, some Web designers discourage the use of a splash page in favor of going directly to a Web site's home page.

Using text and icon hyperlinks

Text and icon hyperlinks can also be positioned in the body of a Web page to help viewers navigate a Web site, or to allow viewers to access Web pages at a different Web site. Traditionally, text hyperlinks have been underlined and formatted with a dark color, generally blue. Today, many Web sites do not use underlining for text hyperlinks and may use different accent colors to indicate text hyperlinks. Some Web sites use animated hyperlinks that appear or disappear as the viewer moves the mouse pointer over the hyperlink. When designing a Web site, remember that the whole point of hyperlinks is for viewers to quickly get to information they want that is located elsewhere. Hyperlinks that are pretty, animated, but hard to locate and use may frustrate viewers and drive them to competitors' Web sites.

Often, Web sites use graphic hyperlinks instead of text hyperlinks. Certain graphic images can effectively communicate the purpose of a hyperlink, especially to global viewers who speak different languages. For example, an e-tailer might use a shopping cart icon in addition to the text Shopping Cart to indicate the hyperlink that displays the contents of the customer's shopping cart. Also, an envelope icon is often used to indicate a hyperlink viewers can click to send an e-mail message. However, graphic hyperlinks can be confusing if the graphic image doesn't clearly indicate where the hyperlink takes the viewer. A well-designed Web site uses a combination of text and graphic hyperlinks. Figures 7-9 and 7-10 illustrate the use of a shopping cart icon hyperlink and shopping cart text hyperlink on the Amazon.com home page.

Figure 7-9
Shopping cart icon hyperlink

Figure 7-10
Shopping cart text hyperlink

Using color

Choosing the right colors for a Web site can be critical, because color is one of the first things a viewer sees as a Web page loads in his or her browser. Color can quickly set the tone for the viewer's experience—for good or bad. In general, bright colors (blue, red, yellow, orange, green) are happy, loud colors used sparingly on conservative corporate Web sites. However, bright colors such as red can be used effectively to call attention to specific Web page elements. Pastel colors (colors that contain a large proportion of white) are more relaxing and undemanding. Earth tones (brown, beige, tan) are unobtrusive and tend to contrast well with primary colors. A good rule to follow for an e-business Web site is not to use too many different colors in the color scheme. Also, the background color should either be a very light and neutral color (possibly white, which evokes a no-nonsense businesslike attitude) or black.

Color choices should reflect the values of the Web site's audience. Because the Internet is global, remember that people around the world respond differently to colors. For example, in the United States, blue is a color that represents trust; however, in Korea, the idea of trust is enforced with pastel colors, especially pink. Understanding the emotional effect of different colors on a Web site's audience can enhance the Web site's design.

Also important in selecting a color scheme for a Web site is the issue of available colors. Web pages use the RGB (red, green, blue) color model that uses three numbers from 0 to 255 to represent each of the three colors. Web browsers are capable of displaying only a small set of the 16 million possible RGB color combinations, and will substitute a different color if the one used on a Web page is not available in its recognizable RGB color set. Therefore, most Web designers use a **browser-safe palette** of colors when designing a color scheme. There are a number of Web sites that illustrate and discuss browser-safe colors. Figure 7-11 illustrates browser-safe RGB colors.

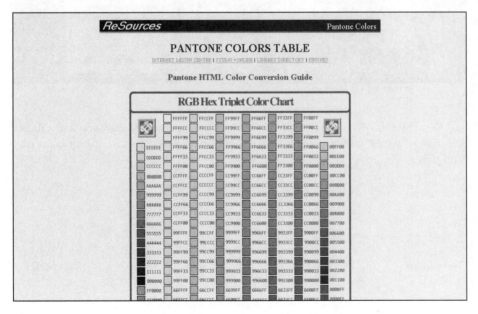

Figure 7-11
Pantone Colors

When eLUXURY.com (Figure 7-12), the upscale fashion site, was launched in mid-June 2000, Alain Lorenzo and his staff paid very close attention to customer feedback. What they found surprised them. Customer feedback indicated that the Web site's audience was younger than anticipated, with more customers falling into the 25–35 age group, instead of the typical luxe consumer found in brick-and-mortar stores. By October, a scant 3½ months after launch, a redesign of the Web site was underway to exploit that trend.

The redesign included replacing the dominant black-and-white color scheme with bolder colors and using bolder typefaces to suggest a fresher, younger attitude. More product illustrations and photographs of a wider product assortment were included, to place a greater emphasis on a fashion-forward approach. Fashion show video clips were added to the Web site's online magazine, and the Web site's navigational structure was modified to add hyperlinks to the online magazine from any page at the Web. The immediate result of the redesign was a doubling of pages viewed by visitors and an increase in sales.

Figure 7-12
eLUXURY.com

Cisco Systems' Web site (Figure 7-13), which uses a white background and a muted selection of accent colors, provides a good example of the effective use of color on a Web page.

Figure 7-13
Cisco Systems

A professional Web designer can help a startup e-business sort through color issues and select a message-appropriate color scheme for its Web site, using a browser-safe color palette. It is also important to carefully design the use of text at a Web site.

Using text

There are important differences between writing for the Web and writing for the printed page. Although the same basic grammar and spelling rules apply, the way a viewer reads text online is different. Online readers scan text instead of reading it word for word. They try to pick out a few words or phrases to get the information they want. Instead of long scrollable pages of dense text, online readers prefer Web pages to be short and to the point. Finally, online readers prefer concise factual information to marketing "fluff" or overly hyped language. Because online readers want to scan text quickly, use frequent paragraph breaks, headings, bulleted lists, and ample white space in Web page text. Online readers also prefer to read narrow columns rather than lines of text that go from margin to margin.

A **font** is a collection of characters that have the same appearance. A **serif** is the small tail at the end of a line in characters such as "I," "M," and "N." A serif leads the viewer's eye smoothly to the next character. Fonts containing characters without a serif are called **sans serif** fonts. For the sake of consistency and easy reading, some Web designers suggest that Web page text should use only one or possibly two different fonts. A common combination in both print media and Web pages is to use a sans serif font, such as Arial, for headings and a serif font, such as Times New Roman, for body text. Web browsers display text in a font that a viewer has on his or her computer system. If a Web page contains fonts not available on the viewer's system, the Web browser will substitute a different font. For this reason, it is usually preferable to use common fonts such as Arial or Times New Roman for Web page text.

Font styles, or attributes, such as bold or italic can be used to call attention to a word or phrase in a Web page body text. Underlining is often used for emphasis on a printed page, but underlining is generally not appropriate for emphasis on a Web page because most viewers associate underlining with hyperlinks. Figure 7-14 illustrates text from a printed document saved as a Web page without formatting changes. The text is dense and difficult to read. Figure 7-15 illustrates how a portion of that text could be revised for easy online reading, by editing the text to be more concise and by adding different font headings, bulleted lists, bolded words or phrases, and ample white space.

Figure 7-14
Original text

Millions of computer users access the Internet each day to shop, listen to music, view museum exhibits, manage their investments, follow current events, and send electronic mail to other computer users. Additionally, thousands of people are using the Internet at work and at home to view and download to their local computers computer files containing graphics, sound, video, and text. The World Wide Web (or WWW), a subset of the Internet, uses computers called Web servers to store these multimedia files. These files are called Web pages.

A network is a group of two or more computers linked by communication media like cable or telephone lines. The Internet is a worldwide collection of computer networks connected by communication media that allow users to view and transfer information between computer locations. For example, an Internet user in New Mexico can access a computer in Canada, Australia, or Europe to view the contents of files stored there or download files to their computer quickly and easily. The Internet is not a single organization or entity but a cooperative effort by multiple organizations managing a variety computers and different operating systems.

In the late 1960s, the United States Department of Defense developed an internet of dissimilar military computers called the ARPAnet. Computers on this internet communicated with new standard communication rules called the Transmission Control Protocol/Internet Protocol or TCP/IP. Additionally, a new technology called "packet switching" was developed for this internet that allowed data transmitted between computers to be broken up into smaller "packets" before being sent to its destination over a variety of communication routes. The data was then reassembled at its destination. These changes in communication technology enabled data to be communicated more efficiently between the different types computers and operating systems.

Soon, scientists and researchers at colleges and universities began using this internet to share data. In the 1980s the military portion of this internet became a separate network called the MILNET and the National Science Foundation began overseeing the remaining non-military portions now called the NSFnet. Thousands of other government, academic, and business computer networks began connecting to the NSFnet. By the late 1980s, the term Internet became widely used to describe this huge worldwide "network of networks."

There are a wide variety of services available to users on the Internet: electronic mail, online discussion groups, e-mail discussion groups, file access, news, weather, traffic reports, product information and support, and much more.

To access the Internet you need a physical communication medium connected to your computer like network cable or a dial-up modem. You will also need the Internet communication protocol software TCP/IP. If you are using a dial-up modem, you will need a version of TCP/IP Stack for telephone lines like Serial Line Internet Protocol (SLIP) or Point-to-Point Protocol (PPP). You should also have front-end browser software like Explorer Navigator to help you use the Internet resources easily. All the communication software you need is generally included when you purchase your browser software.

After setting up your computer hardware and installing the communication and browser software, you must make arrangements to connect to a computer already on the Internet called a "host." Usually, you will connect to a host computer via a commercial Internet Service Provider. An Internet Service Provider (ISP) maintains the host computer, provides a gateway to the Internet, and provides an electronic "mail box" with facilities for sending and receiving e-mail. A Commercial ISP usually charges a flat monthly fee for unlimited access to the Internet and e-mail services. Often, a commercial ISP will supply the communication protocols and front-end tools you will need to access the Internet.

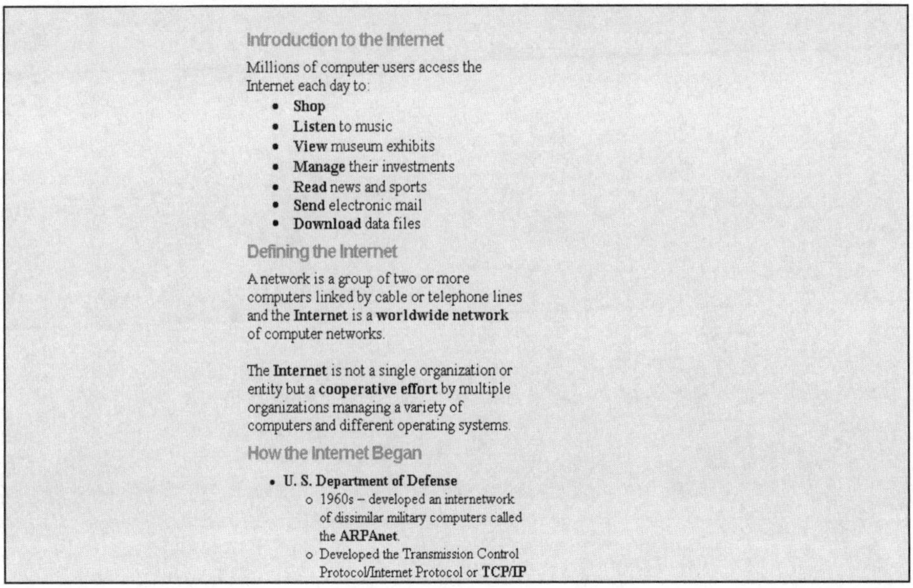

Figure 7-15
Revised text

Finally, it is critical that all the text at a Web site be checked for spelling errors and professionally proofread and edited for grammatical and stylistic errors. Next comes a careful consideration of the use of background images on pages at the site.

Background images

Large background images that fill the whole screen can cause problems for viewers using different monitor resolutions. A small image that repeats over and over, called a **tiled image**, is preferable. Unfortunately, a background image may obscure the text on a Web page, making it difficult for viewers to read the text. Many Web designers think that it is better to forgo using a background image in favor of a light colored background. Others suggest using only a white or black background color. Figure 7-16 illustrates the ineffective use of an unattractive background image that detracts from the Web page text.

Images and other multimedia options can, if used carefully, effectively communicate a Web site's message.

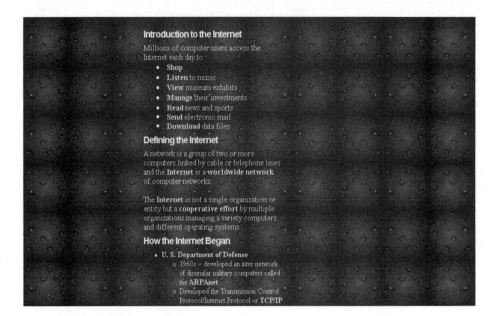

Figure 7-16
Ineffective
background
image

Using images and multimedia

Almost every Web site uses images to enhance the site's design and to effectively communicate information. Some Web sites also make use of animation, sound, or video to provide information. All multimedia used should support the purpose of the Web site and shouldn't be included simply because it looks or sounds great.

Image file size is important when designing Web pages because the larger the image file size, the longer a Web page takes to download. To improve download times, Web page image files should be compressed. There are two primary types of compressed images used on Web pages, **GIF images** (usually pronounced "jif," as in "jiffy") and **JPEG images** (pronounced "jay-peg"), each with their own properties and uses. GIFs are often used for simpler images such as logos and icons, but are usually not used for photographs because photographs compress better as JPEG files. GIFs have some interesting features that JPEGs do not. GIFs can be animated, like a very short movie or flipbook animation, showing a sequence of frames rapidly. GIFs can also allow for a transparent color—a color that will appear transparent on a Web page, allowing the background to show through. In order to maximize Web page loading, it may be necessary to optimize the Web page image file sizes. There are several e-businesses such as Spinwave, GIF Wizard (Figure 7-17), and xat.com that offer image optimization software packages.

There are some techniques that make Web pages containing images load faster. For example, a small version of an image, called a **thumbnail**, can be used to link to the larger version of the image. When the viewer loads a Web page, the smaller thumbnail images load faster. The viewer can click a thumbnail image to see the larger version of the image. Thumbnail images are often used by e-businesses to illustrate their product catalogs. Figure 7-18 illustrates the use of thumbnail images to describe products at a Southwestern Bell Web site.

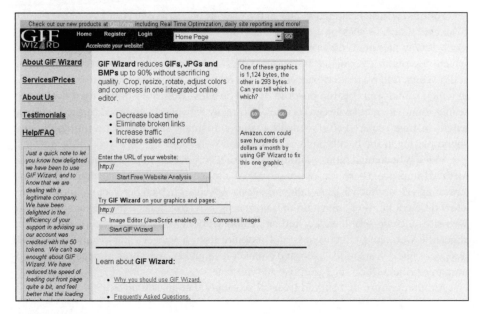

Figure 7-17

GIF Wizard

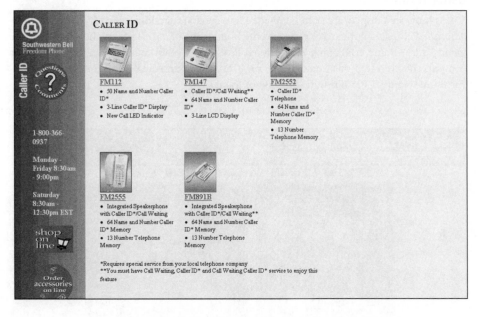

Figure 7-18

Southwestern Bell phone products thumbnail images

Sounds can also be added to play in the background as a page is viewed or to respond to an event, such as clicking an icon. Sound files come in three formats: audio recordings (such as WAV files or AU files), which are like a digital tape recording of sounds; MIDI files, which describe the sequence of musical notes and how they are played; and streaming audio, which is like a radio broadcast over the Internet. In general, there are few situations in which sound supports the purpose of an e-business Web site—unless the e-business is selling sound or creative design services. Also, many viewers are annoyed by the additional length of time required to download a page containing sound, and by the continuous sound playing in the background as they view a Web page.

Video is becoming more common as Internet access becomes faster and compression methods improve. There are two methods of providing video: downloadable video and streaming video, which is like a radio broadcast. Advances in Web browser technology now allow different kinds of animation of text, images, and hyperlinks. Unfortunately, too many Web sites include sound, video, and animation without considering how these features affect the Web site's effectiveness and usability from a viewer's perspective. Most Web designers agree that sound, video, and animation should be used only as necessary to support an e-business Web site's purpose and message.

Another feature that should be used sparingly, if at all, is frames.

Using frames

A Web browser's display area can be divided into separate sections called **frames**, in which different Web pages appear. For example, the home page of a Web site might consist of two frames: a top frame containing navigational links and a main frame containing scrollable content. The Web site of a major Houston, Texas, brick-and-click furniture store, Gallery Furniture (Figure 7-19), illustrates the use of frames.

Figure 7-19
Gallery
Furniture

Unfortunately, frames may make a page look cramped and cluttered, and may, if not correctly designed, cause navigation problems for the viewer. Additionally, frames may cause problems for search engine robot or spider programs trying to index the Web site pages, and may make it difficult for a viewer to add the Web page URL to his or her bookmarks or favorites folder. Because of these potential problems, many Web designers suggest avoiding frames in Web site design.

An important component of any Web site is its use of forms to gather information.

Using forms

Another Web page element that should not be overlooked is forms. **Forms**, consisting of text labels and the related input boxes, option buttons, drop-down lists, or check boxes, are used to collect information from viewers. Forms allow a viewer to enter specific information on the Web page and then send that information to the e-business's e-mail address or Web site database. Forms are used for many things; collecting site feedback, registering for approval to use Web site functions, and ordering products online are just three examples. Figure 7-20 illustrates eBay's registration form.

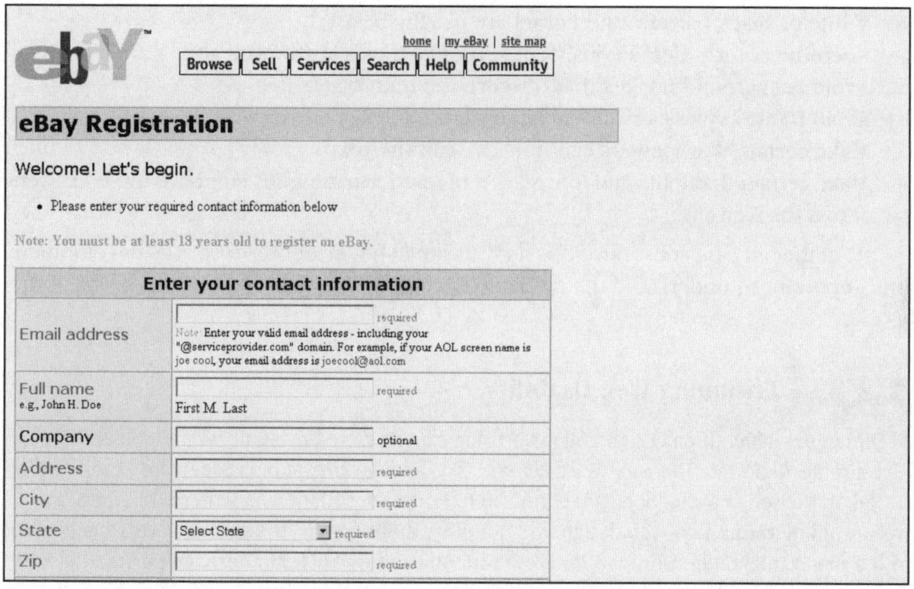

Figure 7-20
eBay
registration
form

Sometimes a viewer is required to complete a paper form. Creating downloadable forms, as well as other documents, is a way to provide these paper forms without having to mail them or fax them to the user. A model form, or form template, is created in a software application such as Microsoft Word. Then either the Word file itself is made available for download (which would require users to have Word installed on their own computers) or the file is converted into a format that any computer can read, such as plain text, rich text (RTF), or Adobe's PDF format. A PDF reader named Adobe Acrobat, which can be freely downloaded from the Adobe Web site, is used to read and print PDF documents. The advantage of a PDF file is that it displays and prints on any computer the way it was designed, regardless of the word-processing programs or fonts on a viewer's computer.

Web design tips

When establishing its Web site design, an e-business should consider the following:

- Remember to keep it simple—the "less is more" idea really works for most e-business Web sites.
- Use lots of white space for an uncluttered look.
- Make certain that the colors used in the design are not only browser-safe but also fit the Web site's message.
- White or black background colors are usually best.
- Keep the color scheme consistent across all pages at the Web site.
- Avoid background images that obscure the text.
- Avoid frames unless absolutely necessary.
- Make certain that viewers can quickly scan the text.
- Make certain that fonts and font sizes and text formatting for emphasis are consistent across the Web site.

Whether an e-business develops its Web site in-house or outsources its development, it is important to understand the available development tools.

E-CASE Critiquing Web Usability

In September 2000, three design and advertising experts were invited to take a look at three high-profile Web sites: *The New York Times*, ESPN, and *Seventeen* magazine. The experts were: Dr. Jakob Nielsen, a usability engineer and Web design consultant who writes books and a weekly online column on Web design and usability; Brian Collins, the executive creative director of the Brand Integration Group at the New York office of Ogilvy & Mather; and Michael Grossman, creative director for *Saveur* and *Garden Design* magazines.

As might be expected, there was a wide range of opinions about what worked and what didn't work at the three Web sites. However, the criticisms of these three Web sites seem to reflect what many other viewers are finding on the Web: big blocks of text, unnavigable Web sites, and hard-to-use features. Some of the individual complaints included misplaced search areas, inconsistent labeling of Web page areas, misplaced links and action buttons, too much text crammed on a page, annoying animation (scrolling and blinking banners), inappropriate writing styles for target audiences, and poor navigational elements. It seems that even some of the big guys still have a difficult time getting it right when it comes to Web site usability.

Identifying Web Development Tools

There are many kinds of Web development software available, from Web page design software to entire Web application environments. Some of the most common Web page design software includes Microsoft FrontPage and NetObjects Fusion, but there are many other programs, which range from freeware and shareware to other, more expensive applications.

Web development software

For simple Web site design, many people use Microsoft FrontPage or a comparable software application because these applications are easy to learn, provide a lot of functionality, and are relatively inexpensive. The cost of many Web design software applications ranges from around $50 to $150. Other, more expensive Web design software applications, such as NetObjects Fusion (Figure 7-21), have a more comprehensive selection of design features and provide more control over the design process. However, these more complex Web design software applications are also more difficult to use, and many cost more than twice as much as an application like FrontPage. (See the Appendix to this book for a brief tutorial on creating a Web site with Front Page.)

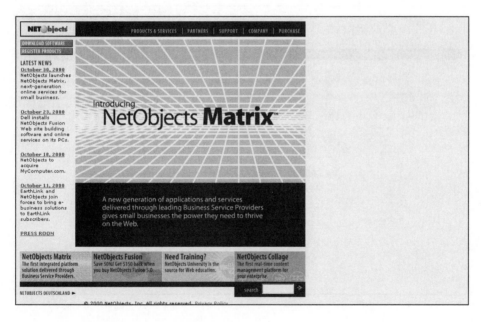

Figure 7-21
NetObjects, Inc.

For more advanced Web site development, there are other, much more sophisticated packages that are used for managing Web site development in addition to creating Web pages. Some of these software applications, such as Allaire's ColdFusion and Microsoft Active Server Pages (part of Microsoft Internet Information Server, IIS), control connections from the Web server to a database, allowing dynamic content to be included instead of just static pages. Other software, such as StoryServer from Vignette (Figure 7-22), provides content management tools. StoryServer allows nontechnical users to enter the information to be displayed on the Web site via online forms and templates.

Web development turnkey packages

Many consulting and software companies sell what is called a **turnkey package**, a package that already has a Web site including a storefront ready to publish, except for any custom content to be input. Many companies that sell turnkey packages also provide, for an additional charge, the data entry necessary to input the custom content. Most of the framework for these turnkey sites is the same from site to site. Businesses that provide turnkey Web site development include Eplex, Maden Tech, and Zeus Technology (Figure 7-23).

Figure 7-22
Vignette

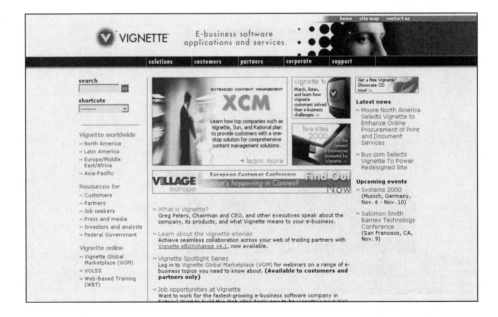

Figure 7-23
Zeus
Technology

There are also prepackaged software solutions, which provide a series of templates used to create Web pages. An e-business simply selects the Web page templates appropriate for its Web site. The advantages of using a prepackaged solution are lower cost and less time to get the Web site operational. Also, there is usually no hardware or software to purchase, just a monthly usage fee. These prepackaged solutions can be used to create a simple one-page Web site or a complex e-tail site with thousands of products. A disadvantage of using prepackaged solutions is the limited range of customization. A good example of a prepackaged solution is the Yahoo! Store (Figure 7-24). An e-business can quickly build an e-tail Web site using the Yahoo! Store for a very low cost by entering the e-business's information into predefined Web page templates in a matter of minutes.

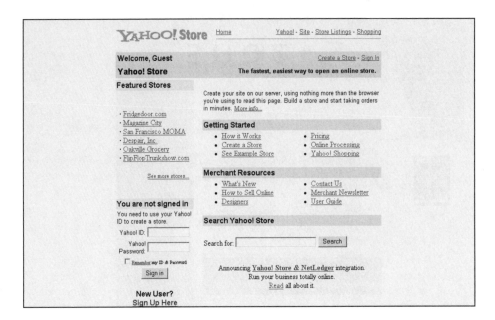

Figure 7-24
Yahoo! Store

Outsourcing web design

Outsourcing Web design work can save a startup e-business time and money by saving the cost of recruiting and hiring in-house Web design professionals. Outsourcing Web design also can enable an e-business to access experienced design specialists who are familiar with best practices and current technological changes. Before agreeing to an outsourcing contract, an e-business should thoroughly review several outsourcing candidates to get answers to the following questions:

- What services do they provide?
- What are their staff capabilities, and what portion of the design work, if any, will they subcontract?
- Can they provide references and examples of their work?
- What is their track record for completing projects on schedule?

Any contract to outsource Web design must also address the important issues of: (1) who is responsible for updating and maintaining the site, (2) what happens if updates are not made on a timely basis, and (3) who owns the Web site content. An e-business must be sure it is not giving away copyright ownership in the resulting Web site to the designer. If in-house employees do the Web design and work within the scope of their duties, the copyright falls to the e-business. However, if the Web design is outsourced to an independent contractor, the copyright ownership may remain with the creator of the design unless transferred in writing to the e-business. It is important to clarify copyright ownership in writing as part of a Web design agreement with an outside designer or developer. And, as with all other contractual relationships, it is a good idea to have an attorney review the terms of the written agreement.

An e-business that outsources its Web design may be able to take advantage of a usability analysis before the Web design is completed. For example, many Web design firms—such as Cyberplex, which has offices in both the U.S. and Canada—employ usability analysts, human-computer interaction specialists who work directly with clients to fine-tune the clients' Web site plans into usage scenarios and process flow diagrams. These scenarios and diagrams are then passed on to Web designers and technicians who use them as the basis for the Web site's design. Usability analysts look for ways to make certain that the appearance, layout, and interaction offered by a Web site reinforce the viewers' understanding of where they are at the Web site, what they can do at the Web site, and where to go next.

After a Web site is constructed, it should be thoroughly tested.

Testing and Maintaining a Web Site

After your e-business has a completed Web site, you should test the site thoroughly before publishing it to its final destination Web server and making it available to the public. Test it for usability, and make certain that everything at the Web site works correctly. One way to do this is to publish the Web site to a temporary location, called a **staging server**, and then to perform a variety of tests on the Web site. For example, all of the hyperlinks should work correctly, data submitted by viewers on forms should update the appropriate database or Web page, and all dynamic or active elements should work as expected. Additionally, the Web site should undergo a "stress test" to ensure that it can handle a heavy load of customer activity.

Unfortunately, many startup e-businesses lack the in-house capability to do a thorough testing process. When this is the case, it is a good idea for an e-business to use the products or services of a professional online Web site testing company, such as Envive Corporation. Other companies, such as Segue Software, BMC Software, and RSW Software (Figure 7-25), sell Web site testing software applications and services.

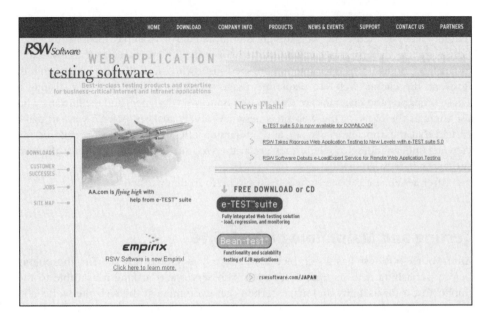

Figure 7-25
RSW Software

After a new Web site is thoroughly tested, it is published to its final destination server, but this doesn't mean that all the work at the Web site is finished. An e-business must carefully monitor and evaluate its Web site activities, to determine overall traffic at the site and the success or failure of the Web site to achieve its overall purpose. Viewer feedback must be solicited and evaluated. Your Web site should be retested at intervals as it evolves over time.

A successful Web site cannot be static but must continue to evolve because of the dynamic nature of the Web, the constant advancement in Web technologies, and the growth in e-business possibilities. The overall purpose of a Web site may be modified as new e-business opportunities arise. An e-business may need to restate its goals and objectives in light of changing customer expectations and available technologies. Viewer feedback may require an evaluation of the overall Web design as well as specific usability issues.

E-CASE It's Better the Second Time Around

What do online customers really want? According to its consumer research, Nordstrom.com customers want three things: the ability to shop and search for items easily, the ability to locate popular brands quickly, and a fast checkout process. With this in mind, the Nordstrom.com site was redesigned to speed up download times, improve navigation, and streamline sales transactions less than a year after it was launched.

Following the lead of its brick-and-mortar stores, Nordstrom.com (see Figure 7-26) can now point customers directly to popular merchandise brands such as Calvin Klein, Kenneth Cole, DKNY, and Liz Claiborne. On the old version of its Web site, Nordstrom.com customers could search only by merchandise category. At the redesigned site, customers can now search by more than one parameter—including brand, color, size, and gender. Other new components at the Nordstrom.com Web site include a simplified user registration process, the ability to chat live with a customer-service representative, and an express checkout option.

Figure 7-26
Nordstrom.
com

It is highly likely that the entire process of identifying the Web site purpose, setting goals and objectives, reviewing Web design issues, creating and publishing the Web site, and monitoring its effectiveness will take place many times in the life of a successful e-business Web site.

. . . SELLING THOSE SPECIALTY ITEMS

In order to meet the budget goals for foodlocker.com's Web site, Bates and Wagner decided to save money by creating their demo Web site in-house, using a prepackaged Web development tool from Yahoo! called Yahoo! Store. This enabled them to quickly create a simple Web store to demonstrate the foodlocker.com e-business idea to suppliers and investors. Wagner and Bates also took advantage of the ease of a prepackaged Web development product to try different design approaches before embarking on the actual Web site design.

After creating a successful demo Web site, the next step was to design the complete foodlocker.com Web site. To achieve the primary goal of enabling customers to easily find and purchase specialty food items from the foodlocker.com Web site, Bates and Wagner decided on a Web site navigational structure that would give customers four ways to locate the food items. Customers could use hyperlinks to locate products by state, by type of product, and by product brand name; they could also search the Web site using keywords to find products. Bates and Wagner decided to use pictures of the available products to create the navigational hyperlinks.

(Continued on the next page)

Bates and Wagner carefully chose the Web site colors, fonts, background, product images, and foodlocker.com logo with an eye to using all these design elements both to enable a positive viewer experience and to promote branding of the foodlocker.com concept. Because it was important to be certain that search engines could index all pages at the site, and that all viewers could bookmark or create a favorite for each of the Web site pages, Bates and Wagner chose not to use frames in the foodlocker.com Web design.

After familiarizing themselves with the prepackaged Web store product from Yahoo!, Bates and Wagner decided to create the foodlocker.com Web site in-house, using a combination of tools such as Microsoft FrontPage 2000 and direct HTML coding to get just the right look and operation at the site.

Before publishing the Web site to its final destination server, Bates and Wagner carefully tested all the features at the foodlocker.com Web site. They discovered that the graphic image hyperlinks and the size of the product images caused the Web pages to load too slowly. They also discovered that viewers had to use too many mouse clicks to get to the product and purchasing information.

To fix these problems, Bates and Wagner modified the Web site design to include more text hyperlinks. They modified the Web site structure to reduce the number of page levels. Finally, they used an image optimization program to minimize the file size of the product images. Are Bates and Wagner finished with the Web site design? Not quite. They continue to monitor the Web site usage and to try to find ways to modify the design in order to increase traffic and the number of return customers at the Web site.

Summary

- Before beginning to create a Web site, an e-business should determine the Web site's overall purpose and the goals and objectives that support that overall purpose.
- An e-business Web site should be designed from the perspective of the potential viewer, rather than from the perspective of the e-business.
- The design of an e-business Web site should take into consideration the technical sophistication of the target audience and the tools that audience is likely to use to view the Web site.
- The cost of developing an e-business Web site will vary with the complexity of the site, from $1,000–$5,000 for a simple site up to $100,000 or more for a complex site.
- An e-business Web site should be structured so that a viewer can find useful information with three to five mouse clicks.
- The storyboarding process is useful for designing the structure of an e-business Web site.
- Good Web design supports an e-business Web site's message rather than detracting from it.

- Web design elements include navigational links, text and icon hyperlinks, color, fonts, font size, font styles, graphic images, animation, sound, and video.
- Certain design elements such as splash pages, frames, sound, video, and animation should be avoided or used sparingly.
- An e-business can create a simple Web site including a storefront easily, using Web design software and prepackaged Web site development software.
- Usability analysts with a background in human-computer interfaces can help an e-business determine how easily a viewer can use the Web site, and resolve any problems before the design is complete.
- Before publishing its Web site to the destination server, an e-business should thoroughly test the site for usability, functioning hyperlinks, working dynamic elements, and activity loads.
- After publishing its Web site and making it available to the public, an e-business must constantly monitor, update, and retest the site as it evolves.

CHECKLIST ✓ Does Your E-Business Web Site Pass the Test?

❏ Does the overall Web site design support its goals and objectives?

❏ Does the Web site design use text, color, and graphics in a way that provides viewers with a clear message about the e-business?

❏ Do the pages at the Web site download quickly using slower modem speeds?

❏ Is the Web site content useful to viewers?

❏ Can viewers quickly scan the Web site text?

❏ Is the Web site text consistently formatted, edited by a professional editor, and spell checked?

❏ Are navigation links, buttons, and bars easy to find on each page?

❏ Can viewers find important information quickly without having to scroll Web pages at the site vertically or horizontally?

❏ Does the Web site provide interactive features that viewers can play with?

❏ Does everything at the Web site work properly when tested on older machines and in older Web browser versions?

❏ Does the Web site provide useful links to other Web sites that can also help viewers get information or solve their problems?

❏ Do viewers have a way to provide feedback?

❏ Has the Web site been thoroughly tested?

❏ Can the Web site handle unusually large volumes of customer transactions?

Key Terms

browser-safe palette
entry page
font
font styles
forms
frames
GIF images

internal hyperlink
JPEG images
navigation bar
navigational outline
sans serif
serif
site map

splash page
staging server
storyboard
thumbnail
tiled image
top-of-page hyperlink
turnkey package

Review Questions

1. When considering a Web site, the first thing an e-business entrepreneur should do is to:
   a. Quickly create the Web site.
   b. Determine the purpose, goals, and objectives of the e-business Web site.
   c. Find a Web-safe color palette.
   d. Decide to use all the latest design "bells and whistles" available.

2. Which of the following items should not appear on every page at an e-business Web site?
   a. The Web designer's name and phone number
   b. The Web page's update date
   c. A copyright notice
   d. Navigation elements

3. A splash page is:
   a. Mandatory for all e-business Web sites.
   b. A Web site entry page usually containing flashy graphic images.
   c. A Web page that shows each page at a Web site and how all the pages are linked together.
   d. A miniature version of a larger page.

4. Frames are used to:
   a. Create a showy entrance into a Web site.
   b. Read and print PDF documents.
   c. Divide a Web browser's display area into sections that display different Web pages.
   d. Connect two Web pages.

5. Viewers are likely to:
   a. Appreciate background images that obscure Web page text.
   b. Be pleased by clicking the mouse seven or eight times to locate useful information at a Web site.
   c. Enjoy constant sounds that play while they are viewing a Web page.
   d. Scan Web page text.

6. Too often Web sites are designed around the needs of an e-business rather than the needs of the Web site's target audience. **True or False?**

7. Web sites designed to work successfully using high-speed connections and the latest Web browser technology pose no problems for viewers using older technology and slower connection speeds. **True or False?**

8. The storyboarding process can help Web authors to get a good picture of an optimum Web site structure. **True or False?**

9. A navigation bar must use icons and be placed at the bottom of a Web page. **True or False?**

10. Graphic images or icons should never be used to represent navigational hyperlinks. **True or False?**

Exercises

1. Using the Splash Page example page at www.csc.calpoly.edu/~ebrunner/WebDesign/SplashExamples.html or other sources, review several Web sites that use a splash page as the entry page to the Web site. Then write a one-page paper describing one of the splash pages and answering the following questions:
   a. Was the name of the e-business prominently displayed on the splash page?
   b. Was the purpose of the Web site clear to you from viewing the splash page?
   c. Do you think that the splash page helps build the e-business's brand?
   d. Did the splash page enhance or detract from your experience at the Web site?
2. Using Internet search tools or other relevant sources, locate several lists of best and worst Web sites. Check out the e-business sites on the lists, and then select one "worst" and one "best" site for review. Review each selected Web site and create a one-page paper stating why you agree or disagree with the Web site's "best" or "worst" designation.
3. In your Web browser load the Web Site Garage Web page (websitegarage.netscape.com/) and run a free Web site analysis on the home page of your choice by entering the URL of the home page. Then review and print the Web Site Garage analysis.
4. Using Internet search tools or other relevant sources, locate two Web sites that provide sample browser-safe palettes and Web design tips for using color. Print the Web pages.
5. Using Internet search tools or other relevant sources, locate a Web site that uses several different navigational elements, including navigation bars and text or icon hyperlinks. Write a one-page paper analyzing the effectiveness of the Web site's navigational elements from a viewer's perspective.

CASE PROJECTS

◆ 1 ◆

You are the assistant to the human resources manager for a Web design firm. You have been asked to prepare a job posting for a usability analyst. This is a new position at your e-business and you want to know more about usability analyst positions being listed by other Web technology firms. Using Internet search tools or other relevant sources, research the kind of work done by a usability analyst (may also be known as a usability engineer, information architect, or information designer). Locate information on current usability analyst job openings, including educational requirements and salary ranges. Then write a one-page job posting description for the new position.

◆ 2 ◆

You and your partner are starting a new B2C e-business selling custom-designed educational toys. You want to use a light-colored background with bright primary color accents, and your partner wants to use a black background with red accents. The two of you are meeting for lunch tomorrow to make the final decision on the color scheme. Using Internet search tools or other relevant sources, review design issues relating to the use of color on a Web site. Then create a one-page outline of topics and issues supporting your color scheme choice.

The executive assistant to the vice president of sales has been assigned the task of putting a section of the printed customer support manual on the company's Web site. He asks for your help. Create a one-page checklist he can use to ensure that the text from the printed document is easy to read online and easy to understand.

TEAM PROJECT

You and three classmates are developing the Web site for a B2B e-business startup. Working together, create a description of the e-business and define the purpose of the e-business's Web site, including three goals and objectives necessary to meet that purpose. Then complete the following tasks:

- Using the storyboarding process, design the Web site's structure.
- Develop a color scheme for the Web site.
- Design the layout for a splash page, if appropriate, and a home page, including the use of fonts, graphics, and navigational elements.
- Create a mock-up of a splash page, if appropriate, and a home page in a word processor, graphics application, Web design application, or other available tool.

Use Microsoft PowerPoint (or other presentation tool) to create a 5–10 slide presentation describing the e-business idea, the Web site structure, and the Web site design elements. Give your presentation to a group of classmates, who will critique the mock-up Web site using the following criteria:

- Will the Web site structure enhance the Web site's usability from the viewer's perspective?
- Does the overall design of the Web site support the Web site's functionality and the e-business's message?

Useful Links

AnyBrowser.com
http://www.anybrowser.com/

AskTog
http://www.asktog.com/

BizReport.com
http://www.bizreport.com/

CNETBuilder.com
http://www.builder.com/

Color Matters
http://www.colormatters.com/entercolormatters.html

Color Perception Page
http://www.lcc.gatech.edu/~herrington/WebSite/
    response/hyper/gt7447b/index.html

Design Resource – Web Resources for Web Design
http://www.mvd.com/webguide/

Echo Web: The Web Developers Starting Point
http://echodev.com/

Efuse
http://www.efuse.com/

Good Documents – How to Write for the Intranets or the Internet
http://www.gooddocuments.com/homepage/
    homepage.htm

iBoost Web Design Forums
http://www.webdesignforums.com/

Outsourcing Center
http://www.outsourcing-academics.com/

PageResource.com
http://www.pageresource.com/

Pantone – All About Color
http://www.pantone.com/allaboutcolor/allaboutcolor.asp

SitePoint.com – Webmaster Resources
http://www.sitepoint.com/?

The Max Model: A Standard Web Site User Model
http://www.informatics-review.com/thoughts/lynch.html

Usable Web
http://www.usableweb.com/

VisiBone Webmaster's Color Laboratory
http://www.visibone.com/colorlab/

Web Developer's Journal
http://www.webdevelopersjournal.com/

Web Developer's Virtual Library
http://www.wdvl.com/

Webmonkey
http://www.webmonkey.com/

WebDeveloper.com
http://www.webdeveloper.com/

WebReference.com
http://webreference.com/

Yale Web Style Guide
http://info.med.yale.edu/caim/manual/contents.html

Links to Web Sites Noted in This Chapter

ActivMedia
http://www.activmedia.com

Adobe Acrobat Reader
http://www.adobe.com/products/acrobat/readstep.html

Allaire
http://www.allaire.com/

Amazon.com
www.amazon.com/

BMC Software, Inc.
http://www.bmc.com/

Burr & Forman LLP
http://www.burr.com/csindex.htm

Cisco Systems
http://www.cisco.com/

Cyberplex, Inc.
http://www.cyberplex.com/Home.asp

eBay
http://www.ebay.com/

eLUXURY.com, Inc.
http://www.eluxury.com/

Envive Corporation
http://www.envive.com/

Eplex
http://www.eplex.com/

foodlocker.com
http://www.foodlocker.com/

Gallery Furniture
http://www1.galleryfurniture.com/

GIF Wizard
http://www.gifwizard.com/pn=10629

Gossamer Threads
http://www.gossamer-threads.com/

Infineum
http://www.infineum.com/sitemap.html

Jakob Nielsen
http://www.useit.com/

Lightfoot, Franklin & White, L.L.C.
http://www.lfwlaw.com/

Maden Tech Consulting
http://www.madentech.com/index.cfm

Maynard, Cooper & Gale, P.C.
http://www.mcglaw.com/

Microsoft FrontPage
http://www.microsoft.com/frontpage/

NetObjects, Inc.
http://www.netobjects.com/

Nordstrom, Inc.
http://store.nordstrom.com/

Ogilvy & Mather
http://www.ogilvy.com/o_mather/who_flash.asp

Pantone Colors Table
http://www.sfo.com/~coredata/colors.htm

RSW Software
http://www.rswsoftware.com/index.shtml

Segue Software
http://www.segue.com/

Southwestern Bell Phone Products — Caller ID
http://www.swbfreedomphone.com/products/cid/

Spinwave
http://www.spinwave.com/

Splash Page Examples
http://www.csc.calpoly.edu/~ebrunner/WebDesign/
SplashExamples.html

Vignette
http://www.vignette.com/

Web Pages That Suck
http://www.webpagesthatsuck.com/

Web Site Garage
http://websitegarage.netscape.com/

xat.com
http://www.xat.com/

Yahoo! Store
http://store.yahoo.com/

Zeus Technology
http://www.zeus.com/

For Additional Review

Association Management. 1999. "Putting Users First," 51(12), November, 26.

Bacheldor, Beth. 2000. "The Art of E-Biz – The Good—And Not Good Enough—of Web-Site Design," *Information Week*, February 14, 42.

Baeb, Eddie. 2000. "Web Site Botch Could Blossom into Major Problem for FTD.com; Cash Reserves Also Shrinking at Online Florist," *Crain's Chicago Business*, 23, July 31, 15.

BizReport.com. 1999. "Global Web Development Budget is $10 Billion," November 9. Available online at: http://www.bizreport.com/news/1999/11/991109-1.htm.

Brown, Kalisha. 2000. "Simplicity Makes Web Sites 'Stickier': Experts Say Consistency and Content Will Keep Online Visitors Coming Back for More," *Denver Business Journal*, 51(45), June 23, 29A.

Chan, Gilbert. 2000. "Rancho Cordova, Calif. Tech Firm Helps Little Guys on the Web," *The Sacramento Bee*, October 6.

Downes, Larry. 2000. "Deconstructing the Web: Your E-Business Strategy Is Not the Same as Your Web Site," *TheStandard.com*, August 7. Available online at: http://www.thestandard.com/article/display/0,1151,17374,00.html.

Dreier, Troy. 2000. "Stuck on the Web," *PC Magazine*, November 21, 155.

Ensor, Pat. 2000. "What's Wrong With Cool?" *Library Journal*, 125(7), April 15, S11.

Flanders, Vincent. 2000. "The Top's Gotta Pop," *Web Pages That Suck.com*. Available online at: http://www.webpagesthatsuck.com/topgotta.html.

Ganz, John. 2000. "Get the Most Out of Web Site Evaluations," *Computerworld*, September 4, 32.

Garber, Joseph R. 2000. "Does Your Web Site Sing?" *Forbes*, June 12, 250.

Gillen, Stephen E. 2000. "Outsourcing Web Site Design: Who Owns the Rights?" *Business Courier Serving Cincinnati-Northern Kentucky*, 17(3), May 12, S-6.

Goldsboro, Reid. 2000. "Creating Substantive Web Sites and Learning About the Web," *Poptronics*, 1(11), November, 20.

Guenther, Kim. 2000. "Evidence-Based Web Redesigns," *Online*, 24(5), September, 67.

Halpern, Ann. 2000. "YourAttorney.com: Law Firms Introducing/Redeveloping Web Sites," *Birmingham Business Journal*, 17(37), September 15, 15.

Hercz, Robert. 2000. "Making It Click: About 90% of Web Sites Are Not User-Friendly," *Canadian Business Journal*, 73(1), January 10, 18.

Holzschlag, Molly E. 1999. "Satisfying Customers With Color, Shape, and Type," *Web Techniques*, 4(11), November, 24.

Hoy, Richard. 2000. "Shopping Cart Options for Small Business," *ClickZ*, May 12. Available online at: http://www.clickz.com/cgi-bin/gt/print.html?article=1714.

Janal, Daniel S. 1997. *Online Marketing Handbook: How to Promote, Advertise, and Sell Your Products and Services on the Internet*. New York, NY: Van Norstrand Reinhold.

King, Nelson. 2000. "eStoreManager," *PC Magazine*, October 3, 147.

Klein, Leo Robert. 1999. "Slicing the Image Pie," *Library Journal*, 124(17), October 15, S13.

Lee, Mie-Yun. 2000. "Good Web Site Design Is Not Just Skin Deep," *Puget Sound Business Journal*, 20(49), April 7, 15.

Lewenstein, Marion. 2000. "Eyetracking: A Closer Look," *Stanford University and The Poynter Institute*, July 12. Available online at: http://www.poynter.org/centerpiece/071200.htm.

Mack, Ann M. 2000. "About Face(Lift): The Redesign of People-Powered Portal About.com Goes More Than Skin Deep," *ADWEEK Eastern Edition*, 41(18), May 1, IQ52.

Mack, Ann M. 2000. "Sites by Design: Three Design and Advertising Gurus Take a No-Holds-Barred Look at Three Content Sites," *ADWEEK Eastern Edition*, 41(36), September 4, IQ36.

Mateyaschuk, Jennifer. 2000. "Envive Puts Its Performance-Monitoring Application Online – Application Aimed at Web Sites That Lack Resources for In-House Testing," *Information Week*, April 24, 145.

Miller, Drue. 1998. "Seven Deadly Sins of Information Design," *Netscape Webbuilding*, July 29. Available online at: http://home.netscape.com/computing/webbuilding/studio/feature19980729-1.html.

Miller, Michael J. "Forward Thinking," *PC Magazine*, November 7, 7.

Morkes, John and Nielsen, Jakob. 1997. "Concise, SCANNABLE, and Objective: How to Write for the Web," *Useit.com*. Available online at: http://www.useit.com/papers/webwriting/writing.html.

Morris, Charlie. 1999. "Amateur Web Sites – the Top Ten Signs," *Web Developer's Journal*, October 6. Available online at: http://www.webdevelopersjournal.com/columns/abc_mistakes.html.

Morris, Charlie. 1999. "Navigation 101," *Web Developer's Journal*, April 2. Available online at: http://www.webdevelopersjournal.com/articles/navigation.html.

Morris, Charlie. 1999. "Good Page, Bad Page," *Web Developer's Virtual Library*, February 28. Available online at: http://www.wdvl.com/Authoring/Design/Pages/good_bad.html.

Musich, Paula. 2000. "BMC Goes Onsite With Web Site Testing, Certification," *PC Week*, January 31, 43.

Napier, H. Albert and Judd, Philip J. 2000. *Mastering and Using FrontPage 2000*. Boston, MA: Course Technology.

Navarro, Ann and Khan, Tabinda. 1998. *Effective Web Design*. San Francisco, California: Sybex, Inc.

Nemes, Judith. 1999. "Web Site Design Getting Personal: Relationships With Customers Help Boost Sales," *Crain's Chicago Business*, 22, November 29, SR24.

Neuman, Alica. 2000. "A Better Mousetrap Catalog," *Business 2.0*, 117, February.

Nielsen, Jakob. 2000. *Designing Web Usability*. Indianapolis, Indiana: New Riders Publishing.

Powell, Thomas A. 1999. "Web Site Planning," *symbex.net.au*. Available online at: http://www.symbex.net.au/internet.htm.

Schmeiser, Lisa. 1999. "Dress It Up: Make Your Web Site More Effective and Attractive With These Expert Tips," *Macworld*, 16(11), November, 100.

Seckler, Valerie. 2000. "First Impressions Prompt a Swift ELuxury Makeover," *Women's Wear Daily*, October 2, 1.

Seckler, Valerie. 2000. "Second Time Around: The Relaunched Nordstrom.com Is Offering Consumers Better Service and an Easier Way to Shop," *Footwear News*, October 16, 22.

Shikli, Peter. 1999. "How To Hire a Good Web Developer," *CMPnet*, October 26. Available online at: http://www.techweb.com/wire/story/TWB19991026S0021.

Siegel, David. 1997. *Creating Killer Web Sites*, Second Edition. Indianapolis, Indiana: Hayden Books.

Spyridakis, Jan H. 2000. "Guidelines for Authoring Comprehensible Web Pages and Evaluating Their Success," *Technical Communications*, 47(3), August, 359.

Story, Derrick. 1999. "Usability Checklist for Site Developers," *webreview.com*, October 15. Available online at: http://www.webreview.com/wr/pub/1999/10/15/usability/index.html.

Stoughton, Stephanie. 2000. "Wal-Mart Hires Cambridge, Mass. Firm to Redesign Web Site," *The Boston Globe*, August 7.

Sullivan, Carl. 2000. "The Text Big Thing at 'Providence Journal,'" *Editor and Publisher*, May 22, 12.

Travis, David. 2000. "Ten Tips to Make Your Website Usable," *System Concepts*. Available online at: http://www.system-concepts.com/articles/tenwebtips.html.

Treese, G. Winfield and Stewart, Lawrence C. 1998. *Designing Systems for Internet Commerce*. Reading, Massachusetts: Addison Wesley Longman, Inc.

U. S. Copyright Office. 2000. "Circular 9: Works Made for Hire Under the 1976 Copyright Act." Available online at: http://www.loc.gov/copyright/circs/circ09.pdf.

Vaswani, Vikram and Kamath, Harish. 1999. "Zen and the Art of Website Design," *Times Computing Online*, December 8. Available online at: http://www.timescomputing.com/19991208/wws1.html.

Vicker, Graham. 2000. "Window Dressing," *Management Today*, March, 8.

Weiss, Todd R. 2000. "Walmart.com Shuts Down at Top of Holiday Season," *Computerworld*, October 9, 20.

Williamson, Mickey. 1995. "Outsourcing: Take My Web Site, Please!" *CIO Magazine*, September 15. Available online at: http://www.cio.com/archive/webbusiness/feature_sept1595_content.html.

Wonacott, Laura. 2000. "Site Savvy: Web Site Design Is a Combination of Both Science and Art That Satisfies Many Users," *InfoWorld*, 22(5), January 31, 60.

CHAPTER **8**

Defining Security Issues

In this chapter, you will learn to:

Discuss general e-business security issues

Identify network and Web site security risks

Enhance e-business security

Identify security service providers

Understand e-business risk management issues

When Bank X, located in New England, rolled out its online banking Web sites, it realized that its e-business risks included more than just problems with network software that could be handled by its IT staff. Bank X, like many other e-businesses, understood that it is important to treat e-business risks like any other business risks. Accordingly, the first thing Bank X did was to involve top management in the security process by forming a technology risk committee that included the executive vice president, the senior vice president, the senior vice president of retail operations, the vice president of IT, and the chief financial officer.

The technology risk committee realized that it was important for the bank and its IT department to be able to tolerate outsiders probing their network security. The committee also wanted to know where the bank's network and Web site security were vulnerable. Next, the bank hired a security-consulting firm, DefendNet Solutions, Inc., to perform a security audit. On a Friday morning in July 2000, the IT vice president got the good news...and the bad news.

General E-Business Security Issues

Why does an e-business need to worry about security? Any business, whether it is a traditional brick-and-mortar business, a brick-and-click e-business, or a pure-play e-business, needs to be concerned about network security. The Internet is a public network consisting of thousands of private computer networks connected together. This means that a private computer network system is exposed to potential threats from anywhere on the public network. Protection against these threats requires businesses to have stringent security measures in place. In the physical world, crimes often leave evidence—fingerprints, footprints, witnesses, video on security cameras, etc. Online, a cybercrime also leaves physical, electronic evidence, but unless good security measures are taken, it may be difficult to trace the source of a cybercrime.

Additionally, e-businesses must protect against the unknown. New methods of attacking networks and Web sites and new network security holes are being discovered with disturbing frequency. By carefully planning its network and Web site security system, an e-business can protect itself against many known and as yet unknown threats. An e-business must always be prepared for network and Web site attacks, or risk the loss of assets.

Another very important reason to protect an e-business's network and Web site is to protect the e-business's relationships with its customers. Many Internet users perceive that there is a large risk to their privacy and security when they buy products and services or submit personal information online. Although the perception of risk may be greater than the actual risk, it is still a cause for concern. An e-business must address customers' perceived risks just as much as any actual risks.

An e-business cannot expect to achieve perfect security for its network and Web site. The important issue for an e-business is to have adequate security to protect its assets, revenue stream, customer privacy, and its own reputation. Determining adequate security depends on an individual e-business's situation. For example, a Web site providing information on flavors of dog food may not require the same level of security as an online banking Web site. An e-business must determine its security needs according to the risks involved, the value of the assets at risk, and the cost of implementing a security system.

How does an e-business identify the security issues to be addressed? First, the e-business must thoroughly understand its business and how all its systems, not just its Web servers, are used. Several aspects of e-business computer systems security need to be addressed:

- *How secure is the server software?* Security should be in place to prevent any unauthorized remote logon to the system. It should be extremely difficult to make changes to the server software. The servers themselves should be physically located in a secure environment.
- *How secure are communications?* Customer credit card information and other sensitive data that is being transmitted across the Internet must be protected.
- *How is the data protected once it is delivered to the e-business?* Is it stored in unencrypted text files at the Web site? Is it moved to offline storage?
- *How are credit card transactions authenticated and authorized?*

Besides implementing secure technologies, an e-business should develop security policies and procedures. Everyone working in an e-business should understand his or her responsibilities for keeping the business secure. Also, a plan of action should be ready to deal with any potential security problem.

The biggest potential security problem in an e-business is of human, rather than electronic, origin. The weakest link in any security system is the people using it. The employees of an e-business may not understand the security policy. Sometimes, the security policy is so burdensome that the employees are not able to follow it, or refuse to follow it because it makes it difficult for them to get their work done. For example, employees might resent having to make frequent changes to logon passwords. At times, they may not understand the importance of security measures. Educating employees about the need for security and their role in the security process is essential. Table 8-1 summarizes general security issues that an e-business must consider:

Issue	Comment
Connection to the Internet	Private computer networks are at risk from potential threats from anywhere on the public Internet network.
Unknown risks	New security holes and methods of attacking networks are being discovered with alarming frequency.
Customer privacy and security of customer information	Not only must steps be taken to protect the privacy of customer information, but customers must be made aware of those steps and have confidence in them.
Security consciousness	Management and employees must understand the importance of security policies and procedures.

Table 8-1

General security issues

Part of addressing security risks is to first understand the inherent risks associated with a network and Web sites.

Network and Web Site Security Risks

An entire glossary of words and phrases identifies network and Web security risks, such as hacker, cracker, Trojan horse, and more. As part of planning a startup e-business's security, management should become familiar with network and Web server security risk terminology. Originally, hacker was a term used to describe gifted software programmers. Now, however, **hacker** is a slang term for someone who deliberately gains unauthorized access to individual computers or computer networks. Ethical hackers use their skills to find weaknesses in computer systems and make them known, without regard for personal gain. Malicious hackers, also called **crackers**, gain access to steal valuable information such as credit card numbers, to attempt to disrupt service, or to cause other damage. Because of the wide press coverage of computer system security breaches, the terms "hacker" and "cracker" are now generally used interchangeably for those involved in malicious unauthorized computer system access.

E-CASE Does Fame Bring Fortune?

His story has launched several newspaper and magazine articles, at least two books, a 1999 movie, and an untold number of Web sites. Who is this extraordinary individual: a rock star? a politician? a "rags-to-riches" e-businessman? No, it's Kevin Mitnick, arguably the most famous hacker of them all. Mitnick's hacking career began when, as a teenager, he stole computer manuals from Pacific Bell and later broke into computers at the University of Southern California, Santa Cruz Operations, Digital Equipment Corp., and many other organizations, repeatedly violating terms of an earlier probation. In 1992 Mitnick went underground and in 1995 was tracked down and arrested in North Carolina. Based on a 25-count indictment that included charges of wire fraud and illegal possession of stolen computer files, Mitnick was sentenced to prison for several years. In January 2000 Mitnick was released after plea bargaining and receiving credit for time served.

An e-business must protect itself against unauthorized access to its computer network, denial of service traffic overloads, and the intrusion of destructive viruses.

Denial of service attacks

A **denial of service**, or **DoS**, attack is an attack on a network that is designed to disable the network by flooding it with useless traffic or activity. A **distributed denial of service**, or **DDoS**, attack uses multiple computers to launch a DoS attack. While a DoS attack does not do any technological damage, it can do substantial financial damage to an e-business, because every second an e-business's network or Web site is down may result in lost revenues.

The only reward to a hacker for launching a DoS attack seems to be the opportunity to show off his or her skills. DoS attacks are not limited to high-profile e-businesses such as CNN.com or Amazon.com. On August 10, 2000, linkLINE Communications, a California-based ISP and Web hosting service (Figure 8-1), had its dial-up and high-speed Internet operations interrupted. The attack against linkLINE lasted six hours, although service to customers was out only 53 minutes because of the hard work by linkLINE's technicians. Additionally, while the technicians worked to stop the DoS attack, the hacker gained entry to other systems via an unprotected customer Internet account and swamped other company systems with information in an attempt to bring all of linkLINE's systems down.

Fortunately, linkLINE's technicians had previously installed filters and routers on key servers, to allow them to redirect DoS attacks away from crucial systems. A **filter** is a process or device that screens incoming information and allows only the information meeting certain criteria to pass through to the next area. A **router** is a device that connects two or more networks and forwards information between networks. However, the hacker was able to find a way around one of the routers designed to filter out DoS attacks.

Figure 8-1
linkLINE

Viruses

Viruses are the most common security risk faced by e-businesses today. A **virus** is a small program that inserts itself into other program files that then become "infected," just as a virus in nature embeds itself in normal cells. The virus is then spread when an infected program executes and infects other programs. Examples of virus effects include preventing a computer system from booting, erasing files or entire hard drives, preventing the saving or printing of files, and thousands of other possibilities. A **logic bomb** is a virus whose attack is triggered by some event such as the date on a computer's system clock. A logic bomb may simply release a virus or may be a virus itself. Viruses are generally introduced into a computer system via e-mail or by unauthorized network access. Virus examples include Stoned.Michelangelo and AutoStart 9805.

A **Trojan horse**, which takes its name from a story in Homer's *Iliad*, is a special type of virus that emulates a benign application. It appears to do something useful or entertaining but actually does something else as well, such as destroying files or creating a "back door" entry point to give an intruder access to the system. A Trojan horse can be an e-mail attachment or a downloaded program. Trojan horse examples include BackOrifice, VBS/Freelink, and BackDoor-G.

A **worm** is a special type of virus that doesn't alter program files directly. Instead, a worm replaces a document or application with its own code and then uses that code to replicate itself. Worms are often not noticed until their uncontrolled replication consumes system resources, which slows down or stops the system. Worm examples include VBS/Loveletter.a, VBS/Godzilla.worm, and Happy.99.

A **macro** is a short program written in an application such as Microsoft Word or Excel to accomplish a series of keystrokes. A **macro virus** is a virus that infects Microsoft Word or Excel macros. Examples of macro viruses that infect the Microsoft Word Normal document template are W97M/Ethan.a, W97M/Thus.a, and Melissa. Macro viruses can be introduced into a computer system as part of a Word or Excel document received as an e-mail attachment, or as a file on disk. Opening the e-mail attachment or file triggers the macro virus.

Some so-called viruses are, however, just hoaxes. Several antivirus software vendors maintain up-to-date information on viruses, worms, Trojan horses, and hoaxes, such as the Virus Information Library at McAfee.com, the AntiViral Pro Virus Encyclopedia, or the Symantec AntiVirus Research Center Encyclopedia (Figure 8-2).

E-businesses also face other security issues related to doing business on the Web, such as Web site defacement, information theft, and data spills.

Web site defacement

Web site vandalism or **defacement** can be the result of a hacker breaking into a network, accessing the Web site files, and modifying the HTML to physically change Web pages. Not only do Web site defacements embarrass an e-business, but some Web site defacements can

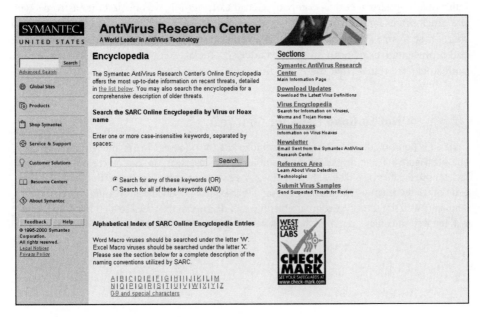

Figure 8-2
Symantec AntiVirus Research Center Encyclopedia

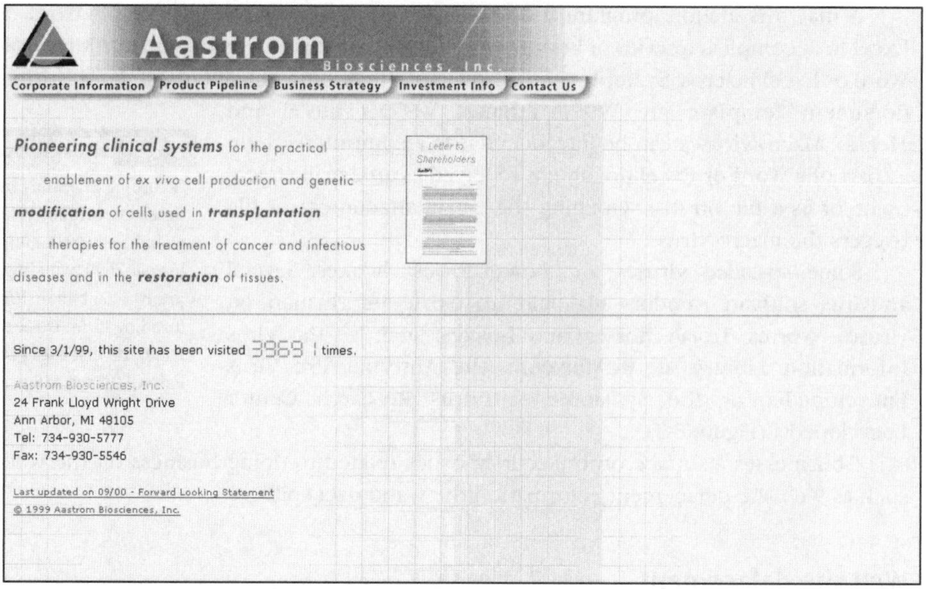

Figure 8-3
Aastrom
Biosciences, Inc.

have serious financial repercussions. Aastrom Biosciences, Inc. (Figure 8-3), a Michigan-based medical products company, experienced a serious defacement created to manipulate its stock price. In February 2000 a bogus news release announcing a merger with a California biopharmaceutical company, Geron Corporation, was posted on Aastrom's Web site. Stock prices for both companies rose: Aastrom shares rose from $4 to $4.41 and Geron shares rose from $47.19 to $51. After discovering the defacement, Aastrom notified Geron, and representatives of both companies advised officials with the NASDAQ index, where both stocks are traded, that there was no merger.

E-CASE Not So Good News for Goodyear

In the early hours of June 24, 2000, a hacker broke into the Goodyear Tire & Rubber Co. corporate Web site (Figure 8-4) and several Goodyear subsidiary sites overseas. On the corporate Web site, the hacker posted a message filled with profanity and a picture of a wrecked auto with the caption "The result of Goodyear tyres." The message questioned the quality of Goodyear tires and ridiculed the company's Web site security. Goodyear's security staff discovered the message after a few hours and took the Web site offline, removed the offensive message, and installed new security measures.

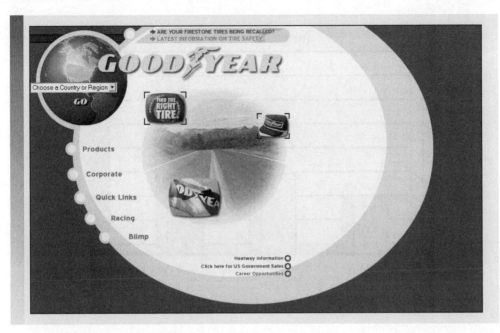

Figure 8-4
Goodyear

Another serious threat is electronic industrial espionage.

Electronic industrial espionage

It is a major risk and a big dollar issue that most companies are reluctant to discuss openly—electronic industrial espionage. According to the American Society for Industrial Security, in 1999 alone Fortune 1000 companies sustained losses of $45 billion from theft of proprietary information. While exactly how much of those lost dollars was the result of electronic industrial espionage is not clear, a survey by the Computer Security Institute indicated that more than half of the 600 companies surveyed felt that their competitors were the likely cause of an attack by hackers and further reported $60 million in losses related to those attacks. Based on data from the Computer Security Institute, Figure 8-5 illustrates that the value of losses from electronic theft of proprietary information is steadily increasing.

TIP

"Spoofing" is a way to give the appearance of a defaced Web page without hacking into a network and actually changing the HTML on a Web page. Hackers can spoof a Web site by creating a fake Web site and then redirecting traffic from the legitimate Web site to the fake Web site by manipulating DNS records (IP addresses).

Often, e-businesses that have been hacked and had business secrets stolen are too embarrassed to admit to the break-in. However, in late October 2000, one very high-profile company—Microsoft—found itself scrambling to deal with first rumors and then published reports of a serious hacking incident with industrial espionage overtones. The apparent culprit was a Trojan horse virus named QAZ Trojan that was first identified in mid-July in China. The QAZ Trojan virus infects a computer system when a user opens an e-mail attachment containing the virus. Then the virus replaces the system's Notepad text editor with its own code, searches for other shared hard drives to infect, and sends the IP addresses of infected computers to an outside e-mail address. This creates a "back door" a hacker can

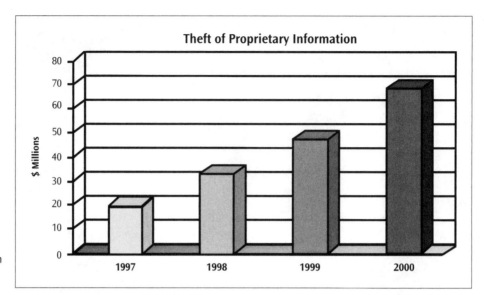

Theft of Proprietary Information

Figure 8-5
Value of stolen information

use to enter a system, search for passwords, and install software programs to allow remote control of the computer. Although by August 2000 all major antivirus software makers had included the QAZ Trojan information in their downloadable virus updates, somehow the QAZ Trojan virus was used to create a "back door" at Microsoft.

On Friday, October 27, *The Wall Street Journal* broke the story of the hack attack on Microsoft with a confirmation by company president Steve Ballmer that the hackers had infiltrated Microsoft's network for up to three months and accessed the source code for major products. Over the next few days, other Microsoft spokespersons began to contradict Ballmer and to indicate that the hack attack was far less lengthy or severe than first reported. By Monday, October 30, Microsoft's story was that the attack originated from an employee's unprotected home computer connected to the network, that the length of the hack attack was only 12 days, that Microsoft was watching and trying to identify the hacker during this time, and that major product source code was not likely seen by the hacker.

Experts were outspoken both about the manner in which Microsoft was vulnerable to a hack attack—failure to follow best practice to prevent such attacks—and about how Microsoft mishandled the distribution of information about the attack to the public. Some experts felt that the changing stories made it appear that Microsoft was unprepared to address the hack attack with the public. Also, because Microsoft minimized the damage of the attack, other experts were concerned that the changing story had damaged Microsoft's brand equity and its customers' trust. Concern was expressed that the attack on Microsoft was the most damaging instance of electronic espionage to date, and the attack raised new and important questions about e-business security issues. Finally, other experts reiterated that, for Microsoft and everyone else, protection against electronic industrial espionage really boils down to familiar advice:

◆ Don't open questionable e-mail attachments.
◆ Keep security software and virus protection software updated and enabled.

◆ Implement corporate-wide best practices for system security.

Not only is an e-business's own proprietary information at risk, its customers' information is also at risk.

Credit card fraud and theft of customer data

Almost all B2C purchase transactions involve credit cards. An e-business that accepts credit cards in payment for goods and services must secure the credit card information in transit to its Web site, and it must secure stored credit card information. Also, systems must be in place for credit card transaction authentication (verifying that the person placing the order really is the holder of the credit card used in the transaction), and credit card authorization (verifying that the charge can be made to the card number).

A hacker can break into a database server and steal thousands of credit card numbers and other information in a matter of moments, and an e-business might not even recognize that the hacker was there. For example, one of the largest reported cases of stolen credit card information took place in January 1999 (but was not reported until much later) when information on 485,000 credit cards, including card numbers, expiration dates, names, and addresses, was stolen from an e-business Web site and stored at a U.S. government agency's Web site, where the agency's Web site administrator discovered the data. There was no reported evidence of fraudulent use, and some of the accounts were not active. But this event highlights the risk to a vulnerable e-business of the theft of sensitive information.

TIP

It's not just consumers at B2C Web sites who have anxiety over the risks inherent in buying products and services online. A Cahners In-Stat survey of 1,000 purchasing professionals found that 66 percent of respondents feared the risks inherent in fraudulent use of payment information and thought those risks were a significant barrier to purchasing online. Fifty-six percent of respondents indicated that they were uncomfortable ordering online for fear of exposing their buying plans to competitors. A primary concern for purchasing professionals is not credit card theft but the protection from competitors of information about purchase orders, inventory, and other sensitive customer/supplier information.

E-CASE IN PROGRESS foodlocker.com

The most important information that foodlocker.com has to protect is its customer credit card information. In order to keep customer credit card data secure, foodlocker.com takes many measures. All ordering is done using a secure server that encrypts the information sent by the customer to the foodlocker.com Web site. Second, all customer information stored on the Web site is protected by more than one password, so even if one password were discovered by a hacker, he or she would still not be able to get to sensitive information. Also, all credit card information is removed from the foodlocker.com servers soon after the transaction, so that data cannot be accessed online, even if all of the passwords were compromised. Finally, all passwords used by foodlocker.com are alphanumeric combinations that are not allowed to consist of easily guessed information such as names, phone numbers, and addresses. These passwords are also changed regularly.

Although the Internet has made global e-business possible, it has also led to an increasing risk of credit card fraud, with up to one-third of all online credit card fraud being perpetrated on U.S. e-businesses by international cybercriminals. U.S. law enforcement is hampered in apprehending international cybercriminals by the differences in white-collar crime laws in other countries, questions of jurisdiction, foreign governmental indifference, and the time and expense of conducting international investigations. Consumers are responsible for only the first $50 of fraudulent online credit card purchases; therefore, in many cases an e-business bears the brunt of losses related to fraudulent credit card purchases.

Another issue is exposing customers to **identity theft**—the theft of sensitive customer information and its use by cybercriminals to obtain new credit or make large credit purchases. In some cases it takes years for consumers to repair the damage to their credit records from identity theft. Protection of sensitive customer data, during transmission to an e-business's Web site and while the data is stored, is paramount. To reduce anxiety over buying products and services online, an e-business must not only have appropriate security measures in place but must also build customer confidence in those measures.

E-CASE Pay Up or Else!

In January 2000, CD Universe, a music e-tailer, realized that its Web site had been hacked when it received an e-mail extortion message from a Russian teenager requesting $100,000 to destroy a list of over 300,000 unencrypted customer credit card numbers the hacker had copied from the CD Universe Web site. When CD Universe (Figure 8-6) refused to pay up, the hacker released about 25,000 of the credit card numbers to a public Web site that the FBI quickly shut down. Although only a handful of fraudulent credit card charges were linked to the theft, thousands of credit cards were reissued as a precaution.

Figure 8-6
CD Universe

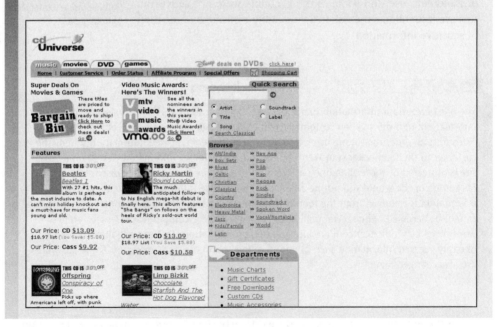

Data spills

Failure to properly protect customer information submitted to a Web site can result in accidental publication, sometimes called a **data spill**, of sensitive information such as customers' names, addresses, phone numbers, and e-mail addresses. Such data spills involving Yahoo!, Seagate, AT&T, Nissan, and Butterball have exposed customers' personal data.

For example, Butterball (Figure 8-7) offered an opportunity to register for a Turkey Mail e-mail newsletter at its Web site, for people interested in turkey recipes and cooking tips. The registration form also requested certain demographic information. Butterball accidentally published the information it gathered from those registration forms to a Web page accessible by the general public. The security breach was found by an employee of a pharmaceutical firm who had read stories of similar data spills and used a popular Internet search tool to look for his and his wife's addresses.

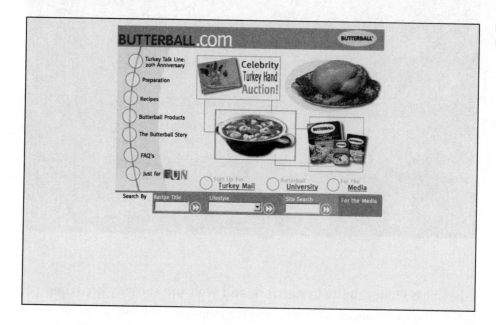

Figure 8-7
Butterball

Are Diamonds Forever?

In April 2000, about 35,000 customer e-mail and home addresses were exposed on the Diamond Information Center Web site (Figure 8-8), an informational Web site sponsored by De Beers Consolidated Mines Limited (generally known as De Beers). How were the names and addresses exposed? Chad Yoshikawa, a San Francisco Bay area consultant, located the names and addresses simply by using the on-site search engine. Yoshikawa's wife had entered a contest at the Diamond Information Center Web site and registered her entry by supplying name and address information. Yoshikawa immediately notified the Web site administrator, who removed the sensitive information from public access.

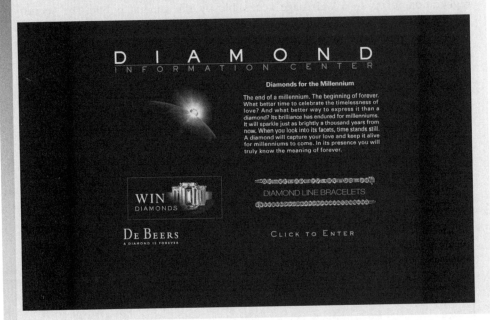

Figure 8-8
Diamond
Information
Center

E-business vulnerability to network and Web site security breaches

In May 2000, the Computer Security Institute (CSI) announced the results of its fifth annual computer crime and security survey of primarily government agencies and large corporations. The survey was conducted with the participation of the Computer Intrusion Squad of the San Francisco office of the Federal Bureau of Investigation (FBI). The survey results emphasize the ongoing vulnerability of organizations to computer security violations.

◆ Ninety percent of survey respondents detected computer security breaches within the past 12 months.

◆ Seventy percent of survey respondents reported a variety of serious security breaches, including theft of proprietary information, financial fraud, system penetration from outsiders, denial of service attacks, and sabotage of data or networks.

- Seventy-four percent of survey respondents admitted to financial losses due to security breaches, and 42 percent quantified those losses at more than $265 million.

The threat to e-businesses of computer crime and other information security breaches continues to grow. The CSI (Figure 8-9) survey also reported separately on the security breaches for the 93 percent of the respondents with Web sites.

- Forty-three percent of survey respondents have e-business Web sites.
- Nineteen percent of survey respondents incurred unauthorized access or misuse of their Web sites within the past 12 months, while 32 percent reported that they didn't know if there had been any unauthorized access or use.
- Thirty-five percent of survey respondents who admitted unauthorized Web site access noted from two to five incidents, with 19 percent reporting 10 or more incidents.
- Sixty-four percent of survey respondents who admitted unauthorized Web site access had their Web sites vandalized, 60 percent reported denial of service incidents, 8 percent reported theft of transaction information, and 3 percent reported financial fraud.

Because of the ever-present and growing risks to its computer networks and Web sites, an e-business should carefully review its network and Web site security needs.

E-Business Security

Network and Web site security issues are too often not a top priority for startup e-businesses, whose primary efforts are directed toward getting the e-business up and running. A startup e-business with limited in-house security expertise and limited resources should consider

Figure 8-9

Computer Security Institute (CSI)

outsourcing its Web hosting needs to an ISP or Web hosting company that has the expertise and resources to handle security issues. E-businesses that elect to host their own Web site should make security issues a top priority.

Network and Web site security

The best way to recognize when a hacker is attempting unauthorized network access is to monitor network performance. Setting up, logging, and monitoring established network reference points, called **benchmarks**, can alert an e-business to security problems. A skilled system administrator and other well-trained technicians, who use these benchmarks to monitor and manage the network and servers, are critical. Additionally, the system administrator should regularly monitor software vendors' Web sites, security-related Web sites such as AntiOnline (Figure 8-10), and security newsgroups to stay abreast of network and Web site security issues.

Other tools such as passwords, firewalls, intrusion detection systems, and virus scanning software should be used to protect an e-business's network and Web site.

A **password** is a code, or more often a common word, used to gain access to a computer network. Passwords are only effective when used properly. Often a computer user chooses a bad password, such as a short common word, a name, or birthday, because he or she wants to be able to remember the password easily. One way hackers penetrate network security is by using software that "guesses" a password by trying millions of common words until one of the words is accepted. Passwords that require a minimum length of six characters in a mix of letters and numbers increase the number of potential passwords into the billions and make it more difficult for a hacker to guess them. A computer user should also change passwords regularly. If a user has access to multiple systems, it is a good idea to require different passwords on each system.

Figure 8-10
AntiOnline

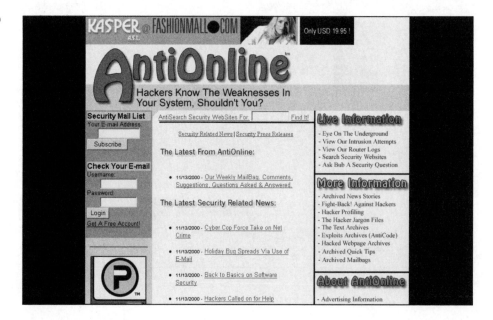

A **firewall** is software or hardware used to isolate a private system or network from the public network. A firewall provides an easy-to-manage entry point to multiple systems behind it. Firewalls can control the type of information that is allowed to pass from the public network to the private network, as well as what services inside the firewall are accessible from the outside. Firewalls can also log activity to provide an audit trail, in case the network is penetrated. Figure 8-11 illustrates the use of a firewall.

Intrusion detection is the ability to analyze real-time data to detect, log, and stop unauthorized network access as it happens. Businesses can install intrusion detection systems that monitor the network for real-time intrusions and respond to intrusions in a variety of user-determined ways. An intrusion detection system can defend a Web site against DoS attacks by adding more servers to increase the traffic the Web site can handle, by using filters and routers to manage traffic, and by having a backup plan to reroute legitimate traffic during an attack. Cisco's Secure Intrusion Detection System (Figure 8-12) and Network ICE's ICEpac Security Suite are two examples of intrusion detection systems.

Virus scanning software, including e-mail virus scanning, should be installed on all network computers. Antivirus software should be kept updated. Two top-rated virus scanning software products are McAfee's VirusScan (Figure 8-13) and Symantec's Norton AntiVirus.

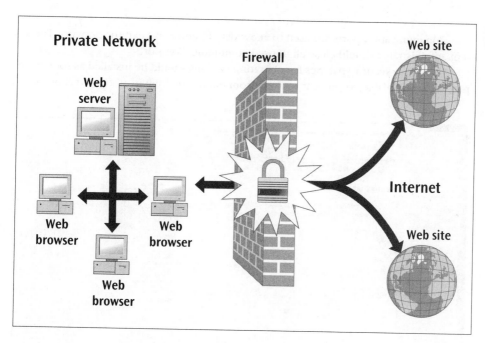

Figure 8-11
Firewall

Figure 8-12

Cisco Secure
Intrusion
Detection
System

Communication ports are used to allow data to enter and exit the network. The system administrator should close all unused communication ports.

Up-to-date security patches for operating systems should be installed as soon as the patches are available, to prevent hackers from exploiting built-in system weaknesses.

Figure 8-13

McAfee
VirusScan

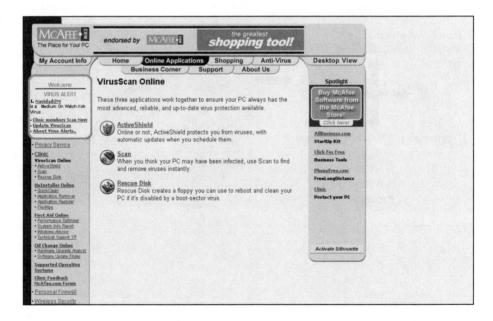

Transaction security and data protection

Transaction security, especially for credit card transactions, and the protection of customer data are as important as Web site and network security. Tools to protect transaction data and customer data include:

- Using a predefined key to encrypt and decrypt the data during transmission
- Using the Secure Sockets Layer (SSL) protocol to protect data transmitted over the Internet. SSL provides encryption of data between the browser on the customer's computer and the software on the Web server, allowing data such as credit card information to be transmitted securely. SSL uses digital certificates so that a Web browser can authenticate the server it is connected to, making sure that credit card data is going to the appropriate server
- Moving sensitive customer information such as credit card numbers offline or encrypting the information if it is to be stored online
- Removing all files and data from storage devices, including disk drives and tapes, before getting rid of the devices
- Shredding all hard-copy documents containing sensitive information before trashing them

Of course, an e-business's security solutions are only as strong as its weakest link—often its employees. An e-business must maintain a security-oriented culture, starting at the top, in order for employees to take security seriously. An e-business should also consider having its security systems tested or audited.

Security audits and penetration testing

Security audits can provide an overall assessment of an e-business's systems and security issues by checking for vulnerabilities in those systems and providing recommendations for fixing those vulnerabilities. Security consultants such as DefendNet Solutions Inc., Internet Security Systems, and Pinkerton Systems Integration (Figure 8-14) offer security auditing services.

Accounting firms, such as Arthur Andersen L.L.P. and Ernst & Young, also offer security auditing services. Some of the Big Five national accounting firms use the American Institute of Certified Public Accountants (AICPA) WebTrust seal and audit criteria. The WebTrust seal indicates to customers that the Web site is verified as being safe and secure by the AICPA. The AICPA audit criteria cover best business practices, site security, and customer information privacy. Some accounting firms use their own audit seal instead of, or in addition to, the AICPA WebTrust seal. An example of an e-business that has passed a Web site audit and displays the WebTrust seal is E*TRADE (Figure 8-15).

Figure 8-14

Pinkerton Systems Integration

Following a full review or audit of security issues, an e-business should have comprehensive network penetration testing done on a regular basis. **Penetration testing** uses real-world hacking tools to test computer security. Regular penetration testing provides an opportunity to see how an e-business's computer systems' security stands up to the

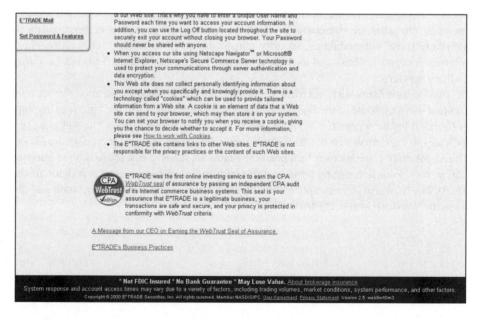

Figure 8-15

E*TRADE displaying WebTrust seal

most current hacking tools and techniques. Penetration testing can measure the effectiveness of intrusion detection measures and response mechanisms. Penetration testing is often used to supplement security audit findings.

When evaluating security consultants who will perform the penetration testing, there are several factors to consider:

◆ Get evidence that the security consultants have insurance to protect against accidental system damage or downtime.

◆ Have everyone on the consultant's penetration team sign a nondisclosure agreement.

◆ Consider requiring a third-party background check on each member of the consultant's penetration team.

◆ Decide whether it makes sense to use a security consultant who employs former hackers.

◆ Determine if the consultant's team is going to use packaged security scanning software that could be employed by the in-house staff, or if they are using custom tools.

◆ Develop a clear scope for the penetration test and a workable time frame.

◆ Determine whether to have a DoS attack done, and if so, when to schedule it to least disrupt customer access.

◆ Make sure the final report from the consultant includes an accounting of all attacks attempted and whether or not they were successful, a return of all the paper or electronic information gathered by the consultant, and recommendations on how to fix any problems discovered during the tests.

E-CASE Renting a Hacker

When Leo Jones (not his real name) realized someone was snooping around in the sensitive information stored on his company's network, he knew he needed help. Jones called in a security consultant, John Klein of Rent-A-Hacker (Figure 8-16), to find out who was hacking into the system and how they were doing it. Rent-A-Hacker employs about 300 freelance computer security consultants—or ethical hackers—from around the world. Klein handpicks a consultant to fit each consulting job his firm gets.

Klein logged on to Jones's network and quickly discovered that an intruder had taken advantage of a commonly known operating system bug. Klein turned the Jones situation over to one of his top specialists, Kelvin Wong. Wong is a former youthful hacker who successfully intruded into Web sites such as nasa.gov, army.mil, and usda.gov. When he turned 18, Wong became a legitimate ethical hacker and Rent-A-Hacker's chief operating officer and top consultant. Wong successfully back-traced the intruder's Internet connection to a Canadian ISP and then to the intruder's cable modem. After launching several DoS attacks to confuse Wong, the intruder finally lost the game and was turned over to the Royal Canadian Mounted Police.

Some traditional security consultants question the advisability of hiring former hackers with questionable backgrounds and trusting them with sensitive information. Klein defends his operation by saying that to catch a hacker, you have to think like one.

(Continued on the next page)

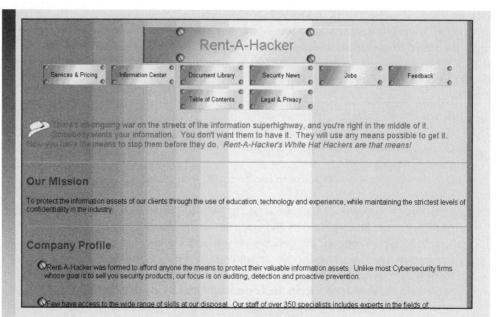

Figure 8-16

Rent-A-Hacker

Individual PC security risks

Often the principals in an e-business use standalone personal computers during the startup phase until funds are available to build and operate a network, or until the e-business can outsource its IT operations. Additionally, some e-businesses offer their employees the opportunity to telecommute—allowing an employee to use his or her home computer or installing a business-owned personal computer in the employee's home. Certainly, business employees often work on business files at home. Because of these factors, it's important for an e-business to understand that individual PCs are also at risk from hackers.

For example, Jim Jarrard, the president of Cinenet, a stock film footage company, didn't think twice about leaving his computer running all night as he tried to download a large file over his DSL Internet connection. During the night a hacker accessed Jarrard's PC and attempted to gain control over the PC, steal valuable film files, and erase the computer's hard drive by installing a software program. Fortunately, the hacker was not completely successful. When Jarrard checked the computer in the morning, he found a frozen system and an error message. Jarrard spent two weeks trying to determine what had happened before he finally realized he had been "hacked" and that the only solution was to back up his data files and reformat the PC's hard drive. During his investigation of the problem, he realized that, because he was connecting to the Internet using an "always on" DSL connection, he needed to install **personal firewall** software to guard against future intrusions.

Surprises on a Home Computer

What are home computer hackers looking for? Some are just playing a game, just trying to see if they can hack the computer. Others are trying to steal information such as credit card numbers. Still others are trying to store files and programs that allow them to take control of the computer, perhaps to use it later in a DoS attack against an e-business's Web site.

Robert Smith (not his real name) will likely never know why his home computer network was hacked. Smith, an IT professional, installed a cable Internet connection to provide his children with high-speed, always-on Internet access. A few weeks later, concerned about security, he began installing personal firewall software on his home computers. When he installed the firewall on his daughter's computer, alarm bells went off. For several days hackers had been logging on to his daughter's computer and storing files and programs. Now the firewall software had locked out the hackers. Over the next several days, Smith watched as hackers tried unsuccessfully to break into the computer. At one point he noted 27 different hackers trying to break in.

Whatever the circumstances, personal computers may be at the same risk of hacking and loss of business-related material as are computer networks. While hacking individual PCs is still relatively rare, the chance of some type of hacking is greater if an uninterrupted, always-on Internet connection such as DSL or cable modem is used. Personal firewall software such as Network ICE's BlackICE Defender and Zone Lab's ZoneAlarm (Figure 8-17) can help protect an individual PC from serious hacker attacks.

After recognizing general network and Web site security issues and specific types of security risks, an e-business should decide whether to outsource its security management or manage it in-house.

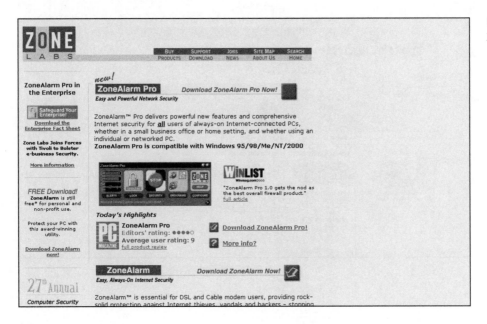

Figure 8-17
Zone Labs

E-Business Security Providers

Unless an e-business maintains an in-house security expert—and few startup e-businesses can afford to do that—it is important to review security needs with an outside security consultant.

Security products

If a startup e-business prefers to keep its security management system in-house, there are a number of security products available. For example, WatchGuard Technologies' (Figure 8-18) LiveSecurity System includes firewall software and other security features and combines them into a device called the WatchGuard Firebox. The device, which has three ports, can be connected to an Internet connection, LAN connection, and e-mail server.

Other companies that provide security products such as firewalls, intrusion detection software, and authentication and encryption products are Check Point Software Technologies Ltd., Internet Security Systems, Inc., Axent Technologies, Inc., Verisign, WebTrends, and Entrust Technologies Inc.

Security services providers

For many e-businesses, hiring, training, and maintaining an in-house staff for systems security monitoring is difficult and expensive. For these e-businesses, outsourcing their

Figure 8-18
WatchGuard

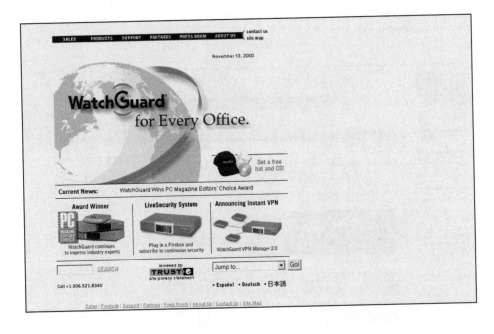

security needs to a technology provider that specializes in information security, including managed services, is a good option.

The advantages to an e-business of outsourcing security services to firms such as Verizon, ManagedFirewall.com, Riptech, myCIO.com, and Counterpane Internet Security, Inc. (Figure 8-19) include $24 \times 7 \times 365$ monitoring of its systems, access to in-depth knowledge about security issues, current expertise in using security tools, and lower costs.

E-businesses outsourcing their security management should look closely at the contract with the security management company to determine whether or not the security management company assumes any liability and covers damages resulting from security breaches. Additionally, the security management company should have errors and omissions insurance coverage that protects against any financial damage to clients resulting from the company's professional advice or actions.

TIP

According to the Gartner Group, the market for security consulting, integration, and managed services was $7.5 billion in 2000 and growing at an annual compounded rate of 40 percent. Although the mainstays of information security technology are firewalls and antivirus software, many experts believe that the next big opportunity for information security professionals is in managing, rather than installing, security technology.

E-Business Risk Management Issues

An e-business should manage its e-business risks as a business issue, not just as a technology issue. An e-business must consider the direct financial impact of immediate loss of revenue, compensatory payments, and future revenue loss from e-business risks such as:

◆ Business interruptions caused by Web site defacement or denial of service attacks

Figure 8-19
Counterpane Internet Security, Inc.

- Litigation and settlement costs over employees' inappropriate use of e-mail and the Internet
- Product or service claims against items advertised and sold via a Web site
- Web-related copyright, trademark, and patent infringement lawsuits
- Natural or weather-related disasters

An e-business should put in place an effective risk management program that includes:

- Network and Web site security and intruder detection programs
- Antivirus protection
- Firewalls
- Sound security policies and procedures
- Employee education

Another important component of a risk management program is the transfer of risk via insurance. Table 8-2 illustrates some of the different kinds of insurance coverage an e-business should consider when developing an effective risk management program. It is a good idea for an e-business's management to consult with a commercial insurance broker that offers e-risk management services, to help develop a risk management plan including insurance coverage.

TIP

In June 2000 Assurex International, in association with the Human Resources Institute of Eckerd College, announced the results of its E-Risks Survey, in which only 13 percent of e-business respondents reported purchasing unauthorized access/use insurance to protect against loss at their Web site.

Table 8-2
Insurance coverage options

E-Risk Insurance	Coverage
Computer Virus Transmission	Protects against losses that occur when employees open infected e-mail attachments or download virus-laden software
Extortion and Reward	Responds to Internet extortion demands and/or pays rewards to help capture saboteurs
Unauthorized Access/ Unauthorized Use	Covers failure to protect against third-party access to data and transactions
Specialized Network Security	Responds to breach of network security and resulting losses
Media Liability	Protects against intellectual property infringement losses
Patent Infringement	Covers defensive and offensive costs when battling over patent infringement issues
Computer Server and Services Errors & Omissions	Protects e-businesses against liability for errors and omissions when their professional advice causes a client's financial loss

Traditional Insurance	Coverage
Employment Practices Liability	Protects employers from workers' claims of discrimination
Director's and Officer's	Protects corporate assets and the personal assets of directors and officers against wrongful acts such as mismanagement, fiscal irresponsibility, or security law violations
Product Liability	Covers risks of third-party bodily injury or property damage from products sold online
Business Interruption	Mitigates revenue losses if computer systems go down
Crisis Communication	Provides funds to hire professional public relations experts and others to handle damage control in a crisis
Crime Loss	Protects against electronic theft of funds
Electronic Data Processing (EDP)	Covers hardware and software replacement and extra expenses related to hiring technical experts and others to recapture lost data

Table 8-2 (continued) Insurance coverage options

Insurance companies such as CIGNA, AIG, The St. Paul Companies, Fidelity and Deposit, and syndicates at Lloyd's of London offer e-risk and other business insurance products through agents and commercial insurance brokers such as Marsh & McLennan (Figure 8-20).

Figure 8-20 Marsh & McLennan

To prepare for the security audit, DefendNet conducted a preliminary interview with the bank's internal auditor and the IT vice president. During this interview the security audit objectives were determined. The bank wanted DefendNet to test the security of its Web sites: the one that handled its customers' online banking and the other that handled static information pages. Two different hosting companies hosted the Web sites. The bank also wanted a security test of its internal network, especially its mail server. DefendNet did not ask the bank for any details about the Web sites or the internal network because it was important that DefendNet use only publicly available information, the same information available to potential hackers.

DefendNet began by using the bank's Web sites to locate every IP address related to the hosted sites. Next, DefendNet set up its testing team, which would attempt to breach the Web sites' security in the early morning hours over the next several days. The early morning hour of 2 a.m. was selected for the tests, since early morning is a more vulnerable time, when system administrators are less likely to be monitoring system logs.

The good news…within 30 minutes of the first attempted intrusion at the bank's online transaction Web site, the hosting company notified the bank by phone message and by e-mail that the bank's site had been scanned, and the scan had the feel of a security test. Within 12 hours of the scan, the hosting company had traced the scan back to DefendNet's IP address, obtained DefendNet's phone number, and called the security consultant to find out what was going on. The IT vice president and the internal auditor were thrilled with the hosting company's quick identification of the attack and its thorough response to the attack. More good news: the audit report on the network and e-mail server security was positive.

The bad news…the second hosting company, which hosted the bank's static information pages, was vulnerable to attack. Additionally, the second hosting company was not aware that the bank's Web site had been scanned until the IT vice president called them with the bad news. While there was no access to sensitive customer information at this Web site, the site was susceptible to Web site defacement by hackers, which could embarrass the bank and adversely affect its reputation. DefendNet (Figure 8-21) recommended switching the Web site containing static pages to the more secure hosting company. However, because of a longstanding relationship between the bank and the second hosting company, the bank elected to make the second hosting company upgrade its security in order to keep the bank's business.

As it turned out, having an independent review was one of the best things to happen to Bank X's security, and the IT vice president felt that the cost of the security audit was money well spent.

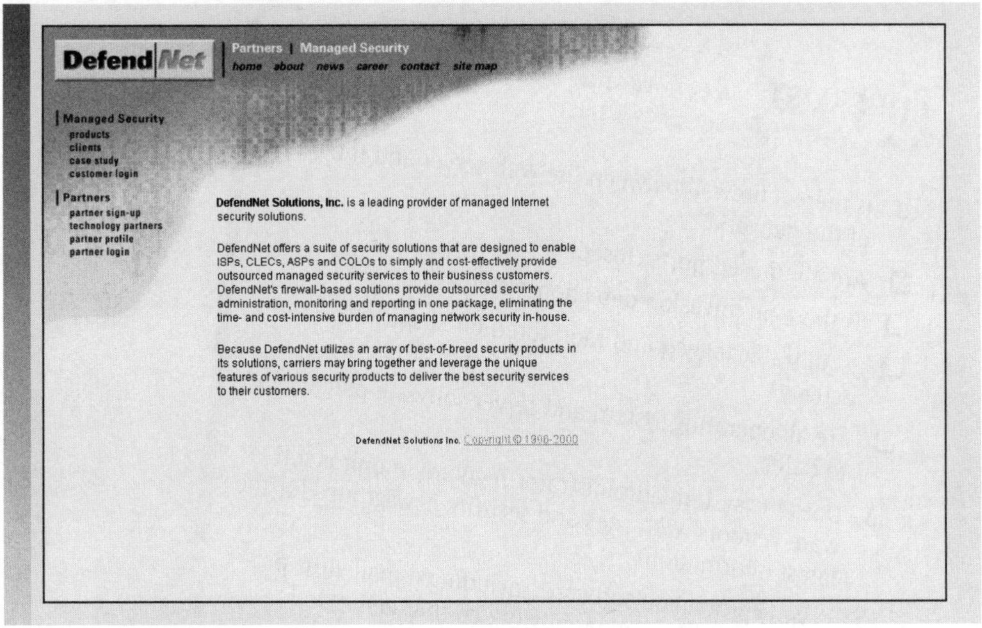
DefendNet
Partners | Managed Security
home about news career contact site map
Managed Security
products
clients
case study
customer login
Partners
partner sign-up
technology partners
partner profile
partner login
DefendNet Solutions, Inc. is a leading provider of managed Internet security solutions.
DefendNet offers a suite of security solutions that are designed to enable ISPs, CLECs, ASPs and COLOs to simply and cost-effectively provide outsourced managed security services to their business customers. DefendNet's firewall-based solutions provide outsourced security administration, monitoring and reporting in one package, eliminating the time- and cost-intensive burden of managing network security in-house.
Because DefendNet utilizes an array of best-of-breed security products in its solutions, carriers may bring together and leverage the unique features of various security products to deliver the best security services to their customers.
DefendNet Solutions Inc. Copyright © 1996-2000

Figure 8-21 DefendNet Solutions, Inc.

Summary

- An e-business must have a security plan and procedures to protect its network and Web site from intruders.
- Another important reason to maintain stringent network and Web site security measures is to protect sensitive information provided by customers, such as credit card numbers.
- The biggest potential security problem facing many e-businesses is a failure to develop a business culture that recognizes security risks and supports security measures.
- "Hacker" and "cracker" are terms generally used interchangeably to identify a person who accesses a network or Web site without authorization, for malicious purposes.
- Denial of service, or DoS, attacks overload a network with information requests.
- A virus is a small program that inserts itself into other program files that then "infect" still other files and programs. Viruses can damage a computer system in many ways, including erasing files or preventing files from being printed or saved. A Trojan horse, a worm, and a macro virus are special types of viruses.
- Web site defacement can embarrass an e-business and may cause financial losses.
- International credit card theft and fraud is an increasing problem for e-businesses.
- A data spill is the accidental publication of sensitive customer information on a publicly accessible Web site.
- An e-business should employ passwords, firewalls, intrusion detection systems, virus scanning software, and up-to-date security patches to operating system software to enhance its network and Web site security.
- An e-business could benefit from a security audit and penetration testing of its network and Web site.
- Individual PCs with always-on, high-speed connections to the Internet may be at risk from hacker attacks.
- An e-business with sufficient experience and resources can manage its security in-house; however, without experience and resources, it may be better for an e-business to outsource its security management.
- An e-business should set up a risk management plan that may include insurance to protect against a variety of potential losses of assets or liability claims.

CHECKLIST — Network and Web Site Security Plan

- ❑ Is there a firewall between the Web server and the rest of the network?
- ❑ Are all unused ports closed?
- ❑ Is there an intrusion detection system?
- ❑ Is all traffic logged and monitored for suspicious activity?
- ❑ Are all operating system and server software patches up to date?
- ❑ Does the system administrator regularly monitor software vendors' Web sites and security newsgroups for the latest information?
- ❑ Is virus protection software, including e-mail virus protection, installed on all network computers?
- ❑ Are filters installed to help deny entry to specific types of requests that look like a possible DoS attack or other intrusion?
- ❑ Are routers installed to help reroute legitimate traffic when defending against a DoS attack?
- ❑ Has the Web site passed stringent security testing?

Key Terms

benchmarks	firewall	penetration testing
cracker	hacker	personal firewall
data spill	identity theft	router
defacement	intrusion detection	Trojan horse
denial of service attacks (DoS)	logic bomb	virus
distributed denial of service attacks (DDoS)	macro	worm
filter	macro virus	
	password	

Review Questions

1. Which of the following is not a virus?
   a. Stoned.Michelangelo
   b. BackDoor-G
   c. Melissa
   d. DDoS

2. A logic bomb is:
   a. A short program written to accomplish a series of keystrokes.
   b. Used to secure credit card information at a Web site.
   c. A virus whose attack is triggered by an event.
   d. A process or device that screens incoming information.

3. A firewall is:
   a. An established network performance reference point.
   b. Software or hardware used to isolate a private network from a public network.
   c. A virus that infects macros.
   d. A predefined encryption key used to encrypt and decrypt data transmissions.

4. A router:
   a. Screens incoming information.

   b. Distributes information between networks.
   c. Clears all viruses from a computer system.
   d. Is a worm virus.

5. Web page defacement:
   a. Enhances the appearance of a Web page.
   b. Secures the Web page from intruders.
   c. Embarrasses Web site owners.
   d. Prevents identity theft.

6. A data spill is the accidental publishing of sensitive information at a publicly accessible Web site. **True or False?**

7. Intrusion detection systems protect against e-mail viruses. **True or False?**

8. Penetration testing uses real-world hacking tools to test the security of a network or Web site. **True or False?**

9. Network and Web site security is always the top priority for e-business employees. **True or False?**

10. Insurance can be an effective part of an e-business's risk management program. **True or False?**

Exercises

1. Using Internet search tools or other relevant sources, research famous (or infamous) network and Web site hacking events over the past three years. Then create a timeline of these events, including a description of each event.

2. Using Internet search tools or other relevant sources, research articles on the differences between ethical hackers and crackers. Then write a one-page paper describing the difference between ethical hackers and crackers, including recent real-world examples of ethical hacking and cracking.

3. Using several security-focused Web sites such as InfoSysSec, Microsoft Security, SecurityFocus.com, and securitywatch.com, review current news

about security issues. Then write a one-page paper describing 3–5 current security issues of particular concern to e-businesses.

4. Using the Virus Information Library at McAfee.com, the AntiViral Pro Virus Encyclopedia, Symantec's Virus Encyclopedia, or other relevant sources, note five recently identified computer viruses. Then write a one-page paper listing the viruses, including name, type of virus, and destructive capability.

5. Using Internet search tools or other relevant sources, research and identify the four basic types of firewalls. Then write a one-page paper summarizing each type.

CASE PROJECTS

◆ 1 ◆

You are the assistant to the president of a startup e-business. The president wants to create a business culture that supports system security, and asks you to draft an announcement to all employees about the importance of network security. Using any relevant resources, draft a one-page announcement explaining the importance of network security and especially the correct use of passwords.

◆ 2 ◆

The mock-up of your new B2C e-business Web site is ready for testing. Your partner, who created the Web site, insists that it is okay to store customer credit card information in a text file at the Web site. He insists that the credit card information will be perfectly safe. However, you are not sure that using a Web site text file is the correct way to securely store the credit card information.

Before you meet with your partner tomorrow, you want to learn more about the best way to safeguard the credit card information. Using Internet search tools or other relevant sources, research secure ways to store credit card information. Then create an outline you can use to either propose a better solution or support your partner's solution.

◆ 3 ◆

You and your B2C e-business partner have decided to hire a security consultant to help you understand network and Web site security issues and to perform a security audit of your e-business's internal network and hosted Web site. You are meeting with the consultant tomorrow and want to be prepared for the meeting. Create an outline of the topics you need to discuss during the meeting before entering into a contract with the consultant.

TEAM PROJECT

You and three classmates are part of the sales team for a B2B e-business. The purchasing manager for a potential major client has declined to order products through your Web site because of security concerns. You and your sales team are meeting tomorrow with the purchasing manager and her supervisor, the vice president of production, to convince them that there are adequate security measures in place.

Assume that your e-business has a risk management plan in place, has adequate network and Web site security in place, has survived a security audit, and schedules regular third-party penetration testing of the network and Web site.

Working with your team, use Microsoft PowerPoint or other presentation software to prepare a 5–10 slide presentation you can use to reassure clients that ordering at your Web site is safe. Then use the presentation to persuade two classmates, selected by your instructor to pose as the target client's purchasing manager and vice president, that ordering products online at your Web site is safe.

Useful Links

Arthur Andersen Case Studies
http://www.arthurandersen.com/WebSite.nsf/
    Content/Resources?

Arthur Andersen Risk Management
http://www.arthurandersen.com/website.nsf/content/
    MarketOfferingsRiskConsultingEWRMPublications?
    OpenDocument

Association for Computing Machinery
http://www.acm.org/

Biometric Consortium – Examples of Biometric Systems
http://www.biometrics.org/html/examples.html

Biometric Digest
http://webusers.anet-stl.com/~wrogers/biometrics/

Center for Secure Information Systems – George Mason University – Security links
http://www.isse.gmu.edu/~csis/links.html

CERT® Coordination Center – Carnegie Mellon Software Engineering Institute
http://www.cert.org/

CNN.com – News Special: Insurgency on the Internet
http://cnn.org/TECH/specials/hackers/

CRN: Research
http://staging.crn.com/sections/research/%5C

CSRC – Computer Security Resource Center – U.S. Department of Commerce
http://csrc.ncsl.nist.gov/

Electronic Frontier Foundation
http://www.eff.org/

Hackers.com
http://www.hackers.com/

Hackers Hall of Fame at Discovery.com
http://www.discovery.com/area/technology/hackers/
    hackers.html

HNN – Hacker News Network
http://www.hackernews.com/

Information Systems Audit and Control Association & Foundation
http://www.isaca.org/

InfoSysSec – Security Portal
http://www.prognosisx.com/infosyssec/

InfoWar – Security Portal
http://www.infowar.com/

International Biometric Group
http://www.biometricgroup.com/

International Computer Security Association
http://www.icsa.net/

Lucent NPS – Network Industry Surveys
http://www.lucentnps.com/surveys/

Microsoft Security
http://www.microsoft.com/security/

National Infrastructure Protection Center (NIPC)
http://www.nipc.gov/

National Security Institute – Security Resource Net
http://www.nsi.org/compsec.html

Packet Storm
http://packetstorm.securify.com/index.shtml

RMIS-Web – Risk Management Information Systems
http://rmisweb.com/

Risk Management – Online Magazine
http://www.rims.org/rmmag/

SANS Institute Online
http://www.sans.org/newlook/home.htm

Securify, Inc. – Security Industry Resources
http://www.securify.com/resources/siresources.html

Security World Wide Web Sites
http://www.alw.nih.gov/Security/security-www.html

SecurityFocus.com
http://www.securityfocus.com/

SecurityGeeks
http://securitygeeks.shmoo.com/

SecurityPortal
http://www.securityportal.com/

SecuritySearch.net
http://www.securitysearch.net/

securitywatch.com
http://www.securitywatch.com/

Smart Card Resource Center
http://www.smart-card.com/

The Jargon File – The New Hacker's Dictionary
http://www.tuxedo.org/~esr/jargon/html/index.html

The WebTrends Network
http://www.webtrends.net/

The WWW Security FAQ
http://www.w3.org/Security/Faq/

TruSecure Corporation
http://trusecure.com/

Virus Protection & Software Primer – Clemson University
http://virtual.clemson.edu/client/repprob/vprimer.htm

Windows IT Security
http://www.ntsecurity.net/Articles/

Wired News
http://www.wired.com/

workz.com
http://www.workz.com/content/default.asp

Links to Web Sites Noted in This Chapter

Aastrom Biosciences, Inc.
http://www.aastrom.com/

AICPA Web Trust
http://www.aicpa.org/webtrust/viewinst.htm

AIG
http://www.aig.com/

Amazon.com
http://www.amazon.com/

American Society for Industrial Security
http://www.asisonline.org/

AntiOnline
http://www.antionline.com/

AntiViral Pro Virus Encyclopedia
http://www.avp.ch/avpve/findex.stm

Arthur Andersen
http://www.arthurandersen.com/WebSite.nsf/
    Content/Homepage?OpenDocument

Assurex International
http://www.assurex.com/

AXENT Technologies, Inc.
http://www.axent.com/Axent/Public/

Butterball Turkey Co.
http://www.butterball.com/

Buy.com
http://www.buy.com/selectcountry.asp

CD Universe
http://www.cduniverse.com/asp/cdu_main.asp

Check Point Software Technologies Ltd.
http://www.checkpoint.com/

CIGNA
http://www.cigna.com/

Cinenet® Stock Footage Library
http://www.cinenet.com/

Cisco Secure Intrusion Detection System
http://www.cisco.com/warp/public/cc/pd/sqsw/sqidsz/

CNN Interactive
http://www.cnn.com/index.html

Computer Security Institute (CSI)
http://www.gocsi.com/

Counterpane Internet Security, Inc.
http://www.counterpane.com/

DefendNet Solutions, Inc.
http://www.defendnet.com/defendsite/default.asp

Diamond Information Center
http://www.adiamondisforever.com/

E*TRADE
http://www.etrade.com/

eBay
http://www.ebay.com/

Entrust Technologies Inc.
http://www.entrust.com/

Ernst & Young
http://www.ey.com/global/gcr.nsf/US/US_Home

Fidelity and Deposit Companies
http://www.fidelityanddeposit.com/

Geron Corporation
http://www.geron.com/

Goodyear Tire & Rubber Co.
http://www.goodyear.com/

Information Week
http://informationweek.com

Internet Security Systems, Inc.
http://www.iss.net/

linkLINE Communications
http://www.linkline.com/default.asp

Lloyd's of London
http://www.lloydsoflondon.co.uk/

ManagedFirewall.com
http://www.salinasgroup.com/mf/index.asp

Marsh & McLennan
http://www.mmc.com/

McAfee Virus Information Library
http://vil.mcafee.com/

myCIO.com
http://www.mycio.com/

Network ICE
http://www.networkice.com/

Pinkerton Systems Integration
http://www.psi.pinkertons.com/

Rent-A-Hacker
http://www.rent-a-hacker.com/

Riptech, Inc.
http://www.riptech.com/

Symantec AntiVirus Research Center
http://www.symantec.com/avcenter/vinfodb.html

The St. Paul Companies, Inc.
http://www.stpaul.com/wwwstpaul/static/index.htm

VeriSign
http://www.verisign.com/

Verizon – Genuity Network Security
http://www.gte.com/products/prods/sitepatrol.html

WatchGuard
http://www.watchguard.com/

WebTrends
http://www.webtrends.com/

Yahoo!
http://www.yahoo.com/

Zone Labs
http://www.zonelabs.com/

For Additional Review

Andress, Mandy. 2000. "AppShield Repels Hack Attacks—Speeds E-business Applications to Market While Keeping Web Servers Safe," *InfoWorld*, 22(20), May 15, 45.

ArrowPoint Communications: White Papers. 2000. "Web Site Security and Denial of Service Protection." Available online at: http://www.arrowpoint.com/solutions/white_papers/Web_Site_Security.html.

Assurex International. 2000. "Business Not Prepared for E-Risks, Survey Reveals," June 1. Available online at: http://www.assurex.com/otbtemp/newspecial.asp.

Baltazar, Henry and Dyck, Timothy. 2000. "Openhack: Lessons Learned — In-Depth, Ongoing, and Consistent Security Planning and Execution are Organizations' Only Defense in Today's E-Business Economy," *eWeek*, August 7, 1.

Banham, Russ. 2000. "Hacking It," *CFO, The Magazine for Senior Financial Executives*, 16(9), August, 115.

Berinato, Scott. 2000. "After Hack, Microsoft Mistakes Linger," *eWeek*, November 5. Available online at: http://www.zdnet.com/zdnn/stories/news/0,4586,2650237,00.html.

Berinato, Scott. 2000. "A UL-Type Seal for Security? Don't Bet on It," *eWeek*, October 15. Available online at: http://www.zdnet.com/filters/printer-friendly/0,6061,2640597-2,00.html.

Berinato, Scott. 2000. "Industry Reaction to Microsoft Hack: It Will Only Get Worse," *eWeek*, October 27. Available online at: http://www.zdnet.com/eweek/stories/general/0,11011,2646167,00.html.

Bowman, Lisa M. 2000. "MS Attack Takes Hacking to New Levels," *ZDNN*, October 28. Available online at: http://www.zdnet.com.

Brunker, Mike. 2000. "Vast Online Credit Card Theft Revealed: And Related Stories," *MSNBC News*, March 17. Available online at: http://www.msnbc.com/news/382561.asp#BODY.

Business Week. 2000. "Cyber Crime," (i3669), February 21, 36.

Campanelli, Melissa. 2000. "A Wall of Fire," *Entrepreneur*, 28(2), February, 48.

Chain Store Age Executive with Shopping Center Age. 2000. "Look, Don't Touch," 76(7), July, 82.

Cisco Systems. 2000. "Intrusion Detection Planning Guide." Available online at: http://www.cisco.com/univercd/cc/td/doc/product/iaabu/idpg/.

CNN.com. 2000. "Cyber-attacks Batter Web Heavyweights," February 9. Available online at: http://www.cnn.com/2000/TECH/computing/02/09/cyber.attacks.01/index.html.

CNN.com. 2000. "E*TRADE, ZDNet Latest Targets in Wave of Cyber-attacks," February 9. Available online at: http://www.cnn.com/2000/TECH/computing/02/09/cyber.attacks.02/.

CNN.com. 2000. "Legendary Computer Hacker Released from Prison," January 21. Available online at: http://www.cnn.com/2000/TECH/computing/01/21/mitnick.release.01/.

CNN In-Depth Specials. 1999. "Two Views of Hacking." Available online at: http://cnn.org/TECH/specials/hackers/qandas/.

Coffee, Peter. 2000. "The Microsoft Hack: Welcome to the Paper House," eWeek, October 27. Available online at: http://www.zdnet.com/zdnn/stories/comment/0,5859,2646203,00.html.

Coffee, Peter. 2000. "Trojan Horse, Virus, or Worm?" eWeek, February 9. Available online at: http://www.zdnet.com/zdhelp/stories/main/0,5594,2435378,00.html.

Coles, Robert. 2000. "Safety Net," The Banker, 150(895), September, 7.

Computer Security Institute. 2000. "2000 Computer Crime and Security Survey," May 22. http://www.gocsi.com/.

Daudelin, Art. 2000. "E-security Advances for Everyday Banking," Bank Technology News, 13(2), February 1, 21.

Davis, Bruce. 2000. "Goodyear Victim of Hack Attack; Violated Site Now Fixed," Crain's Cleveland Business, 21(30), July 17, 24.

Demers, Marie Eve. 2000. "Not Buying E-Commerce: The Purchasing Community Suffers from Online Transaction Anxiety," Electronic News, 46(32), August 7, 36.

Drury, Tracey. 2000. "Safe and Secure," Business First of Buffalo, 16(33), May 8, 21.

Edwards, Mark Joseph. 2000. "Something Old, Something New: DNS Hijacking," Windows IT Security, February 16. Available online at: http://www.ntsecurity.net/Articles/Index.cfrm?ArticleID=8170.

Felten, Edward W. et al. 1997. "Web Spoofing: An Internet Con Game," Department of Computer Science, Princeton University. Available online at: http://www.cs.princeton.edu/sip/pub/spoofing.pdf.

Fennelly, Carole. 2000. "Hacker's Toolchest: Techniques and Tools for Penetration Testing," SunWorld, May. Available online at: http://www.sunworld.com/sunworldonline/swol-05-2000/swol-05-security_p.html.

Figg, J. 2000. "Cyber Insurance to Cover E-Business," Internal Auditor, 57(4), August, 13.

Freeman, Emilly Q. 2000. "E-Merging Risks," Risk Management, 47(7), July, 12.

Gips, Michael. 1999. "Is Your Web Site a Hacker's Delight?" Security Management, 43(8), August, 64-6.

Gray, Lisa Waterman. 2000. "More Firms Considering Benefits of Internet Insurance," Memphis Business Journal, 22(17), August 25, 21.

Greenemeier, Larry. 2000. "IBM Offers Web-Site Checkup—Three Levels of Global Services' Web Security Scan Augment Firewalls and Encryption," Information Week, June 5, 139.

Harris, Donna. 2000. "Security Expert Offers Tips to Stop Web Site Defacement," Automotive News, 74(5865), March 13, 1.

Harrison, Ann. 2000. "Stopping Attacks at Their Source," Computerworld, October 2, 78.

Harvey, Thomas W. 2000. "Asleep at the Packet Switch," CFO, The Magazine for Senior Financial Executives, 16(11), Fall, 31.

Hayes, Frank. 2000. "Wanted: Security Champion; Find Yourself A Security Champion—Or Else," Computerworld, January 17, 94(1).

Hopper, D. Ian. 2000. "Denial of Service Hackers Take on New Targets," CNN.com, February 9. Available online at: http://www.cnn.com/2000/TECH/computing/02/09/denial.of.service/.

Hurwitz Report. 2000. "Web Application Security: Protecting e-Business from Attack," Sanctum, Inc. Available online at: http://www.sanctuminc.com/security/more/index.html.

Impellizzeri, Laura. 2000. "Keynote Finds Perfect Pitch in E-Commerce Chaos," San Francisco Business Times, 14(43), May 26, 4.

Inam, A. 2000. "Companies Now Confront Cyber Risk," Global Finance, 14(6), June 1, 85–86.

International Biometric Group (IBG). 2000. "Biometric Market Report." Sample data available online at: http://www.biometricgroup.com/.

Jopeck, Edward J. 2000. "Five Steps to Risk Reduction," Security Management, 44(8), August, 97.

Kang, Leslie. 2000. "Vyou Gives Web Site Security," Publishers Weekly, 247(25), June 19, 30.

Kunkel, Julie. 2000. "Evaluating the Risks of B2B Trading Exchanges," Chain Store Age Executive with Shopping Center Age, 76(9), September, 4.

Larsen, Amy K. 1999. "Global Security Survey: Virus Attack," *Informationweek.com*, July 12. Available online at: http://www.informationweek.com/743/security.htm.

Larsen, Eric and Stephens, Brian. 2000. *Web Servers, Security, & Maintenance*. Upper Saddle Rivers, NJ: Prentice-Hall, Inc.

Lemos, Robert. 2000. "Microsoft—Burned by Anti-Virus Tools?" *ZDNN*, October 27. Available online at: http://www.zdnet.com/zdnn/stories/news/0,4586, 2646200,00.html.

Lemos, Robert. 2000. "MS Intruder May Elude Authorities," *ZDNN*, October 27. Available online at: http://www.zdnet.com/zdnn/stories/news/ 0%2C4586%2C2646331%2C00.html.

Lucent Technologies. 2000. "Achieving Network Security Through a Managed Service," July 19. Audio and downloadable presentation available online at: http://www.lucent-netcare.com/ news/events/sec/.

Machrone, Bill. 2000. "A Security State of Mind," *PC Magazine*, November 21, 91.

Mandeville, David. 1999. "Hackers, Crackers, and Trojan Horses: A Primer," *CNN.com*, March 29. Available online at: http://cnn.org/TECH/specials/hackers/.

Masland, Molly. 2000. "The Dark Side of Online Shopping: Trail of Fraud Leads From Amazon.com to Thailand," *MSNBC*, June 24. Available online at: http://www.msnbc.com/news/283239.asp.

McClure, Stuart and Scambray, Joel. 1999. "Security Watch: Scanned Your Web Applications Lately for Security Holes? Try These Free Audit Tools," *InfoWorld*, 21(48), November 29, 58.

McMillan, Dan and Goldfield, Robert. 2000. "Internet Security Firms Profit from Hack Attack," *Business Journal-Portland*, 16(52), February 18, 8.

Moran, John M. and Halloran, Liz. 2000. "Hackers Jam More Internet Doors—Concern Rises over Web Site Security, Privacy," *The Hartford Courant: Statewide*, Issue PSA-2532, Main (A) Section.

MSNBC News. 2000. "New Account of Microsoft Attack," October 29. Available online at: http://www.msnbc.com/news/482011.asp.

Mullins, Robert. 2000. "Safety Seal," *The Business Journal-Milwaukee*, 17(28), April 7, 1.

Nemzow, Martin. 1997. *Building CyberStores: Installation, Transaction Processing, and Management*. New York, New York: McGraw-Hill.

Null, Christopher. 2000. "How to Hire a Hacker," *Ziff Davis Smart Business for the New Economy*, July 1, 112.

Olson, Scott. 2000. "Protection From Hackers Available for E-tailers," *Indianapolis Business Journal*, 20(50), February 21, 5.

Ostrow, Nicole. 2000. "Clamor for Data Resources Industrial Espionage, Security Technology," *Sun-Sentinel: Computers: Computer Industry*, September 19. Available online at Northern Light Special Collection documents: http://www.northernlight.com/.

Pappalardo, Denise. 2000. "Avoiding Future Denial-of-Service Attacks," *NetworkWorld* as reported by *CNN.com*, February 23. Available online at: http://www.cnn.com/2000/TECH/computing/ 02/23/isp.block.idg/index.html.

Penenberg, Adam L. 1999. "The Troubled Path of Kevin Mitnick," *Forbes*, April 19. http://www.forbes.com/ forbes/99/0419/6308050s1.htm.

Pfister, Nancy. 2000. "How Do Firms Weave a Secure Web Site?" *Orlando Business Journal*, 16(39), February 18, 23.

Power, Richard. 2000. *Tangled Web: Tales of Digital Crime from the Shadows of Cyberspace*. Indianapolis, IN: Que/Macmillan Computer Publishing.

Quinn, Michelle. 2000. "How Internet Users React to Hacking Incidents," *San Jose Mercury News*, February 12. Available online at: http://www.mercurycenter.com/svtech/news/ indepth/docs/hkpoll021300.htm.

Radcliff, Deborah. 2000. "Keep Hackers Out of Your Web Site," *Computerworld*, 34(1), January 3, S28.

Rapoza, Jim. 2000. "Locking Up Content—Lockstep Security Application Stops Web Attacks," *eWeek*, July 3, 57.

Raymond, Eric S. 2000. *The New Hacker's Dictionary— 3rd Edition*. Cambridge, MA: The MIT Press.

Reshef, Eran. 1999. "Internet Application Security," Perfecto Technologies. Available online at: http://www.sanctuminc.com/

Savage, Marcia. 2000. "Locking the Doors—Denial of Service Attacks and Viruses Prime the Market for Security Solutions and Services," *Computer Reseller News*, September 25, 72.

Savage, Marcia. 2000. "Case Study—Security Audit Yields Action Items—Crescendo Lands Long Term Client by Taking a Different Approach," *Computer Reseller News*, September 25, 76.

Savino, Lenny. 2000. "Teen Hacker Charged in Canada May Face Matching Charges in U.S.," *Knight-Ridder/Tribune News Service*, August 4, pK2185.

Scheraga, Dan. 2000. "Hack Attack," *Chain Store Age Executive with Shopping Center Age*, 76(3), March, 204.

Schultz, Eugene. 1999. "A Strategic View of Penetration Testing," *Information Security*, September. Available online at: http://www.infosecuritymag.com/sept99/pen_test.htm.

Sengstack, Jeff. 2000. "Make Your PC Hacker-Proof," *PC World*, 18(9), September, 169.

Shimmin, Bradley F. 2000. "Deconstructing Denial of Service Attacks," *ZDNet Help*, February 8. Available online at: http://www.zdnet.com/zdhelp/stories/main/0%2C5594%2C2434548%2C00.html.

Shur, Jim. 2000. "Hackers Sabotage Company's Web Site," *AP Online:Detroit*, February 18.

Stefanova, Kristina. 2000. "Computer Virus Toll Tops $1.5 Trillion," *Insight on the News*, 16(33), September 4, 31.

Stevens, Michael. 1998. "How Secure Is Your Computer System? Tips for Accountants," *The Practical Accountant*, 31(1), January, 24(8).

Stone, Martin. 2000. "Data Spill Blamed for De Beers Web Site Security Leak," *Newsbytes*, April 5.

Stonely, Dorothy. 2000. "Hack Job," *The Business Journal*, 18(2), May 5, 25.

Sullivan, Bob. 2000. "How Did It Happen?" *MSNBC*, October 27. Available online at: http://www.msnbc.com/news/ 481998.asp.

Sullivan, Bob. 2000. "'Netspionage' Costs Firms Millions: High-Priced Hacking Has Companies Worried," *MSNBC*, September 11. Available online at: http://www.msnbc.com/news/457161.asp.

Sundaram, Aurobindo. 2000. "An Introduction to Intrusion Detection," *Association for Computing Machinery*. Available online at: http://www.acm.org/crossroads/xrds2-4/intrus.html.

Sweeney, Paul. 1999. "Cyber-crime's Looming Threat," *Banking Strategies*, 75(4), July-August, 54(5).

Swift, Sean. 2000. "Effective Penetration Testing Requires Security Review, On-Target Test Team," *SunServer*, 14(9), September 1, 11.

Thomas, Pierre and Hopper, D. Ian. 2000. "Canadian Juvenile Charged in Connection with February 'Denial-of-Service' Attacks," *CNN.com*, April 18. Available online at: http://www.cnn.com/2000/TECH/computing/04/18/hacker.arrest.01/.

Thompson, Clive. 2000. "When Hackers Make House Calls," *Fortune*, 142(8), October 9, 60.

Treese, G. Winfield and Stewart, Lawrence C. 1998. *Designing Systems for Internet Commerce*. Reading, Massachusetts: Addison Wesley Longman, Inc.

Tucker, Darla Martin. 2000. "California-Based Internet Service Companies Fall Victim to Hacker Attacks," *Knight-Ridder/Tribune Business News*, August 28.

Vaas, Lisa. 2000. "Security Checkup—One Bank's Experience at Having Its E-Biz Links Poked, Prodded, Scanned," *eWeek*, August 14, 49.

Vamosi, Robert. 2000. "QAZ.Trojan Infects Networks," *ZDNN*, August 14. Available online at: http://www.zdnet.com/zdnn/stories/news/0,4586,2605063-1,00.html.

Varcoe, Bill. 2000. "Three Ways Your Site is Vulnerable to Attack," *workz.com*. Available online at: http://www.workz.com/content/1187.asp.

Vaughan-Nichols, Steven J. 2000. "Microsoft Can't Spin This Worm," *Sm@rtPartner, ZDNet,* Reported by MSNBC, October 27. Available online at: http://www.msnbc.com/news/482181.asp.

Violino, Bob and Larsen, Amy K. 1999. "Security: An E-Biz Asset," *Information Week*, February 15, 44.

Whitney, Sally. 2000. "Risky Business in Cyberspace," *Best's Review*, 101(2), June, 143.

Williams, Cathy. 2000. "Preparing Your Business for Secure e-Commerce," *Strategic Finance*, 82(3), September, 21.

Williams, Jason. 2000. "Locking Them Out," *Editor and Publisher*, April 17, 20.

Williams, Martyn. 2000. "'Immense' Network Assault Takes Down Yahoo," *ComputerWorld* as reported by *CNN.com*, February 8. Available online at: http://www.cnn.com/2000/TECH/computing/02/08/yahoo.assault.idg/index.html.

Winkler, Ira S. 1997. "Anatomy of an Industrial Espionage Attack," Excerpted from *Corporate Espionage* (Prima Publishing, 1997). Available online at: http://www.smdc.army.mil/SecurityGuide/v1comput/Case1.htm.

Wolverton, Troy. 1999. "Butterball's Data Security For the Birds," *CNET News.com*, May 4. Available online at: http://news.cnet.com/news/0-1005-200-342072.html.

Wolverton, Troy. 1999. "Nissan Privacy Goof Exposes Email Addresses," *CNET News.com*. Available online at: http://news.cnet.com/news/0-1007-200-341234html.

Understanding Back-End Systems

In this chapter, you will learn to:

Define front-end and back-end e-business systems

Discuss integrating front-end Web-based systems with back-end systems

Describe business records maintenance

Discuss the importance of backup procedures and disaster recovery plans

Understand order fulfillment processes

Discuss outsourcing fulfillment management

The Palo Alto, California wine seller, Wine.com, knew it was ready for the 1999 holiday season and the season's anticipated tenfold increase in sales. When Wine.com launched its Web site in 1995, it had a very labor-intensive manual warehousing and order fulfillment system that required the e-business to "staff up" during peak business periods. Mindful that order fulfillment and delivery are critical for making a retail e-business successful, in the fall of 1999 Wine.com installed a warehouse-management system to automate shipping, receiving, and inventory replenishment in its 32,000-square-foot warehouse.

The new warehouse-management system provided Wine.com with more information about the movement of its products through the warehouse and helped improve warehouse employee productivity. Wine.com planned to integrate its new warehouse-management system with its shipping partners' systems, its internal accounting systems, and its customer support and order-tracking systems by the year's end. The VP of technology believed that the new automated warehouse-management system gave Wine.com enough capacity to handle the expected volume of holiday orders.

As the holiday season approached, Wine.com used a $10 million television, print, and online marketing campaign to drive customers to its Web site. It worked. December sales began to skyrocket. Then the holiday rush hit, bringing the Wine.com management down to earth. Phones weren't being answered quickly enough, and order picking (selecting products from warehouse shelves), packing, and delivery slowed down. At one point one of Wine.com's lead investors was in the parking lot packing boxes of wine for shipment. Of course, customer service suffered. After all that planning, what could possibly have gone wrong?

Defining Front-End and Back-End E-Business Systems

Front-end systems are those processes with which a user interfaces, and over which a customer can exert some control. For an e-business, front-end systems are the Web site processes that customers use to view information and purchase products and services. **Back-end systems** are those processes that are not directly accessed by customers. For an e-business, as well as for a traditional brick-and-mortar business, back-end systems include the business's ERP and CRM systems that handle the accounting and budgeting, manufacturing, marketing, inventory management, distribution, order-tracking, and customer support processes.

An e-business's front-end systems require much of the same data already stored in its back-end systems, such as product availability and pricing. Additionally, new data being gathered by the front-end systems—for example, order information—must be made available to the back-end systems for internal business processes such as accounting, billing, payment processing, and order fulfillment. Integrating front-end and back-end systems not only provides an e-business with more useful information about its operations, but also reduces costs by allowing common data to be shared across front-end and back-end applications. Figure 9-1 illustrates the integration of an e-business's front-end and back-end systems.

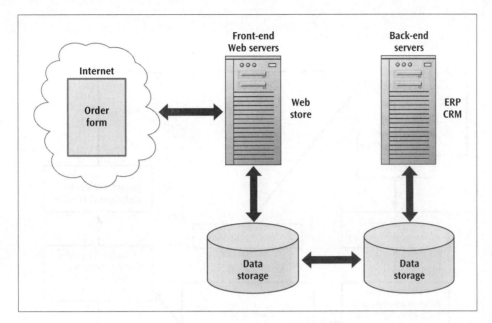

Front-end
Web servers

Back-end
servers

Internet

Order
form

Web
store

ERP
CRM

Data
storage

Data
storage

Figure 9-1
Front-end
and back-end
systems

The process of integrating front-end and back-end systems begins with data generated from online transactions.

Integrating Front-End Web-Based Systems with Back-End Systems

Online sales transactions are the heart of an e-business's operations. The sales and payment information gathered from online sales transactions must be integrated with other back-end processes. For example, in order to authenticate and authorize credit card payments, the credit card information must be either transferred to a credit card processing provider, or processed internally using credit card authorization software.

Orders must be routed to designated suppliers or warehouses for fulfillment. The transaction must be recorded in the accounting system. Inventory records must be updated. It may be necessary to transmit the order transaction to systems outside of the e-business, such as those of suppliers or shipping agents. Figure 9-2 illustrates an example of order transaction processing for an e-tailer such as foodlocker.com.

TIP

Some e-businesses integrate traditional back-end processes such as inventory management and order-tracking into their front-end Web site processes in order to increase customer satisfaction by sharing product availability or order status information.

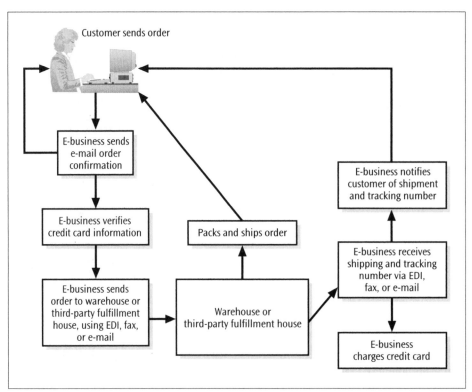

Figure 9-2

Online transaction processing

E-CASE Just What the Doctor Ordered

When Patrick Kelly and three associates founded Physician Sales and Service (PSS) in 1983, their mission was to fulfill the inventory needs of office-based physicians. Today PSS, located in Jacksonville, FL, is a leading provider of medical supplies with 51 service centers distributing medical supplies to approximately 110,000 physician offices in all 50 states. In the belief that "the customer is everything," PSS installed the first Web-enabled order-processing system in the medical supply industry, changing the way its customers order supplies and access information.

Today, PSS customers can order supplies, review billing information and total dollars spent, and access information about product usage and availability by logging on to the PSS Web site. By integrating the order-processing data into its ERP systems, PSS is also able to provide product and order information directly to its remote sales consultants via the Web. Although many customers choose to enter and monitor their own orders at the PSS Web site, this front-end and back-end system integration allows sales consultants to monitor all their customers' activities. PSS (Figure 9-3) effectively uses its distribution facilities, integrated front-end and back-end systems, and more than 700 sales consultants to process and deliver any customer order within approximately three hours.

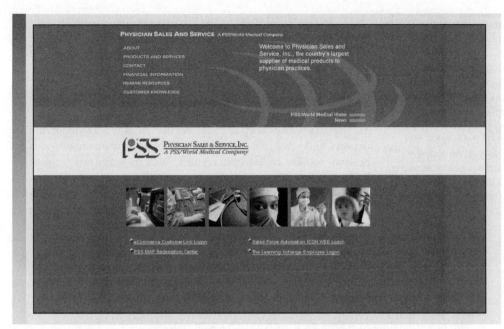

Figure 9-3
Physician
Sales and
Service

There are several issues involved with integrating Web site systems with existing systems, also called **legacy systems**, including:

♦ *Real-time requirements:* Do the various systems require real-time update of data, such as checking the inventory system before allowing the customer to order an item? If there is no need for real-time data exchange, or it is not feasible, then the system must store the data to be updated in a queued list to be sent at a specified time. Many e-businesses do this with their credit card authorization when their systems are not capable of real-time authorization. They store all of the credit card numbers to be authorized during the day, and then authorize them in a batch at the end of the day.

♦ *Security:* Because there are several interconnecting systems, each system must be secure. It is possible that the e-business will exchange data with an outside business such as a supplier or shipping agent. Such connections need to be secure as well. One method of securing these connections is by using **trust protections**, which accept only data sent from the e-business servers, and not data sent from any other computer address. There also may need to be some form of validation code within the data being sent to the third-party systems.

♦ *Technology issues:* Technology is always a consideration when integrating systems not originally designed to communicate with each other. When selecting software components for online transaction processing, it is important to make sure that the components can share data with the other components. To create the exchange of data, it may be necessary to write special data translation programs, or to purchase specialized software, often called **middleware**, for making connections between two systems.

TIP

For a brick-and-click company, it is important to integrate Web site operations with the overall business. The e-business portion of a brick-and-click business should not be a separate operation from the brick-and-mortar part of the business. A Web site customer should be able to call the brick-and-mortar store for customer service, and customer service personnel at the physical store should be able to answer questions about an order placed online.

Data generated from front-end systems and integrated with back-end systems is used to prepare important records of an e-business's activities.

Business Records Maintenance

All businesses must keep records. Records of orders, payment and delivery, and customer data are necessary, and various regulations or laws require transaction records, such as sales tax records. **Primary records** include those records that provide supporting documentation for the key activities of an e-business, including sales, order fulfillment, and payment activities. **Secondary records** include important information generated from e-business activities but not used in daily operations. The primary business records used in day-to-day operations can include:

♦ *Orders:* Data on the actual online order placed, including name, shipping and billing addresses, payment information, items ordered, requested shipping method, prices, and order totals.

♦ *Invoices:* Fulfillment information such as items shipped.

♦ *Payments:* Receipts of payment received.

Secondary records can include:

♦ *Customer data:* The customer database is one of an e-business's most crucial assets. Customer data includes data identifying the customer, but also can include data on when customers visit the Web site, what items they buy, and other information collected from customers, such as product preferences, birthdays, ages of children, etc. This information can be used for many purposes, such as data mining and customer research. The same data is also used by many e-businesses to personalize or customize the Web site for each customer. For example, Amazon.com welcomes repeat customers by name and lists recommended books based on each customer's previous purchases.

♦ *Data on Web site activity:* E-businesses often track visitors as they travel through a Web site, including information about the Web site from which they linked to get to the e-business's site. This information is very useful in studying how the Web site is used, from which Web sites customers are linking, what problems the customers may be experiencing, and the effectiveness of advertising.

♦ *Records needed to meet regulatory requirements:* It is also necessary to keep certain records to meet federal, state, and local government requirements. Most businesses are required to report sales and sales tax records to the state (and perhaps municipality) in which they operate. Sales data are also used in tax reporting, whether it is for personal income taxes for the sole proprietor or partner, or for corporate or franchise tax reporting for the corporation. It may also be necessary to keep other records to meet other regulatory requirements that are specific to a particular type of business or category of product.

With both primary and secondary records, there are other issues, such as how long the records should be kept, which depends on reporting requirements, regulatory requirements, and the organization's record storage capability. However, because these records are initially stored online electronically, it is easy to move them to some form of electronic offline storage, such as CD-ROM or magnetic tape. Additionally, records can be stored in a printed format if necessary.

Another issue is accessibility. It is important that a business be able to access stored records. For example, some records, particularly accounting records and government records, may be required when a business undergoes an audit. An **audit** is an independent review that the business records are accurate and have conformed to accepted practices. Large businesses, particularly public companies, must undergo an accounting audit to verify to investors that the company is accurately reporting accounting information and is complying with accepted accounting practices. Because state and federal government regulations and legal liability issues can affect an e-business's records retention and destruction policies, it is a good idea for an e-business to consult with a certified records management professional, relevant government agencies, and attorneys when developing its records retention and destruction plan.

Another important issue for an e-business is the need to plan for potential disruptions to business caused by computer hardware or software problems or by a natural disaster.

Backup Procedures and Disaster Recovery Plans

Much of the time, e-business systems run smoothly. However, there are times when an e-business must deal with operational problems. For example, what happens if electrical power is interrupted, or if the data communication lines go down? Planning ahead to solve these kinds of problems is particularly critical for e-businesses because most of their business is conducted electronically. Every e-business should have computer system backup procedures and a disaster recovery plan.

Backing up system and data files

One of the most important procedures an e-business can implement is a regular **backup procedure** to copy its critical computer system files and its data files. Creating a backup copy of its computer system files and data can protect an e-business from losses resulting from processing errors, from hardware failure such as a disk crash, or from a catastrophic event such as a building fire. Manual recovery of data lost as the result of processing errors, hardware failure, or a catastrophic event could cost an e-business anywhere from a few hundred to several thousand dollars or more.

TIP

Although the chance of a disk drive crashing may be slim, some reports indicate that as many as four hard drives crash every minute in the United States.

A regularly scheduled backup of a small number of files can be as easy as copying the files to a Zip disk, CD-ROM, or magnetic tape. For large network systems and databases, there are many different types of backup software, such as LiveVault, used to create regular system backups. Additionally, many ERP and CRM providers, such as Oracle, also include backup utilities in their applications. Finally, it is possible to contract with a Web-based backup provider such as @Backup (Figure 9-4) to back up critical files to a server via the Internet.

Figure 9-4
@Backup

E-businesses should take a copy of the backup materials to an off-site location, so that in the event of a building fire or catastrophe there is a safe copy of the computer system files and data. E-businesses with more than one location connected by a network may copy the backup data from one location to the other electronically to create an off-site backup. If an e-business uses a hosting service to host its Web site, the hosting service should make backup copies of the server data on a regular schedule. The e-business can keep copies of the information stored on the hosting company's server at its own location, creating an additional backup.

One critical issue that an e-business must resolve when scheduling regular backups is the need to continue to process online order transactions $24 \times 7 \times 365$, and still back up critical files. To resolve this problem, many e-businesses, especially those with a large volume of transactions, continuously back up critical data, often in real time.

Scheduling and conducting regular computer system and data file backups is only one step. An e-business should be prepared for some unexpected and uncontrollable outside event that can disrupt normal business operations.

Disaster recovery plans

Disaster recovery is the process of recovering from hardware failures, loss of power, communications failures, fires, natural disasters, or any other unexpected and catastrophic event that can interrupt an e-business's normal activities. Following regular backup procedures is an important component of a disaster recovery plan, but not the only component.

TIP

A good source of information about enterprise risk management issues is the International Risk Management Institute (IRMI) Web site.

To create an effective disaster recovery plan, an e-business's management must review the business's minimal requirements in order to not only survive but maintain some semblance of productivity. These requirements include the data-processing tools, information, and personnel required to survive a disaster, such as:

◆ Access to telephones and communication lines
◆ Scaled-down functional servers
◆ Networking software and hardware
◆ Relevant data and databases
◆ Network configuration information
◆ Emergency duty rosters
◆ Procedure for notifying employees where to report following a disaster
◆ Contact information and building blueprints and specifications provided to police and fire departments
◆ Emergency service agreements with outside electrical, telephone, and Internet service providers

E-CASE Ready or Not: Here Comes Mother Nature

Today most healthcare transactions are automated, making it critical that healthcare organizations have a way to quickly bring systems and data back online if a disaster should strike. Ernie Weber, vice president and CIO of St. Joseph's Hospital and Medical Center in Paterson, NJ (Figure 9-5), was well aware of this need. In the late fall of 1999, St. Joseph's was in the middle of a three-year effort to implement a comprehensive information systems disaster recovery plan when Mother Nature decided She couldn't wait.

In November 1999 Tropical Storm Floyd struck New Jersey, causing severe problems for Bell Atlantic Central, St. Joseph's telecommunications provider, including knocking out the two T1 communication lines that connected St. Joseph's computer systems with its offsite recovery service. Unable to access its information systems or the offsite recovery service, St. Joseph's immediately implemented its disaster recovery plan, which required employees to manually record medical test orders and other patient transactions. It took two days for St. Joseph's telecommunications and information systems to be brought online and an additional eight hours to bring all applications back online and integrate the manually collected data.

After operations were back to normal, Weber and his staff reviewed the weaknesses in the St. Joseph's disaster recovery plan and immediately went to work to accelerate changes to the plan and update its telecommunications network. Following Tropical Storm Floyd, St. Joseph's spent approximately $250,000 to make several changes, including rerouting the two T1 lines to protect against both lines being knocked out at the same time, enabling St. Joseph's computer systems to operate over existing copper wires at its data center, and adding fiber-optic cables where there was limited wire capacity. Additionally, St. Joseph's purchased several cellular telephones and CB radios to maintain communications with its data center and offsite recovery service when regular phone lines go down. Having to actually go through a natural disaster enabled Weber and his staff to test St. Joseph's disaster recovery plan, identify its weaknesses, and see just how quickly everyone could react to a disaster. Of course, an e-business shouldn't wait for a real disaster to test its disaster recovery plan! All facets of its plan should be tested under a disaster simulation.

(Continued on the next page)

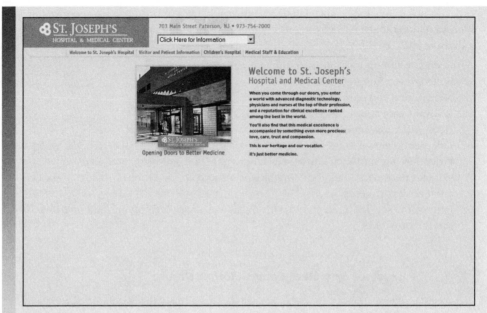

Figure 9-5

St. Joseph's Hospital and Medical Center

There are several ways an e-business can structure its systems recovery in case of hardware or software malfunction or a natural disaster.

Disaster recovery options

An e-business must have its computer and communications systems available at all times and cannot halt operations as the result of a hardware failure or power loss. Failed hardware components such as disk drives or power supplies must be replaced without losing data or shutting down systems for any length of time. Unlike a brick-and-mortar business, an e-business can't make repairs and adjustments at night when the store is closed—an e-business store is never closed!

Since an e-business's computer systems must be available at all times, it should plan to have spare computer system components ready in case of a failure. This may include components that can be installed while the computer system is running, called **hot-swappable** components, such as hard drives, a redundant array of independent disks to provide the ability to maintain data if one of the disks crashes, spare servers, uninterruptible power supplies (UPS), generators, and redundant network circuits to separate carriers.

It is possible to have a system that will automatically switch to a spare system (**fail-over system**) in the event of a failure in the primary system. It is even possible to have the spare system in a different location. This can solve problems of much greater magnitude than system component failure. For example, many Silicon Valley companies keep spare systems in data centers elsewhere in the U.S., to be used in the event that an earthquake disrupts operations in California. It is impossible to completely protect against an earthquake, but an e-business can ensure that its operations can continue after one. In fact, since an e-business is not tied to a particular location, it could easily continue its operations from another location, as long as it can fill orders.

Often, e-businesses replicate their systems on multiple servers. In the event of a failure of one server, network traffic is simply shifted to another server that is still running. If one server is too busy, traffic can be shifted to another server. Often these servers are **mirrored servers** (backup servers that duplicate the processes and data of a primary server) in more than one location, serving as possible remote fail-over servers. Having multiple servers in different locations also puts servers closer to the users connecting to them, reducing lag time between the user and the server.

To replicate the components of a system, the system must be designed from the beginning with scalability in mind. Scalability makes it easy to expand the system as traffic grows or to replace components when they break down. It is also potentially less costly to use many less expensive servers, rather than a single, very expensive and fast server, to handle all of the traffic and transactions. In fact, many e-businesses assign only one part of a back-end system, such as the accounting system, to a given server. An e-business may use separate servers for storing Web pages, for processing transactions, and for maintaining databases, all servers communicating with each other.

> **TIP**
>
> Network operating systems such as Windows 2000 and NetWare include features to protect data and ensure network stability, such as replication of data and network information on many servers to provide redundancy in the event that one server goes down.

Another important—perhaps in the end the most important—back-end system for an e-business that sells products online is its order fulfillment system.

Order Fulfillment Processes

Order fulfillment, getting the ordered product into the hands of the customer, is likely the least exciting—and yet most critical—part of an e-business's operations. Failure to handle the order fulfillment process well can result in unhappy customers who not only won't be back but will also tell other potential customers about their unhappy experience. Fulfillment issues include inventory management, order picking and packaging, and shipping.

Until about 1999, aspects of back-end support in order processing and fulfillment were designed to meet the demands of the typical catalog or direct-mail customer. With the direct-mail model, inventory planning is based on historic patterns of sales and purchasing, providing enough lead time to forecast needs, buy products, move them to the fulfillment center, and fulfill customer orders. This process also typically can include back orders, delayed shipping, and high customer call-to-order ratios. E-tailing has changed the order processing and fulfillment model. Now both brick-and-click and pure-play e-businesses must operate in "Internet time," which has a much shorter procurement cycle and which requires access to real-time information about inventory, order tracking, and customer service.

Additionally, in traditional catalog fulfillment centers, order picking and packaging processes were designed to combine multiple items in one shipment rather than to handle the high-volume, single-item orders generated by e-businesses. The often inflexible components of traditional catalog fulfillment centers, such as limited docking areas, staging space, and unloading or reloading capacity, can constrain efficient order fulfillment in Internet time.

Inventory management

Some e-businesses elect to handle all order fulfillment processes in-house by maintaining their own warehouse space, product inventories, order picking, packaging, and shipping. Large e-tailers such as Amazon.com, BarnesandNoble.com, Webvan, and GroceryWorks.com fall into this category. These e-businesses run their own multimillion-dollar distribution centers, using extensive warehouse and logistics technology to manage their inventories, pick and package products, and ship or deliver those products.

Smaller e-businesses that choose to maintain their own warehouses must also be concerned with inventory management issues. There are several issues that must be dealt with when managing inventory:

◆ How the inventory is stored
◆ How the inventory is arranged in order to find specific items when they are ordered
◆ How the inventory movement (sales and replenishment) is tracked

Tracking inventory is the most important part of managing inventory. Tracking inventory allows the e-business to inform customers whether a product is available or not, and also indicates when products need to be reordered. Information from an inventory tracking system can also help determine which items are not being sold, so the e-business can consider discontinuing those items or at least reducing the inventory that they carry of those items. Tracking inventory is also important in understanding the true cost of the goods being sold, including the holding costs for those items.

In addition to inventory management, e-businesses that manage their own fulfillment processes also have to deal with order-picking issues.

Order picking

Order picking, selecting products from warehouse shelves and bins, is the heart of fulfillment operations, and it affects both the order fulfillment and the inventory replenishment processes. Cahners Research reports that 38 percent of the fulfillment facilities they surveyed reported a mispick error costing more than $100. Fifty-two percent reported mispick errors costing more than $60 per error. A traditional accuracy goal for picking operations in fulfillment centers has been 90 percent or better. However, many fulfillment centers are now striving for a 99 percent accuracy rate, because even a 1 percent error rate—at $100 per error—is extremely expensive. For example, a fulfillment operation with 50 mispicks per day at $100 each can incur additional costs of $5,000 per day!

Because e-business customers tend to place smaller, more frequent orders, it is imperative that the order-picking process be designed efficiently. Fulfillment center managers who design and monitor the order-picking process must be concerned with profiling products into "fast-moving," "medium-moving," and "slow-moving" categories and then arranging the physical layout so that products can be picked, packaged, and replenished more efficiently. Several companies such as Diamond Phoenix, White Systems, and Remstar (Figure 9-6) are developing new carousel-based order-picking systems for the high-speed and fast-flow order picking required by e-businesses.

An example of e-tailing in which the order-picking process is particularly challenging is the online grocery and delivery business. To meet the challenge, Webvan, Albertson's, GroceryWorks.com, and others have invested heavily in their fulfillment systems, including automating the order-picking process. These e-businesses realize that order fulfillment is the heart of their business and that picking the highest quality product in the shortest time means everything. For example, although customers might not get too upset if they personally select poor produce at the market, they might not be so forgiving if a third party picked poor-quality produce for them. Out-of-date perishable items or missing grocery items can quickly turn off a customer, who likely will not continue to use the e-tailer's services.

After an order is picked and packaged, the next step is to get the order to the customer in a timely manner.

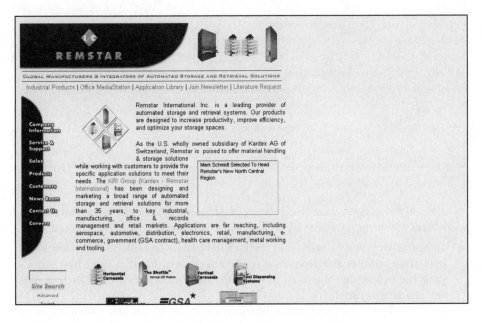

Figure 9-6
Remstar

Shipping and delivery

Unless the product offered is information or another digital good such as downloadable software, e-businesses have to deliver their products to the customer through the U.S. Postal Service (USPS) or a commercial shipping company such as United Parcel Service (UPS), Federal Express (FedEx), Airborne Express (Figure 9-7), or DHL. Each shipping company has its own area of specialty, and no one shipping company will necessarily be the exclusive choice for an e-business. In fact, many e-businesses allow their customers to choose the form of shipment that they desire. Also, there can be benefits to using more than one shipping partner. If there is a problem with shipment through one shipping partner, it is easy to transfer the shipping process to another.

Figure 9-7
Airborne
Express

One of the biggest concerns relating to shipping is how the e-business charges customers for shipping. There are several ways that shipping can be charged.

- *Shipping included in price:* The price of the item being sold includes any shipping charges. Some e-businesses give the impression that they offer free shipping, but in actuality, the price of the shipping is incorporated into the price of the goods being sold.

- *Flat rate:* There is a fixed charge added to all orders. This is generally used by e-businesses that offer only a small variety of similar goods.

- *Weight:* Weight-based metered shipping charges provide more accurate shipping costs for heavier goods.

- *Number of items:* Basing shipping charges on the number of items sold provides more accurate charges when all items are of a similar weight, such as CDs or

videos. For example, shipping charges for one CD might be
$3.50 plus $2.00 for each additional CD.

- *Order total:* Shipping charges based on a percentage of the
dollar value of the order are easy for the customer to understand and can be used to encourage the customer to increase
the order size, by having the percentage decrease as the order
size increases. For example, shipping charges may be $8.50
on orders totaling $0–$50, $10.00 on orders totaling
$51–$100, or $12.00 on orders totaling more than $100.

- *Actual rate calculation:* The shipping charge is calculated from actual shipping costs
based on the weight of the order, the location it is being shipped from, and its destination. This requires either copies of the shipping company's rate tables or the ability to
look up rates in real time on the shipping company's server.

When deciding how to charge for shipping, an e-business must consider several factors.
It is important to understand how the shipping charges are levied and how changes in order
size and destination address will affect those charges. If items are being shipped from multiple locations, the calculation becomes more complex. Also, in addition to shipping costs,
handling costs must be considered. The box and packing materials cost money, as well as
the time of the employee who packs the order.

There are several e-businesses that offer comparison-shopping for shipping services.
For example, freightquote.com, Stamps.com, and SmartShip.com can provide information
on the best shipping options based on the origin of a package and its destination, taking
into consideration the package's weight and dimensions. Stamps.com and SmartShip.com
(Figure 9-8) also provide a tool that e-businesses can integrate into their Web sites to allow
customers to review shipping estimates and select their own shipping method.

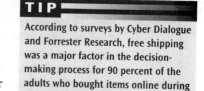

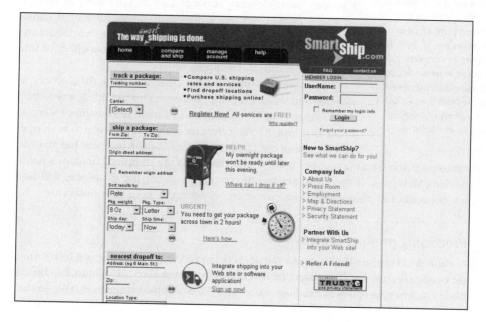

Figure 9-8
SmartShip.com

It is also very important to consider the effect of shipping costs on customer behavior. An e-business's shipping policy must be made very clear to the customer, and the customer must be able to easily determine shipping charges for an order. Problems arise when customers are not informed of shipping charges until they reach the checkout page of the Web site, or find that shipping costs are higher than expected. If shipping costs are disproportionately high compared to the cost of the item, customers may be discouraged from buying that item.

How an e-business transports and tracks orders has a direct impact on customer satisfaction. An e-business must prepare the orders for shipment, arrange for pickup by or delivery to the shipping company, create the necessary paperwork and labeling, and assign tracking numbers to the packages. Often, automated software and/or hardware, such as FedEx PowerShip, generates a tracking number, prints a shipping label or waybill, and schedules a pickup. Most delivery services can be contacted to pick up packages, or will pick up at the shipping location on a regular schedule. Also, most shipping companies have representatives who work with e-businesses to figure out the most effective solution for preparing and shipping orders.

The e-business must inform customers how their orders are shipped and when they ship, and must also provide a way to track orders. Most e-businesses send an e-mail to the customer when an order is shipped, letting the customer know how it was shipped and providing the tracking number. If the e-business has an order status page, the tracking number may be hyperlinked to the shipping company's online tracking Web page. If the customer has the tracking number, the e-business will have fewer e-mails and phone calls from customers wondering where their orders are.

Processing product returns

According to Forrester Research, one out of every ten products sold online will be returned. An e-business must not only state a clear return policy on its Web site, it must also have in place efficient returned-item handling procedures. Many shipping companies have services that e-businesses can integrate into their Web sites to make returns more convenient for customers by allowing customers to download or print postage-paid return labels directly from the Web site. Some of these tools allow the e-business to arrange for a pickup of the returned item from the customer's location, and provide online return package tracking.

Table 9-1 illustrates the return services offered by three shipping companies as of this writing.

Shipping Company	Name of Service	Description of Services
U.S. Postal Service	Merchandise API	Customer downloads return address label from e-business Web site and takes package to post office
Federal Express	FedEx NetReturn	E-business can schedule a package return from the customer's location and electronically submit all return information
United Parcel Service	Online Worldship	Two services: Online Call Tag allows UPS drivers to bring label to package pickup location on user-specified date and pick up the package. With Print Return Label, the e-business generates a return label and sends it to the customer with instructions to take the package to a UPS drop-off location.

Table 9-1
Return services

Because of the global nature of e-business, it is important to consider issues related to international shipping.

International shipping issues

International shipping creates a whole new set of problems for an e-business. Many businesses use different shipping methods for international and domestic shipping, because different shipping companies are better suited to each task. An e-business should consider where it expects to ship orders when deciding which shipper to use for international orders. The options may be more limited than with domestic shipping.

The additional paperwork required for international shipping differs, depending on the country the package is being shipped to and the contents of the package. Some countries may require only a standard commercial invoice and customs form for certain items, while different items or shipments to other countries may

TIP

According to a 1999 survey by BizRate.com, about 90 percent of survey respondents consider an e-tailer's return policy to be the determining factor in whether they will be repeat customers. Most survey respondents also indicated that having a return policy was not enough—only a 100 percent money-back guarantee would be acceptable. Respondents also opposed being charged a restocking fee for returned items. Additionally, 85 percent of respondents were unhappy with e-tailers that did not immediately credit a refund to their credit or debit card.

require certificates of origin, bills of lading, consular invoices, insurance certificates, or international letters of credit.

There may also be import or export restrictions on certain items going into certain countries. For example, encryption software shipped from the U.S. is subject to special export controls, and many European countries do not allow the importation of meat products. Export licenses may be required to ship some items or to ship to certain countries.

TIP

A good source of information on the specific requirements for shipping to any individual international location is a shipping company, such as DHL, that specializes in international shipping.

Taxes or duties may be due in the destination country. Though these vary widely, the adoption of the Harmonized Tariff System Classification has somewhat simplified the confusion surrounding taxes and duties on items shipped internationally. It provides for a standardized numbering system for traded goods. The number assigned to a specific class of goods is used by customs agencies across the globe to calculate duties or tariffs, as well as to determine the regulations that apply to those goods.

To summarize some important ways to keep customers happy with the order fulfillment process:

- *Deliver on time:* Better yet, deliver ahead of time. Make sure the customer understands when delivery is expected, and deliver on that promise.
- *Use effective packaging:* Not only does effective packaging protect the goods being shipped, it greatly affects the impression a customer has of an e-business. Also, packaging is a great opportunity for branding. An e-business should never use a plain box (or worse yet a reused box) when it can instead use a box that proudly displays its logo. It should also be noted that excessive packaging can also create a negative impression.
- *Get it right:* Make sure that all items, and the correct items, are included in the order. Check it again. Most customers find that getting the wrong item is worse than not receiving their order, because they must then go through the hassle of returning the wrong item.
- *Handle returns quickly:* Post a clearly defined return policy at the Web site that makes it easy for customers to return items. Credit refunds to a customer's charge or credit card in a timely manner. If possible, do not charge a restocking fee.

Because of the complexities and costs involved in warehousing an inventory of products, many e-businesses find that it is more cost-effective to hire another company to manage their inventory and fulfillment processes.

Outsourcing Fulfillment Management

Some e-businesses outsource their fulfillment processes to third-party **fulfillment houses**, which are independent companies that warehouse products and pick, package, and ship orders for multiple clients. Fulfillment houses were developed to provide expertise in warehousing (Figure 9-9), packing, shipping, and tracking for both manufacturers and brick-and-mortar retailers. Manufacturers use fulfillment houses for catalog direct marketing. Brick-and-mortar retailers have traditionally used fulfillment houses to maintain a large inventory, reduce inventory costs, and ship orders directly to customers.

Unless an e-business ships thousands of orders each day, it may be more cost-effective to outsource its fulfillment processes. By outsourcing its order fulfillment processes an e-business may be able to reduce costs by not having to pay for warehouse space, including warehouse heating, cooling, and insurance. Additionally, an e-business that outsources its order fulfillment processes can save on employee costs. And finally, partnering with a fulfillment house may provide other benefits, such as increased customer service and support; additional marketing options, such as promotional gifts and gift wrapping; ease of customizing pricing, tagging, packaging, and shipping methods; and better shipping rates.

ShipMax.com, Fingerhut Business Services, USCO Logistics (Figure 9-10), A. N. Deringer, and Ryder Integrated Logistics are examples of fulfillment houses. Fulfillment houses that provide warehouse and shipping logistics services are often called **3PLs** (third-party logistics).

According to a study by Jupiter Media Metrix, order fulfillment problems were high on the list of online consumer disappointments during the 1999 holiday season. Sixty-two percent of consumers surveyed said they had trouble receiving information about delayed deliveries. Thirty-three percent said that they were not notified that out-of-stock items wouldn't be delivered on time. While more than 50 percent of consumers surveyed expected an e-tailer to send them order status e-mails within six hours of an order placement, only 29 percent of the e-businesses involved in those consumers' orders were able to comply with that expectation. In order to alleviate these delivery problems and meet consumer expectations for the 2000 holiday season, more than 50 percent of the e-businesses described in the study have invested more money into back-end operations, have brought fulfillment in-house, or have found better fulfillment partners.

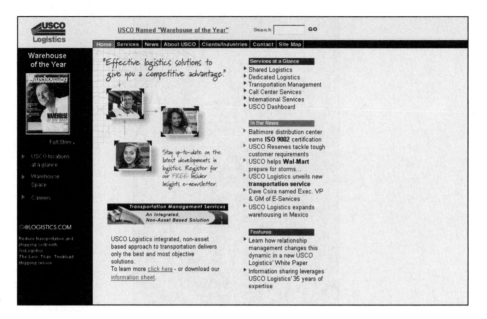

Figure 9-10
USCO Logistics

According to International Data Corporation (IDC), the market for third-party fulfillment services will grow to $8.6 billion by 2004. The growth of e-business has also spurred the growth of a new type of fulfillment house that targets e-businesses, sometimes called **3FLs** (third-party fulfillment logistics). 3FLs such as ShipMax.com, iFulFillment.com, e-fulfillment.com, Electron Economy Inc., ReturnCentral, and Fulfillment PLUS offer a full range of back-end fulfillment services such as payment, shipping, and customer service. Other 3FLs, such as 3PF, offer distribution services including freight and transportation, returns and defective product processing, warehouse services, and inventory management. 3FLs are developing fulfillment centers that have the flexibility to accommodate the extreme highs and lows of single-item, prepackaged customer orders for e-business clients. Figure 9-11 illustrates a generic order fulfillment process for a third-party fulfillment house similar to one an e-tailer like foodlocker.com might use.

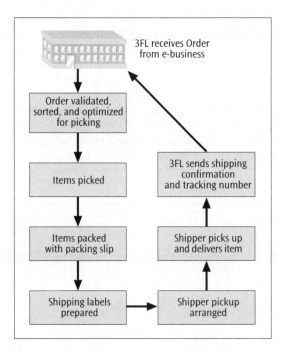

Figure 9-11

Independent fulfillment process

E-CASE The Last Outpost

Founded in 1995 to serve consumers and small office or home office businesses, Outpost.com today is a Web site superstore that has about 750,000 customers and about four million monthly Web site visitors. Outpost.com offers over 170,000 products, including hardware, software, computer accessories, electronics, and other items.

Outpost.com knows that most consumers shop online for convenience and that many of their target customers are "cash rich and time poor." To attract and keep those customers, Outpost.com promises free next-day delivery of any item ordered before midnight. In order to keep that promise, Outpost.com needed a fulfillment partner that was up to the task. Outpost.com looked for three things in a fulfillment partner: total focus on the customer, speed, and cost. They found all three in Airborne Express.

To speed up the shipping process, Outpost.com moved its inventory to the massive Airborne Express sorting center in Wilmington, Ohio, where Airborne Express handles Outpost.com's fulfillment processes. Once a customer's credit card is authorized, Outpost.com sends the order information to Airborne, where the orders are temporarily stored electronically until a sufficient number of orders are received to begin the picking process. By 2 A.M., the last of Outpost.com's orders are picked, packaged, and sent to the central sorting area with other shipments to be placed on outbound planes. By 4 A.M., all the orders are on their way for delivery the next morning. The average arrival time is 9:45 A.M.

(Continued on the next page)

The partnership between Outpost.com (Figure 9-12) and Airborne Express has not been without its challenges. For example, people who order from traditional catalogs generally submit those orders throughout the day, while many people ordering from Outpost.com's Web site do so after 5 P.M., creating a peak order period from 5 P.M. to midnight. Additionally, because overnight shipping was created for business letters and small packages rather than e-business orders, forecasting need and getting the fulfillment process to work well for overnight delivery have been difficult. The relationship was certainly tested during the 1999 holiday season, when orders began to flood into the Airborne Express center. As of this writing, Airborne was expecting at least 30 days with 25,000 to 30,000 orders per day for the 2000 holiday season.

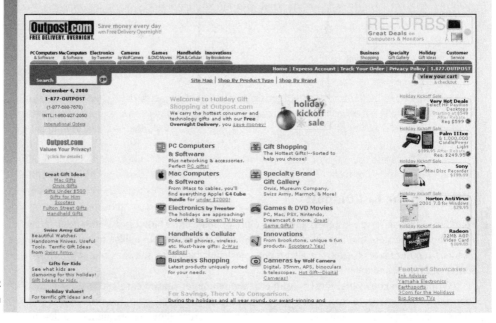

Figure 9-12
Outpost.com

Choosing a fulfillment house

Outsourcing critical functions such as fulfillment means that an e-business loses a measure of control over the processes that lead directly to customer satisfaction. In order to establish an effective e-business partnership with a fulfillment house, an e-business should consider several factors. A fulfillment house should:

- Maintain adequate inventory levels to satisfy anticipated demand
- Maintain a high order-fill rate (the percentage of orders that get filled on the same day)
- Have a reasonable cutoff time for orders to be filled the same day
- Use a variety of well-known shipping companies
- Support real-time communication to the e-business on the status of inventories and orders, instead of simple batch processing
- Add additional value by providing the option of credit card processing and customer service on behalf of the e-business

Because of problems with outsourced order fulfillment during the 1999 holiday season, many e-businesses are now taking control of their order fulfillment processes. At the top of this list is eToys, the clear winner in the 1999 online toy-selling season with $107 million in sales. Despite winning the online toy sales race in 1999, eToys still had a somewhat rocky Christmas season. In 1999 eToys was focused on increasing sales and energizing its brand, and it was so successful at accomplishing these two goals that it had to turn over about half its packing and shipping to a fulfillment house.

Unfortunately, eToys' experience with outsourced order fulfillment was less than stellar. While eToys did not experience the major holiday fulfillment problems experienced by other e-businesses—more than 99 percent of eToys' orders were delivered by Christmas—it still ended the year with excess inventories and a spotty fulfillment record.

In the belief that an e-business must handle its own fulfillment in order to maintain control over customer satisfaction, eToys decided to in-house 100 percent of its order fulfillment during the 2000 holiday season. eToys invested heavily in building its order and fulfillment systems from the ground up, including proprietary software used to run its computer network and two new order fulfillment facilities: a 1.2 million square foot distribution center in Blairs, VA, and a 764,000 square foot distribution center in Ontario, CA. Then during the weeks preceding the 2000 holiday season eToys ran several stress tests of its fulfillment systems—for example, by holding back thousands of routine orders for up to 24 hours and then flooding the system to see how it would cope with a sudden increase in orders, and by shutting down warehouse power to test backup generators.

The 2000 holiday season was particularly critical for eToys (Figure 9-13) and its chance to hit its financial targets. On December 19, 2000, eToys announced that fourth-quarter sales would be about half of earlier expectations, causing its stock price to tumble.

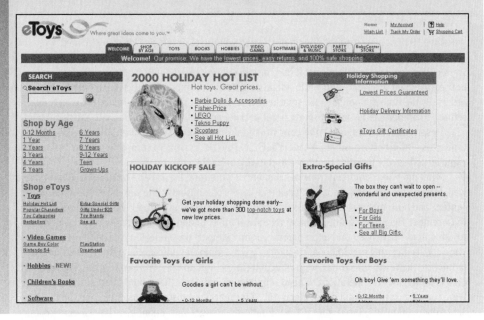

Figure 9-13
eToys

For many businesses, having products warehoused and shipped by its e-business partners makes sense.

Virtual inventories

Still other e-businesses choose to maintain a **virtual inventory**. A virtual inventory of products does not belong to an e-business, but to its partners, who own, warehouse, and ship the products. These partners are third-party manufacturers or distributors who are under contract to the e-business. This arrangement allows the e-business to avoid tying up its cash in inventory. The advantage to e-businesses of maintaining a virtual inventory is its infinite scalability. The e-business can add products without having to build warehouses and manage additional inventory. A disadvantage for an e-business of maintaining a virtual inventory is the loss of control over the fulfillment process. When an e-business maintains a virtual inventory, its partners control the fulfillment process—while the e-business is still responsible to its customers for the effectiveness and timeliness of the order fulfillment processes.

Another reason for maintaining a virtual inventory can be the type of product being sold. For example, in order to ensure product freshness, foodlocker.com maintains a virtual inventory. The producers of the food products foodlocker.com sells, such as New York's H & H Bagels, ship the product directly to foodlocker.com's customers, ensuring the orders' freshness.

The manner in which an e-business handles its fulfillment management is critical and can mean its success or failure.

. . . GHOST OF CHRISTMAS PAST

When the 1999 holiday rush hit, the unanticipated volume of orders quickly overwhelmed Wine.com's understaffed 80-person customer service department, and ringing phones went unanswered. Wine.com also realized that it had misjudged the capacity of its 32,000-square foot warehouse, making it extremely difficult to adequately stock and ship its products for all the holiday orders. By the time the holiday rush smacked Wine.com, it was too late to change the warehouse setup. To compensate for lack of space, order packing and delivery overflowed into the parking lot.

Despite these problems, Wine.com struggled through the season without a major disaster, unlike several other prominent e-tailers such as Toys 'R' Us, macys.com, and CDnow Online, who experienced severe logistics problems and many unhappy customers.

What did Wine.com learn from its experience? Wine.com, like other e-tailers who did not fare as well, learned greater respect for the "back-end" facets of e-business: the warehouses, distribution centers, and customer service call centers, upon which the ultimate success of the e-business hinges. In December 1999, last-minute hiring to try to beef up its understaffed customer service call center resulted in inexperienced and unprepared customer support personnel. To counter these problems, Wine.com is investing in additional customer service employees and special training programs to familiarize them with the Wine.com product inventory. Additionally, Wine.com is keeping close tabs on its entire back-end operations to head off potential problems that would affect customer service. Wine.com also spent $5 million on a new 100,000-square foot distribution center. And, finally, in September 2000 Wine.com (Figure 9-14) announced a merger with online rival WineShopper.com, which has comprehensive back-end systems in place. Perhaps Wine.com will be better prepared for future holiday seasons!

Figure 9-14
Wine.com

Summary

- Users directly access front-end systems, such as Web site ordering systems. Back-end systems, such as accounting and order fulfillment, can be integrated with front-end systems to provide better information and reduce costs.
- Integrating Web site operations with other back-end systems is critical for smooth operation and high-quality customer service.
- Having a solid records retention plan is important for an e-business to be able to provide data to customers, federal, state, and local government agencies, and the legal system.
- An e-business should have procedures for regular system backups and a disaster recovery plan to protect the business and its data from unexpected events.

- Order fulfillment is likely the most important facet of operations for an e-business that sells products online.
- How an e-business charges customers for shipping is an important part of customer satisfaction and customer service.
- Having an easy to understand and use product return policy is critical for customer satisfaction.
- Import restrictions, payment of taxes and duties, and additional paperwork can affect an e-business's international sales.
- Third-party fulfillment houses can be an effective source of fulfillment management for an e-business.

Key Terms

3FLs	front-end e-business systems	order picking
3PLs	fulfillment houses	primary records
audit	hot-swappable hard drives	secondary records
back-end e-business systems	legacy systems	trust protections
backup procedure	middleware	virtual inventory
disaster recovery	mirrored server	
fail-over system	order fulfillment	

Review Questions

1. A regularly scheduled backup process:
   a. Accounts for customer returns.
   b. Tracks inventory.
   c. Protects data from unexpected events.
   d. Provides customer satisfaction.

2. Order fulfillment is:
   a. Not important to a customer.
   b. A process that must be done in-house.
   c. Critical to the success of an e-business that sells products.
   d. A method of inventory replenishment.

3. Fulfillment houses are companies that:
   a. Provide systems backup and disaster recovery.
   b. Provide auditing services.
   c. Provide comparison-shopping for shipping companies.
   d. Manage inventory, order picking, packaging, and delivery for multiple clients.

4. In order to avoid tying up cash in inventory, some e-businesses maintain a:
   a. Virtual inventory.
   b. Invisible inventory.
   c. Improbable inventory.
   d. Vital inventory.

5. Which of the following is critical to the success of an online grocery e-tailer?
   a. A great looking Web page
   b. Outsourcing its fulfillment processes
   c. A high-quality and efficient order-picking process
   d. Establishing a partnership with a shipping company such as FedEx

6. An e-business need not be concerned with how long it maintains its business records. **True or False?**

7. Prior to 1999 most order-processing and fulfillment processes were designed to meet the demands of the typical catalog or direct mail customer. **True or False?**

8. Poor order-picking processes can affect both fulfillment and inventory replenishment operations. **True or False?**

9. Many e-businesses establish relationships with more than one shipping partner. **True or False?**

10. Most customers who buy online are only interested in convenience and are indifferent to the cost of shipping their orders. **True or False?**

Exercises

1. Using Internet search tools or other relevant sources, research two 3FL companies not noted in this chapter. Write a one- or two-page paper describing the companies and their services.

2. Using Internet search tools or other relevant sources, research how various e-tailers performed during the most recent holiday season. Select an e-tailer and write a brief explanation of the e-tailer's reported performance, including suggestions for improvement on that performance.

3. Using Internet search tools or other relevant sources, research different warehouse management or order fulfillment software vendors.

Write a one-page paper describing three of the vendors and the services they offer.

4. Using Internet search tools or other relevant sources, research RAID systems. Write a one-page paper describing the three most common RAID types: RAID 0, RAID 3, and RAID 5.

5. Using Internet search tools or other relevant sources, identify at least 10 major e-tailing sites selling different products. Note whether each site offers free shipping or delivery. List the sites and their shipping charge policies.

◆ 1 ◆

It's only early fall, but you are anticipating your first holiday buying season at your B2C e-business Web site. You anticipate some customer returns, and you want to make the process as simple and easy for your customers as possible. You want to add customer return tools to your Web site, but you are not certain which shipping companies' tools to add. Using Internet search tools or other relevant sources, research the return services provided by the United States Postal Service, Federal Express, and UPS that can be integrated into an e-business Web site. Then write a one- to two-page paper describing each service. Select one of the services for your e-business Web site and indicate the reasons for your choice.

◆ 2 ◆

You and your partner are in the process of developing a new B2C e-business idea. Your partner wants to maintain control over customer satisfaction by having the e-business warehouse and manage its own inventory and fulfillment processes. You think using a virtual inventory is the better choice, and now you have to convince your partner. Create a one- or two-page paper describing your e-business idea and outlining the reasons why following the virtual inventory model is best.

◆ 3 ◆

As the vice president of administration for a B2B e-business, you are responsible for developing a records retention policy. Using Internet search tools and other relevant resources, research issues related to records retention and destruction. Then draft a one- or two-page outline for a records retention plan for the e-business.

You and two classmates are creating a new B2C e-business and have decided to outsource your order fulfillment processes. You are meeting with your board of advisors next week to discuss your outsourcing decision. Working together as a team, define your e-business and its order fulfillment requirements. Then research the services offered by several third-party logistics companies. Select one of the companies to handle your order fulfillment processes.

Create a 5–10 slide presentation, using Microsoft PowerPoint or another presentation tool, describing your e-business, your order fulfillment needs, the company you selected to handle your order fulfillment, and your reasons for your selection. Then make the presentation to a group of classmates, selected by your instructor, who will act as your board of advisors and who will critique your outsourcing decision.

Useful Links

3PLResources
http://www.inboundlogistics.com/resources/
3plhome.html

@brint.com E-commerce Resources
http://www.brint.com/

Armstrong & Associates, Inc. – Logistics
http://www.3plogistics.com/Links.htm

CMPnet – NetBusiness
http://www.techweb.com/netbiz/

DataSafe – Free Records Retention and Destruction Guide
http://www.datasafe-sf.com/files/rrdg.pdf

Disaster Recovery Forum
http://www.rothstein.com/wwwboard/welcome.html

Disaster Recovery Journal
http://www.drj.com/

Disaster Resource Guide
http://www.disaster-resource.com/

DRI International
http://www.dr.org/

eCommerce Info Center
http://www.ecominfocenter.com/index.html

Global Sports
http://www.globalsports.com/

MIT Business Continuity Sample Recovery Plan
http://web.mit.edu/security/www/pubplan.htm

National Retail Federation
http://www.nrf.com/home.asp

Online Logistics Bibliography
http://www.clm1.org/bibliography/

Purchasing Magazine Online
http://www.manufacturing.net/magazine/purchasing/

RAID Advisory Board
http://www.raid-advisory.com/index.html

Small Business Administration
http://www.sba.gov/

The Bureau of Export Administration, U.S. Department of Commerce
http:// www.bxa.doc.gov/

The U.S. Census Bureau, Foreign Trade Statistics
http://www.census.gov/foreign-trade/www/

The Rothstein Catalog on Disaster Recovery
http://www.rothstein.com/catalog.html

TradeSpeak.com – Business, IT, and Science White Papers
http://www.tradespeak.com/

Links to Web Sites Noted in This Chapter

3PF
http://www.3pf.com/

@Backup
http://www.backup.com/

A. N. Deringer
http://www.anderinger.com/mainsite/index.asp

Airborne Express
http://www.airborne.com/

Albertson's, Inc.
http://www.albertsons.com/signin.asp

Amazon.com
http://www.amazon.com/

BarnesandNoble.com
http://www.bn.com/

BizRate.com
http://www.bizrate.com/

CDnow Online, Inc.
http://www.cdnow.com/

Cyber Dialogue
http://www.cyberdialogue.com/

DHL
http://www.dhl.com/

Diamond Phoenix
http://www.diamondphoenix.com/home.html

e-fulfillment.com
http://www.e-fulfillment.com/

Electron Economy
http://www.electroneconomy.com/

eToys
http://www.etoys.com

ExpressFulfillment.com
http://www.expressfulfillment.com/Fulfillment.htm

Federal Trade Commission
http://www.ftc.gov/

FedEx
http://www.fedex.com/

Fingerhut Business Services, Inc.
http://www.4bsi.com

foodlocker.com
http://www.foodlocker.com/

Forrester Research, Inc.
http://www.forrester.com/

freightquote.com
http://www.freightquote.com/

Fulfillment PLUS
http://www.fulfillmentplus.com/

GroceryWorks.com
http://www.groceryworks2.com/

iFulfillment, Inc.
http://www.ifulfillment.com/

International Risk Management Institute
http://www.irmi.com/

Jupiter Media Metrix
http://www.jmm.com/

KBkids.com
http://www.kbkids.com/index.html

Kozmo.com
http://www.kozmo.com/index.html

LiveVault Corporation
http://www.netint.com/products/lv/Default.asp

macys.com
http://www.macys.com/index.html

Mimeo.com
http://www.mimeo.com/

Outpost.com
http://shop.outpost.com/

Physician Sales & Service, Inc.
http://www.pssd.com/index.htm

PlanetRx.com
http://www.planetrx.com/

Remstar International, Inc.
http://www.remstar.com/

ReturnCentral
http://www.returncentral.com/

Ryder Integrated Logistics
http://www.ryder.com/ril/index.shtml

ShipMax.com
http://www.shipmax.com/

SkyMall.com
http://www.skymall.com

SmartShip.com
http://www.smartship.com/

Stamps.com
http://www.stamps.com/

St. Joseph's Hospital and Medical Center
http://www.sjhmc.org/index.html

SubmitOrder.com, Inc.
http://www.submitorder.com/

Toys 'R' Us
http://www.amazon.com/

UPS
http://www.ups.com/

USCO Logistics
http://www.usco.com/

Webvan
http://www000108.webvan.com/central/default.asp?/

White Systems, Inc.
http://www.whitesystems.com/

wine.com
http://www.wine.com/

WineShopper.com
http://wineshopper.com

For Additional Review

Adams, Nancy. 2000. "Integrating E-Commerce: How B-to-B CIOs Are Connecting E-commerce Front Ends and Legacy Back Ends. Hint: It's Not Easy (Part 1 of Series)," *Planet IT*, April 28. Available online at: http://www.planetit.com/techcenters/docs/executive_strategies/features/PIT20000427S0007.

Adams, Nancy. 2000. "Execs Contend With Integration Obstacles: CIOs See a Downside to Linking an E-commerce Front End and Legacy Back End—Complexity (Part 2 of Series)," *Planet IT*, May 2. Available online at: http://www.planetit.com/techcenters/docs/executive_strategies/features/PIT20000501S0023.

Adams, Nancy. 2000. "E-Integration's Future: Experts Foresee Hybrid ERP and E-commerce ASPs, Greater Technical Expertise (Part 3 of Series)," *Planet IT*, May 9. Available online at: http://www.planetit.com/techcenters/docs/management_issues-executive_strategies/features/PIT20000509S0018.

Alaimo, Dan. 2000. "Logistics Key to E-Grocery Success," *Supermarket News*, August 7, 21.

Anthes, Gary H. 1999. "The Quest for Quality: Too Many Online Merchants Are Driving Customers Away in Frustration," *ComputerWorld*, December 13, 46(1).

Bacheldor, Beth and Konicki, Steve. 2000. "Retailers Upgrade Systems to Fix Their Problems," *InformationWeek*, August 7, 30.

Berry, John. 2000. "Dotcom Start Ups Get Physical – Watch for Dotcoms to Launch Stores in Tandem with Web Sites," *InternetWeek*, April 3, 39.

Braunstein, Peter. 2000. "E-Commerce to Do Battle for Holiday Dollars," *Women's Wear Daily*, August 14, 2.

Business 2.0. 2000. "Is It Do or Die for E-tailers This Holiday Season?, " September 29. Available online at: http://www.business2.com/content/insights/filter/2000/09/29/19023.

Cahill, James M. 2000. "Virtual Supply Chains Drive E-Business," *Transportation & Distribution*, 41(5), May, S30.

Campanelli, Melissa. 1999. "Delivering the Goods," *Entrepreneur*, March. Available online at: http://www.entrepreneur.com/Your_Business/YB_SegArticle/0,1314,229968,00.html?CALLER_NODE_ID=634.

Campbell, Scott. 2000. "Distributors Add More Consumer Fulfillment – Delivery Expertise Brings Opportunities," *Computer Reseller News*, June 5, 54.

Casper, Carol. 2000. "New Developments In Hardware and Software," *ID: The Information Source for Managers and DSRs*, 36(9), September, 33.

Chain Store Age Executive With Shopping Center Age. 2000. "Deliverance," 76(8), August, 15B.

Chowdhry, Pankaj. 2000. "For Many, Dot-Com Was Not-Com at Christmas," *PC Week*, January 17, 76.

Christie, James. 2000. "FedEX and UPS to Wrangle Online," *Red Herring*, June 16. Available online at: http://www.redherring.com/industries/2000/0616/ind-fedex061600.html.

Cleary, Mike. 2000. "FTC Warns E-Tailers to Deliver," *Interactive Week as reported by MSNBC*, November 17. Available online at: http: //www.msnbc.com/news/491435.aspw.

Cohen, Mitchell. 1998. "Integrating Front-End Web Systems with Back-End Systems," *INET '98*. Available online at: http://tango.cetp.ipsl.fr/~porteneu/inet98/1f/1f_3.htm.

Cooke, James Aaron. 2000. "3PLs: Riding the Wave," *Logistics Management & Distribution Report*, 39(7), July 1, 69.

Couzin, Jennifer. 2000. "Getting Down to Holiday Business," *TheStandard*, November 20. Available online at: http://www.thestandard.com/article/display/0,1151,20198,00.html.

Cruz, Mike. 2000. "E-Commerce Market's Sell-Fulfillment Prophecy," *Computer Reseller News*, June 12, 7.

Dembeck, Chet. 1999. "Return Policies Top E-Shoppers' Concerns," *E-Commerce Times*, December 29. Available online at: http://www.ecommercetimes.com/news/articles/991229-2.shtml.

Electron Economy. 2000. "The Web's Great Order Disorder," *Business 2.0*, 173, October 10.

Ellis, Juanita. 2000. "Accepting Credit Cards Online," *Entrepreneur*, September 25. Available online at: http://www.entrepreneurmag.com/Your_Business/YB_SegArticle/0,1314,278944,00.html.

Ellis, Juanita. 2000. "Handling Returns Online," *Entrepreneur*, August 28. Available online at: http://www.entrepreneurmag.com/Your_Business/YB_SegArticle/0,1314,278951,00.html.

Ellis, Juanita. 2000. "Regulating Your Online Inventory," *Entrepreneur*, June 26. Available online at: http://www.entrepreneurmag.com/Your_Business/YB_SegArticle/0,1314,277019,00.htm.

Ellis, Juanita. 2000. "Service with a Smile," *Entrepreneur*, April 21. Available online at: http://www.entrepreneurmag.com/Your_Business/YB_SegArticle/0,1314,271960,00.htm.

Ellis, Juanita. 2000. "Special Delivery: E-Commerce Fulfillment," *Entrepreneur*, May 19. Available online at: http://www.entrepreneurmag.com/Your_Business/YB_SegArticle/0,1314,275184,00.html.

Ellis, Juanita. 2000. "Fulfillment Companies at Your Service," *Entrepreneur*, June 24. Available online at: http://www.entrepreneurmag.com/Your_Business/YB_SegArticle/0,1314,278012,00.html.

Fitzpatrick, Eileen. 2000. "Study Says Web Retailers Are Prepared For Holidays," *Billboard*, 112(46), November 11, 12.

Foster, Thomas. 1999. "Dot-Com Retailers Give 3PLs Their Big Chance," *Logistics Management & Distribution Report*, 38(10), October 31, 38.

Giangrande, John T. and Lackmeyer, Pamela J. 1998. "Which Pick/Pack Technologies Are Right for You?" *Operations and Fulfillment*, May-June. Available online at: http://www.opsandfulfillment.com/back/may-jun98/WhichTechnologiesareRightforYou.html.

Gibert, Alorie. 1999. "Rule No. 1: Don't Annoy Your Customers – Online Retailers Are Learning to Use Improved Logistics to Keep Customers Happy," *InformationWeek*, December 13, 134.

Goff, John. 2000. "The Dotcom Before the Storm," *CFO, The Magazine for Senior Financial Executives*, 16(11), Fall, 56.

Golway, Michael W. and Palmer, Thomas O. 1999. "Simulating Carton Flow Pick-to-Light Systems: A Simulation Analysis of Batch Versus Single Order Pick Methods," *AutoSimulations Symposium '99*. Available online at: http://www.asapauto.com/news/published/plantservices/index.htm.

Haskin, David. 2000. "Warehouses: When to Own, When to Rent," *Business 2.0*, September 29. Available online at: http://www.business2.com/content/magazine/indepth/2000/09/29/20131.

Hicks, Ed. 2000. "Memphis a Growing Player in Dot Com World," *Memphis Business Journal*, 22(15), August 11, 2A.

Jastrow, David and Campbell, Scott. 1999. "Distributors Gas Up Web Reseller Engine," *Computer Reseller News*, October 4, 1.

Jastrow, David and Cruz, Mike. 2000. "Happy Holidays for Shippers – Retailers Turn to FedEx, UPS to Avoid Last Year's Delivery Snafus," *Computer Reseller News*, October 2, 8.

Kalakota, Ravi. 1999. *e-Business: Roadmap for Success*. Reading, MA: Addison Wesley Longman, Inc.

Kelly, Beckie. 2000. "CIO's Best Strategies to Prepare for Disaster," *Health Data Management*, 8(7), July 1, 86-91.

Kelly, Patrick. 1996. "Physician Sales & Service – Keys to Hypergrowth," *fed.org*, September 27. Available online at: http://www.fed.org/onlinemag/nov96/casestudy.html.

King, Julia. 2000. "Filling Orders a Hot E-Business – Companies Race to Offer Logistics Services," *Computerworld*, June 12, 3 (1).

Koller, Mike. 2000. "CRM Being Built Into E-Biz Back-Office Apps," *Internet Week*, August 21, 14.

Kosiur, David. 1997. *Understanding Electronic Commerce*. Redmond, WA: Microsoft Press.

Krizner, Ken. 2000. "3PL Providers Will Continue to Add Customers and Services," *Frontline Solutions*, April. Available online at: http://www.findarticles.com/cf_0/m0DIS/4_1/61908254/p1/article.jhtml.

LaHood, Lila. 1999. "For Struggling Dot-Coms to Succeed, They'll Have to Deliver the Goods," *Fort Worth Star Telegram*, December 29.

Lawrence, Stacy. 2000. "From Bricks to Clicks," *TheStandard*, March 27. Available online at: http://www.thestandard.com/research/metrics/display/0,2799,13256,00.html.

Logistics Management & Distribution Report. 2000. "Clicks vs. Mortar: E-tail Shipping Strategies Diverge," 39(7), July 1.

Logistics Management & Distribution Report. 2000. "Up Front," 39(5), May 1.

Los Angeles Business Journal. 2000. "When to Consider Outsourcing Product Fulfillment," 22(35), August 28, 55.

Mann, Jennifer. 2000. "Holiday Season Will Separate Dependable E-tailers from Imposters," *The Kansas City Star*, November 24, K4102.

Mesrobian, Edmond and Ringer, Brian. 2000. "Toward Successful E-Fulfillment," *Web Techniques*, 5(1), January, 57.

Mullins, Robert. 2000. "Careful What You Click For," *The Business Journal-Milwaukee*, 17(10), December 3, 15.

Nicholson, Gilbert. 2000. "Internet Explodes with Web Sites for Booking Truck Freight," *Birmingham Business Journal*, 17(31), August 4, 15.

Ohlson, Kathleen. 2000. "Redundant Arrays of Independent Disks," *Computerworld*, 34(22), May 29, 67.

Press Release. 2000. "Princeton Softech's SyncPoint Transforms ERP Data Into eData," Princeton Softech. Available online at: http://www.princetonsoftech.com/news/press/physicians.htm.

Rafter, Michelle V. 1999. "The Art of Fulfillment," *TheStandard*, September 13. Available online at: http://www.thestandard.com/article/display/0,1151,6283,00.html.

Ramsey, Geoffrey. 2000. "Holiday Hangover: Online Sales Surged over the Holidays, but Researchers Agree – Improvement Is Needed," *Business 2.0*, March 1. Available online at: http://www.business2.com/content/magazine/numbers/2000/03/01/20706.

Razzi, Elizabeth and Burt, Erin. 2000. "Present Not Accounted For," *Kiplinger's Personal Finance Magazine*, 54(12), December, 140.

Rist, Oliver. 2000. "Planning for Disaster Recovery," *Planet IT*, February 14. Available online at: http://www.planetit.com/techcenters/docs/systems_management/expert/PIT20000210S0057.

Romaine, Ed. 2000. "E-Commerce Realities in Warehousing and Distribution," *IIE Solutions*, 32(7), July, 39.

Ryan, Thomas J. 2000. "Penney's to Handle Fulfillment for Planned Ann Taylor Web Site," *Women's Wear Daily*, June 14, 10.

Saenz, Norman. 2000. "It's in the Pick," *IIE Solutions*, 32(7), July, 36.

Saracevic, Alan T. 2000. "Hope for the Holidays: To Avoid Last Year's Dot-Com Debacle, E-tailers Scramble to Bolster Their Back Ends," *Business 2.0*, September 29. Available online at: http://www.business2.com/content/magazine/indepth/2000/09/29/20144.

Scheraga, Dan. 2000. "Paging Rudolph," *Chain Store Age Executive with Shopping Center Age*, 76(10), October, 102.

Scheraga, Dan. 2000. "The Nightmare Before Christmas," *Chain Store Age Executive with Shopping Center Age*, 76(3), March, 126.

Schrage, Michael. 2000. "Toby Lenk (Chief Executive of eToys)," *Brandweek*, 41(42), October 30, IQ36.

Seckler, Valerie. 2000. "Always in Style Gets New Partners," *Women's Wear Daily*, August 16, 15.

Shern, Stephanie. 2000. "Retailing in the Multi-Channel Age," *Chain Store Age Executive with Shopping Center Age*, 76(5), May, 1.

Stoughton, Stephanie. 2000. "eToys Cranks Up Holiday Sales As Traditional Rivals Muscle In," *The Boston Globe*, November 27.

The Christian Science Monitor. 2000. "Rush Job: The Rise of E-tailing and a Culture's Heightened Need for Speed Are Driving a Revolution in Shipping," August 7, 15.

Treese, G. Winfield and Stewart, Lawrence C. 1998. *Designing Systems for Internet Commerce*. Reading, Massachusetts: Addison Wesley Longman, Inc.

Trunk, Christopher. 2000. "Carousels Spin for E-Commerce," *Material Handling Management*, 55(6), June 1, 85–88.

Wilder, Clinton. 2000. "The Complete Package—With the Novelty of the Internet Worn Off, Shoppers Will Expect a Lot More Than Fast Web Sites and Timely Delivery This Holiday Season, "*Information Week*, October 16.

Williams, Mina. 2000. "Logistics Logic: A Clear-Cut Model for On-Line Shopping Has Not Yet Emerged," July 24, 21.

Wilson, Bill. 2000. "E-Commerce: Changing the Rules for Back-End Operations," *Target Marketing*, 23(6), June, 56.

Wilson, Tim. 2000. "An Exchange for Small Biz – Web Service Links Mom-and-Pop Businesses with EDI Marketplaces," *InternetWeek*, August 14, 36.

Wingfield, Nick. 2000. "Amazon Hurt by e-Toys Stumble," *MSNBC*, December 19. Available online at: http://www.msnbc.com/news/505314.asp.

Young, Eric. 2000. "Online Retailers Want to Be Good for Christmas," *TheStandard*, October 10. Available online at: http://www.thestandard.com/article/display/0,1151,19197,00.html.

CHAPTER 10

Launching Your E-Business

In this chapter, you will learn to:

Discuss the importance of a beta launch test

Describe e-business marketing methods

Discuss e-business advertising techniques

Discuss the importance of setting Web site benchmarks

Describe Web site return on investment issues

Discuss Web site metrics

Wine.com (launched as Virtual Vineyards in January of 1995) was founded by Peter Granoff and Robert Olson. Granoff and Olson wanted to use the Web to give potential customers in-depth information about the wines and the wineries that made them. They also wanted to provide that information in a way that was comfortable and not intimidating to their customers. They featured only the best wines from the best vintners, so that customers would get a good bottle every time. They also developed sophisticated software tools to make shopping as easy and convenient as possible.

The two founders believed that their customers were people interested in wine, but were not wine connoisseurs. They believed that their customers would be willing to experiment with bottles of wine costing between $8.00 and $20.00. The ideal target was a woman or man with moderate to high income, who was interested in learning more about wine. These inferences were based on casual observation and communications with individual customers via e-mail.

Olson (the "Propeller Head") and Granoff (the "Cork Dork") had to figure out how to make people aware of their Web site and get them to visit it. There were many options available. There was direct mail, used by many traditional retail businesses, but it was not considered focused enough, and it would not convert recipients into shoppers. Online advertising, particularly banner ads on Web sites visited by their target customers, was also a possibility. Print advertisements, radio ads, and TV spots had the additional hurdle that over half of those who saw them would probably not even have Internet access. How would the two founders bring customers to their Web site?

Testing Your Web Site with a Beta Launch

Software companies allow users to test their software programs, which is called a **beta test**, before selling them to the public, in order to find problems—called bugs—in the software. Hollywood studios preview their movies to a sample audience to get feedback that they then use in the final editing of the movie. For the same reason, it is a good idea for an e-business to test its Web site with a **beta launch**. By allowing visitors to test its Web site before marketing it to the public, an e-business can get critical feedback and make necessary changes before wide public exposure to the Web site.

To get that critical feedback, an e-business can use focus groups composed of beta launch viewers or survey the beta launch viewers to gain insight into potential customers' impressions of the Web site and how easy it is for them to find important information or conduct transactions at the site. Beta launching is also an important opportunity to make sure that everything at the Web site works—that there are no broken links or other problems. A beta launch also allows an e-business to see how its Web servers respond to different volumes of traffic by using the sample traffic to estimate peak traffic loads during busier times.

While a Web site beta launch is being conducted, the marketing plan to launch the Web site for public use must begin. Because there is a lag time between the implementation of marketing initiatives and their effect on the public, the marketing efforts must begin before the official Web site launch.

Marketing an E-Business

One of the fallacies of the marketing approach taken by many early e-businesses was the attitude that "If we build it, they will come." Although this approach may work in the brick-and-mortar world, where hundreds of people drive by every day and see a new business, a Web site is more like a brick-and-mortar store built on a dead-end street with few people driving by. Drive-by Web viewers, those who are simply doing undirected browsing, may never find a specific e-business. Even when viewers use search engines and directories to locate Web sites more closely focused on their interests, the huge volume of available Web sites may make it nearly impossible for any specific e-business's Web site to appear in the viewer's search list. Unless an e-business makes a special effort to let potential viewers know how to find its Web site, it is likely to have few visitors. One way to stand out is to make certain all major search engines and directories have information about the e-business in their indexes.

Search engines and directories

Despite all of the millions of dollars spent on advertising e-businesses, a majority of people find the products or services they are looking for by conducting an online search using a **search engine** or **directory**. For example, a survey by Forrester Research on British Internet users found that 81 percent of those surveyed found Web sites with a search engine or directory, rather than by any other source, including advertising. It is important that an e-business be listed in the indexes of all major search engines or directories, as well as any relevant minor ones.

Search engines and directories use a variety of methods to gather the Web page information that becomes part of their indexing process. Search engines use software programs called **spiders** that automatically update their indexes of Web pages by examining existing pages across the Web. Because of the growing volume of Web pages, changes made to a Web site, such as pages added or deleted, may take some time to be identified by search engines. Eventually, search engines may discover these changes and adjust their indexes. Some examples of search engines are AltaVista, Northern Light, HotBot, GoTo.com, and Google (Figure 10-1).

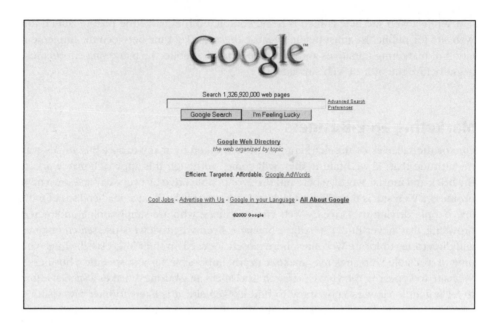

Directories rely on people to generate their indexes or catalogs of information. An e-business must submit a short description of its entire site and a short list of keywords, in order to be listed in a directory's catalog. Sometimes a directory maintains its own editing staff to review and write descriptions of Web sites in its catalog. Changes made to the Web site have no effect on a directory catalog. An e-business must submit any changes it wants to be reflected in a directory's catalog. Yahoo! is a well-known example of a directory. Other directories include Open Directory Project and Magellan Internet Guide (Figure 10-2).

There are some hybrid search engines that have an associated directory. Also, some directories forward searches to a search engine when the directory does not have matching results within its own catalog. For example, if a search at Yahoo! cannot find matching Web site results, Yahoo! uses the Google search engine to attempt to locate matching Web pages. Another directory/search engine hybrid is LookSmart (Figure 10-3).

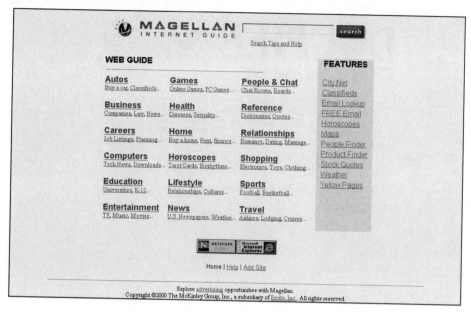

Figure 10-2
Magellan
Internet Guide

There is also a category of search engines called **meta search engines**. Simplistically, all a meta search engine does is search other search engine indexes or multiple directory catalogs and then combine the results of that search action into one report. Meta search engines include MetaCrawler and Dogpile (Figure 10-4).

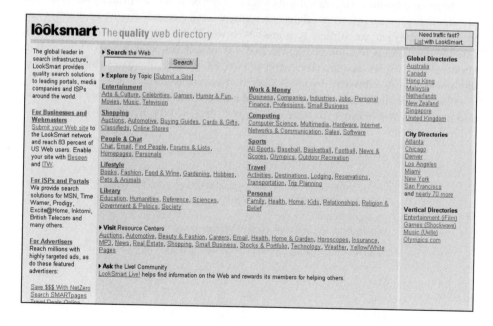

Figure 10-3
LookSmart

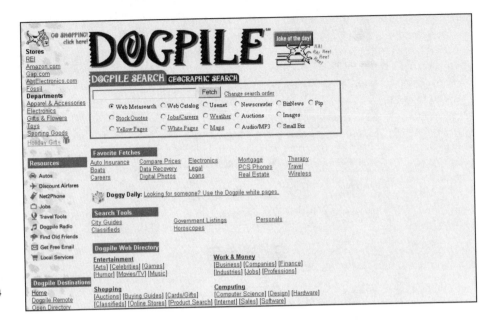

Figure 10-4

Dogpile

So how does an e-business get listed in the index or catalog of a directory? Most directories allow people to submit Web site information using forms at the directory's Web site. The first step in submitting Web site information is to carefully prepare a short description, using effectively chosen words that describe the e-business and capture viewers' interest.

Next, an e-business usually submits a set of approximately 25 primary keywords or phrases that potential viewers might enter when using a search engine or directory to locate Web sites similar to the e-business's Web site. These keywords should not duplicate the description, but complement it. Last, a directory generally requires information about an e-business's designated contact person. Figure 10-5 illustrates the description and the keywords submitted by foodlocker.com to various directories.

Some directories may require additional information beyond the brief description and keywords. For example, to have its Web site listed at Yahoo!, an e-business must

Figure 10-5

Sample description and keywords

Description:
Famous restaurant specialties and regional delicacies. Local food and hard to find food. Food you never forget!

Keywords:
specialty food, regional food, comfort food, local flavors, great food, hometown favorite, home style cooking, food from home, Texas, Louisiana, BBQ, regional gourmet, online food store

identify the specific categories (up to three) in which it wants its Web site to be listed. Some directories may charge a fee to have a submission reviewed and accepted more quickly than usual. For example, foodlocker.com submitted its Web site to Yahoo! for inclusion and waited. And waited. And waited. Apparently, in many categories Yahoo! was running as much as six months behind in reviewing and accepting submissions. foodlocker.com decided to pay a $200 quick review fee, and had its site listed within two weeks instead of six months.

Today search engines typically have a Suggest a Site or Add URL link at the bottom of their home pages that allows users to request that the search engine's spider index a specific site. For an e-business, this is more efficient than waiting for the search engine's spider to find the site on its own. Many of these search engines require only a Web site URL and perhaps the e-mail address of the contact person. Most search engine spiders can "crawl" a site quickly, but may be slow to update the search engine's index with the new information.

Various search engines use different indexing methods. Some use page content and page titles to get their indexing information. Some search engines use meta tags found in the HTML coding of Web pages, which usually contain the site's description and key-words, to index the pages. Many search engines use a combination of methods.

Meta tags are small segments of HTML code that are not visible on the Web page, but are read by the search engines. Meta tags are found between the <head> and </head> HTML tags on a Web page and use the following format:

<META NAME="description" CONTENT="*Description goes here.*">

<META NAME="keywords" CONTENT="*Key words go here, separated by commas.*">

Figure 10-6 illustrates a meta tag included in the HTML codes on the foodlocker.com home page.

In addition to meta tags, it is equally important that each Web page at a Web site have a descriptive title. This descriptive title appears on the title bar of a browser when the Web page is being viewed. Not only do some search engines use this descriptive title for indexing purposes, some also display this title in their list of returned search results to more clearly identify the Web page.

Instead of manually submitting its Web site information to a list of search engines and directories, an e-business can use a search engine submission service such as SubmitIt! (Figure 10-7) to assist in the submission process. Search engine submission services provide a single place for the e-business to complete all of the necessary forms and submit them to all of the major search engines and directories, as well as to some lesser-known search engines and directories.

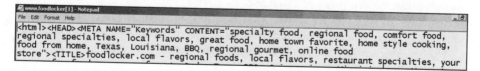

Figure 10-6 foodlocker.com sample meta tag

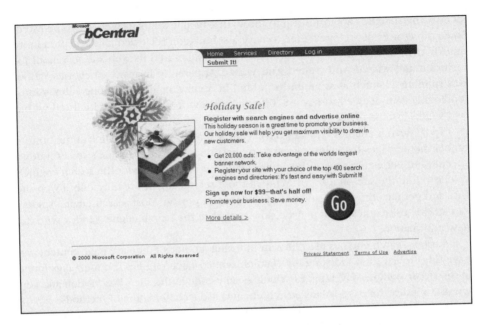

Figure 10-7
SubmitIt!

Search engines and directories use a variety of methods to organize the list of returned Web pages, called **hits**, that a viewer sees when he or she uses the search tool. Depending on the search strategies and keywords or phrases the viewer uses, these hit lists can be very long. Most viewers expect to find useful Web page links near the top of any hit list, certainly within the top 10 or 15 Web pages listed. An issue of concern for an e-business is how to get its Web page listed near the top of a hit list in order to attract a viewer's attention.

Currently there is no guaranteed method of ensuring that an e-business's Web pages will be listed in the top 10–15 hits for every keyword, but there are steps that an e-business can take to increase its chances. Simply by submitting effective descriptions and keywords and by using meta tags and descriptive Web page titles, an e-business has taken several steps in the right direction. Other considerations involve Web site and Web page design:

◆ Write clear but descriptive Web page text that a search engine can index.
◆ Avoid using frames, or have an alternate no-frames site to help search engine spiders locate all Web pages at a Web site.
◆ Use static HTML pages for products that can be indexed by search engines, instead of active pages that are generated by the server from data from a product database.
◆ Use fewer or smaller images to keep page load times down.
◆ Provide links to complementary pages.
◆ Arrange for inbound links from other Web sites, which are used by some search tools to determine a Web page's ranking or relevance.

It is a good idea for an e-business to periodically review search engine results, using the e-business's keyword or meta tag list to make certain its Web site is being listed when those keywords are used by viewers. Checking its Web site's ranking on hit listings also provides feedback to an e-business on whether or not its Web site search optimization techniques are working. Monitoring search engine results can provide other useful information as well, such as the existence of a new competitor or the optimization of a competitor's hit ranking. When the competitor shows up higher in a search engine's list of hits, it is a good idea to examine the competitor's site, including its meta tags, to try to learn why that site ranked higher for those specific keywords.

TIP

You can view a Web page's HTML code, including its meta tags, in a Web browser. For example, if you open a Web page in Internet Explorer and click the Source command on the View menu, the source HTML code opens in the Notepad application window.

Some Web sites try to "cheat" their way into a higher position in a search engine's hit list by "spamming" the search engine with the same keyword repeated over and over in a page. Almost all search engines reject such spamming attempts.

E-CASE Bidding for Placement

One new twist on the pay-for-placement advertising model is the advent of a pay-for-position search engine, called GoTo.com, which allows advertisers to bid for click-throughs. The highest bidder on a given keyword is listed first when that keyword is searched on, followed by the next highest bidder, and so on. This allows advertisers to decide how much a click-through for a specific keyword is worth to them and bid accordingly. Also, advertisers pay only for what is delivered, since they pay for click-throughs, not impressions. GoTo.com benefits because it always sells listings to the highest bidder. The person who conducts a search on GoTo.com (Figure 10-8) should receive very relevant results, because it is not worthwhile for an e-business to bid on keywords that are not relevant to what it offers.

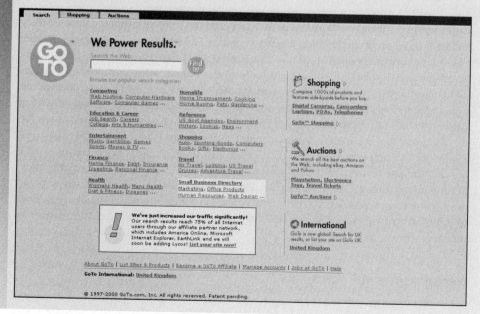

Figure 10-8
GoTo.com

In addition to being indexed by search engines and submitting information to directories, an e-business must consider using traditional marketing tools such as public relations to get its Web site noticed.

Public relations

When it works well, a public relations effort is one of the more cost-effective marketing tools. Public relations, sometimes called simply PR, is the process of establishing and maintaining a company's public image. Public relations activities for an e-business can include media, community, consumer, and governmental relations; informing the general public about the e-business's policies, procedures, activities, and accomplishments; and, if necessary, crisis communications.

Part of the art of a public relations effort is writing effective press releases and sending those releases to the appropriate destinations at just the right time. A press release is a short announcement of a newsworthy item sent to members of the press. An effectively written and timely press release is much more likely to attract the attention of writers and editors, and therefore to be published. Also, sending the release to the right members of the press makes a big difference. It is also important to make sure that the news media's audience matches the e-business's target customers. For example, a news radio station may have audience demographics similar to an e-business offering online stock trading, while a rock 'n roll radio station may not. There are many businesses that specialize in providing public relations services. Additionally, advertising agencies often maintain a public relations department.

Another way an e-business markets itself is with its communications with customers.

Consumer communication

Any communication between an e-business and its customers or potential customers is extremely important. An e-business must appreciate the long-term value of its customers and treat them accordingly. Using e-mail to communicate with customers can help build customer relationships one at a time. An e-business's customer communications should be interactive, instead of one-way. Also, communications with customers should be immediate: customers expect prompt responses to their inquiries. E-mail enables customer communications to be both immediate and interactive. All e-mail should be answered within 24 hours, but quicker responses make an even more positive impression.

One way to use e-mail to build customer relationships is by creating an e-mail mailing list for existing customers and then using it to update customers about new products or Web site features. It is also a good idea for an e-business to obtain a customer's permission to include his or her e-mail address in the list. In this way, customers give permission to an e-business to remind them about its Web site when they haven't been to the Web site recently. Using e-mail mailing lists can help increase the number of repeat customers at a Web site.

Newsgroups and forums

An Internet newsgroup is a topic-specific message-based discussion group. Participants in a newsgroup submit messages and respond to other messages on a specific topic.

Newsgroups may also be called forums, bulletin boards, or special interest groups (SIGs). Because the participants in these special interest groups use the Internet to share information, these groups are a great source of information to an e-business and are an indirect venue for marketing an e-business.

Internet special interest groups can be used to conduct market research by reading about participants' problems or their views of various companies' products and services. Participating in Internet special interest groups can also be an opportunity to build relationships, by getting involved in the group and answering questions posted by participants. This allows an e-business to indirectly market itself to the other participants by developing a positive image. Participation also has an added benefit of portraying the e-business as a leader in the specific industry and an authority on the discussion topic. A side benefit of being involved in the newsgroup is that the e-business can quickly become aware of any negative postings about its products or services and respond to them quickly and effectively.

One important thing for an e-business to remember about participating in Internet special interest groups: It should not post blatant advertisements to the group. Internet special interest groups are forums for information, not for advertising. It may be a good idea for an e-business to simply observe the messages for a while before posting any messages, in order to learn what is acceptable and what is not acceptable to the participants.

Affiliate programs

One marketing method that has been effectively used by many businesses is what is called an affiliate program. In an **affiliate program**, the e-business pays a referral fee (usually a flat amount) or commission (a percentage of the sale) on all sales sent to the e-business from another Web site. When an e-business signs up for the affiliate program, they receive customized links to their Web site that allow them to track the referred visitors into and through the Web site. When those referred visitors make a purchase, the e-business credits the affiliate whose link was used with the purchase.

For example, eToys pays affiliates $10 per new customer referred to their online store. Amazon.com pays its affiliates a 15 percent commission on books that the affiliate links directly to, and 5 percent for all other products. Amazon.com reports that its affiliate program has 430,000 members.

Affiliate programs provide a nice income opportunity for Web sites that offer content but no e-commerce capability. For an e-business, affiliate programs provide a cost-effective method of attracting customers. These programs basically allow other Web sites to do an e-business's marketing. An e-business does not have to pay an affiliate for visitors who buy nothing; the e-business pays a commission or referral fee only if an actual purchase is made. Also, using an affiliate program to attract customers is more cost-effective than other programs such as advertising. Many e-businesses pay rates exceeding $40 per new customer attracted through advertising. It is easy to see that paying an affiliate $10 for each customer referred is a much less expensive option. There are several services such as ClickTrade (Figure 10-9) that run affiliate programs for e-businesses.

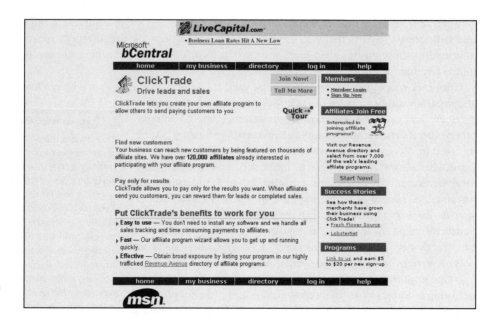

Figure 10-9
ClickTrade

Other Cost-Effective Promotion Methods

There are many other cost-effective ways to promote an e-business. One basic way to promote an e-business is to make sure its Web site URL and e-mail address are on all printed materials such as business cards, letterhead stationery and envelopes, and brochures. Additionally, any building signs, company uniforms, delivery vans, shipping boxes, Yellow Page listings, trade organization directories, or any other venue that can be seen by potential customers should display the Web site URL as well as other pertinent information.

Exchanging links

Many e-businesses have complementary Web sites with which to exchange links. For example, an e-business pet store might exchange links with a veterinary clinic that has a Web site. The first step is for an e-business to locate complementary Web sites. The next step is to send an e-mail message to the Web site suggesting a link exchange. Some e-businesses will not exchange links, or are very selective about the Web sites with which they exchange links. Other e-businesses are happy to add additional links to their Web sites and to have others link to their Web site. Occasionally, an e-business is willing to provide a link without requiring a reciprocal link, in order to provide additional information to their viewers. There are centralized link exchanges such as LinkExchange (Figure 10-10) that an e-business can use to connect with many other e-businesses willing to exchange links.

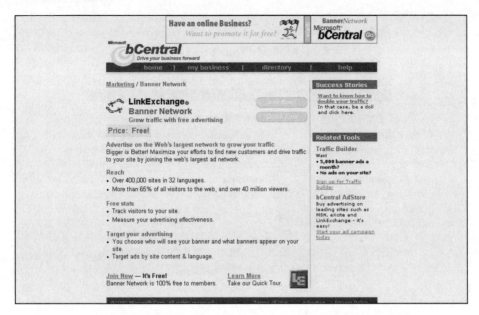

Web rings

Sometimes a group of similar e-businesses will create a "chain" of links among their Web sites that allows a visitor to link through to all of the sites in the chain. This circular chain of links is called a **Web ring**. Participating in a Web ring can increase the number of visitors to an e-business Web site, but will also increase the number of visitors who leave the site to follow the Web ring. The Yahoo! WebRing site (Figure 10-11) is a good place for an e-business to look for or join a Web ring.

Awards

Another way to get recognition for an e-business is to win an award, such as a "cool site of the day" designation. The award-winning e-business will be featured on the award Web site with a link to it. This helps give a fledgling e-business credibility, in addition to the recognition.

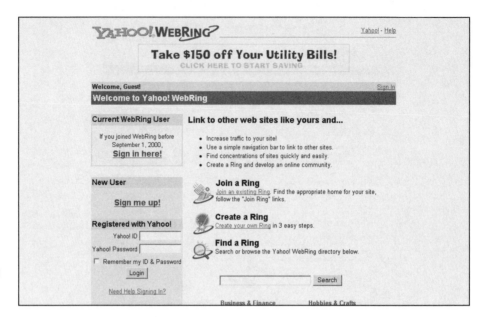

Figure 10-11
Yahoo! WebRing

Old-fashioned word of mouth

It's always a good idea for an e-business to keep its customers happy. This sometimes-overlooked concept is extremely important; some e-businesses generate most of their new customers from referrals by existing customers. Also, while a happy customer may not tell others about his or her experience with an e-business, an unhappy customer is highly likely to do so. In addition to losing that unhappy existing customer, word of mouth about negative customer experiences may prevent other potential customers from shopping with the e-business.

Word of mouth is an important component of "viral marketing," the marketing of an e-business by its customers. When customers tell their friends about an e-business, and those friends tell their friends, and so on, an exponential explosion can occur in the number of visitors who view a Web site and purchase its products or services. One of the best recent examples of viral marketing is Napster.

Napster is a pioneer in person-to-person file sharing over the Internet. Because of word of mouth between college students and music lovers, Napster's popularity in person-to-person music file sharing became so great that Napster became a threat to the recording industry, causing music industry leaders to file suit against Napster (Figure 10-12). In October 2000, German media conglomerate Bertelsmann, whose record group is involved in the Napster litigation, established a strategic partnership with Napster. As of this writing, it is unclear how this partnership will change the basic Napster service. Also, it is unclear whether Napster will create strategic partnerships with other recording companies, and if it does, what will be the impact on Napster's future.

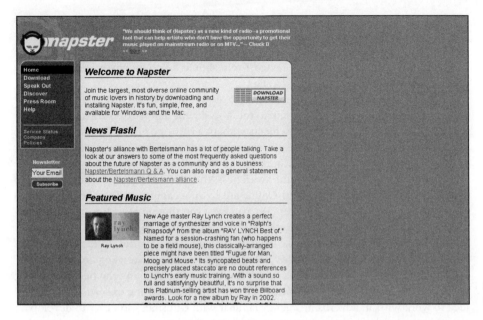

Figure 10-12
Napster

Other viral marketing techniques include providing Web page links that viewers can use to send a copy of the Web page to a friend, or to send a coupon from the Web page to a friend.

Encouraging visitors to create a bookmark or favorite shortcut to a Web page is another no-cost marketing technique. Some e-businesses put text on their home page prompting visitors to remember to create a shortcut to the site so they can easily return to it. Marketers expect that simply asking visitors to create the shortcut will cause many of them to do so. A few e-businesses encourage visitors to use their home page as a browser start page, so that the home page is loaded into the browser automatically when the browser is launched.

Finally, the best marketing idea is to offer an outstanding product or service that customers need. However, even with an outstanding product or service, an e-business may find that it needs to advertise that product or service.

E-Business Advertising

To increase public awareness of the e-business and increase traffic coming to its Web site, it may be necessary to use advertising. Advertising can be purchased directly or through an advertising agency. If your e-business does not have the marketing staff and experience, an advertising agency can design an entire advertising campaign, including both the placement of the ads and the content of the ads. Advertising agencies are very useful in designing an advertising campaign that uses both online and traditional media, partly because they have the ability and time to research the different types of ad placements. Advertising agencies also have the experience and expertise to determine what will work well for an e-business and what won't work as well.

There are many steps an e-business can take to increase the effectiveness of its advertising efforts. However, the most important thing to remember is that all advertising, as well as other marketing, should be integrated. It should all communicate the same message and build the same brand image. There should be no inconsistency among any of the aspects of the advertising campaign.

Another important item to remember is to always include the URL and e-mail in any advertisement (or at least include a link to the Web site for online advertising). It is hard to find magazine or newspaper advertisements without a URL or e-mail address these days, for good reason. It is wise for any business to refer viewers to its Web site for more information.

Besides the goal of driving visitors to the Web site, there are other purposes for advertising. One is increasing brand awareness, the ability of potential customers to recall the company's name or brand. Online advertising campaigns have been shown to raise brand awareness by 5–7 percent on average, even though most businesses are primarily concerned about the traffic that the campaigns drive to their site.

Before buying advertising, an e-business should consider the costs related to an advertising campaign. This includes comparing rates for advertising in comparable media, getting quotes from multiple media sources, and negotiating the best advertising rates. Because many startup e-businesses lack in-house advertising expertise, it is a good idea to work with an advertising agency familiar with e-business, in order to develop an appropriate and timely advertising campaign, including negotiating the best advertising rates, determining when and where ads are to be published, and creating the ads.

An advertising campaign can include many components, one of which is banner ads.

Banner ads

One of the most common advertisements found on the Internet is banner ads. Banner ads are rectangular images, or banners, placed on a Web page that link to the advertiser's Web site. The most typical size for a banner ad is 468 pixels wide by 60 pixels high. Figure 10-13 illustrates a banner ad for foodlocker.com.

Banner ads are generally priced according to the number of impressions, or the number of times the ad is viewed. This pricing is expressed in CPM, or cost per thousand impressions. The pricing can range widely from Web site to Web site, depending on the demographics of the Web site users and demand for ad space. Banners run from $3 to $30 CPM on average, but some sites charge as much as $70 or $80 CPM for an audience that is very attractive to the advertisers.

Figure 10-13
foodlocker.
com
banner ad

It is possible on some Web sites, such as search engines, to target advertising. The ads can be targeted by the demographics of registered users at the Web site. For example, a banner ad could be targeted to Web site viewers living in a specific zip code with a median income greater than $75,000. A banner ad can also be targeted to a specific audience by having it appear only on selected pages at a Web site. For example, a banner ad might be displayed only in the automotive section of a search engine Web site. Another way to target banner ads is by using the keywords a viewer enters in the search process. For example, a banner ad may only be displayed when a particular keyword, such as chocolate or auto, is entered.

The more targeted the banner ad, the higher the cost for the ad. Keyword targeting on search engines costs many times more than untargeted, or "run of site" ads, resulting in prices of $35 to $125 CPM for targeted ads. The higher cost of targeted banner ads is often worth it, because effectively targeted ads can generate high click-through rates. The **click-through rate** is the percentage of viewers who click on a banner ad to view the advertised Web site. Click-through rates for targeted banner ads may be many times higher than the click-through rates for the run of site ads. It is expected that viewers who click through banner ads are likely to be better sales prospects.

Some Web sites charge a monthly fee for banner ads. When this is the case, the e-business must know how many impressions the Web site receives each month, to determine a comparable rate in CPM. A newer model in pricing banner ads is charging per click instead of per impression. This benefits advertisers in that they only pay for performance, being charged only when someone actually clicks on their banner ad.

There is some controversy over the effectiveness of banner ads. Critics of banner ads question their effectiveness. Because viewers see many banner ads while browsing the Web, they often do not pay attention to them. In fact, the average click-through rate for untargeted banner ads has fallen below 0.5 percent. Proponents of banner ads believe that even if viewers do not click through banner ads, simply placing the ads at a Web site increases brand awareness and purchase intent.

Featured placement or sponsorships

It is possible for an e-business to buy featured placement or sponsorship on another Web site. Many Web sites (called **portals**) that offer a wide array of services, including access to search engines and directories, sell placement in their shopping sections. Charges are usually per month, but may have a contract term associated with them. Charges also depend on how the e-business is featured. For example, in AOL's shopping section (Figure 10-14), e-businesses pay much more for the featured position with a large graphic display than for a smaller text link down lower on the page for a given section.

Figure 10-14
Shopping AOL

When purchasing sponsorships or featured placement, it is again important to know how many impressions the placement will get. The quality of the audience is also important. For example, AOL users are known to make many purchases online, while another online mall may have only a small number of viewers who are also buyers. When contracting for the placement, it is a good idea to specify a guaranteed number of impressions so that an e-business gets the coverage for which it is paying.

Opt-in e-mail or newsletters

An e-business can send messages to potential customers who request e-mail from advertisers on a particular topic. Similarly, an e-business can buy advertising space on many e-mail newsletters. It is important to remember that the e-mail or newsletter recipients choose to receive these e-mails. This is called opting in. In other words, these messages are not unsolicited junk e-mail, often referred to as spam.

Opt-in e-mail is much more effective than banner ads or ad placement because the target audience has chosen to receive the e-mail advertising and is a more focused audience. Some opt-in e-mail lists can be targeted even further, based on registration data provided by the recipients, such as state or e-mail domain. E-businesses are starting to recognize the effectiveness of this type of interactive direct marketing. Forrester Research predicts that targeted e-mail advertising will increase to 65 percent of all online advertising by 2003, with a total of $5.5 billion being spent on direct e-mail marketing.

Costs per impression are higher with opt-in e-mail than with banner ads, running in the $275 to $450 range per thousand recipients. However, opt-in e-mail generates a much higher response rate, sometimes exceeding 10 percent for an appealing offer sent to a highly targeted audience. Advertising in newsletters does not generate as high a response rate, but the pricing for advertising in newsletters is less because pricing is determined by many factors such as the number of ads, position in the newsletter, and number of people who receive the newsletter (also referred to as reach).

There are several companies, such as NetCreations and yesmail.com, that maintain and rent lists of opt-in e-mail subscribers. When an e-business wants to rent a particular list, it submits its e-mail advertising message to the company that owns the list, which then sends the e-mail out to all of the members of the list. An e-business never receives a copy of the list members' e-mail addresses.

The time required to execute an opt-in e-mail campaign is very short, just a matter of days. This means that an e-business can execute this portion of its advertising campaign much more quickly, and analyze the results in real time as they come in. This allows an e-business to refine its opt-in e-mail campaign and try it again, all in the same time frame it would take to get its first ad published in a newspaper.

Traditional advertising

E-businesses should consider traditional advertising as part of their advertising campaigns. Traditional advertising includes print media, radio, television, outdoor advertisements, and direct mail. Online advertising has not replaced traditional media. In fact, the two work very well together and complement each other. Costs for traditional media are often measured in CPM as well. Prices vary widely, but are easily compared using the CPM measure.

E-CASE IN PROGRESS foodlocker.com

foodlocker.com faced the challenge of attracting customers to its site. Targeting customers would prove to be a challenge, because the ideal customer for foodlocker.com was someone who had moved away from his or her hometown and missed the food that was available there. There was no print or broadcast medium that corresponded to that particular type of customer. A guess at the closest approximation was airline magazines, because a strong correlation was likely between travelers and the customers that foodlocker.com was targeting. Also, the airline magazines could reach a secondary market, people who have enjoyed foods while traveling that are not available where they currently live.

foodlocker.com's primary method of generating customers was optimization of search engine hit list placement, so that foodlocker.com appeared at the top of a hit listing when someone searched using the names of the regional products that foodlocker.com carries. foodlocker.com also used the pay-for-placement search engine GoTo.com, where foodlocker.com had the highest bid on search terms corresponding to the products they offered. In this way foodlocker.com appeared first in the GoTo.com search hit list. The importance of good public relations was evident, since each mention of foodlocker.com in a metropolitan newspaper notably increased traffic to the Web site. Finally, word of mouth became important as more visitors tried out foodlocker.com products and were satisfied with their purchases.

One important step is to analyze the results of Web site performance in terms of its sales, marketing, and advertising objectives.

Benchmarking a Web Site

Even though the excitement of starting and running an e-business is rewarding for an entrepreneur, there remains a need to measure the e-business's performance against previously set goals. The primary reason to measure performance is to learn how to improve the e-business, not simply to determine success or failure. In fact, success or failure in most businesses, including e-businesses, is determined by the ability to learn from experience and make improvements.

TIP

Sometimes there can be a seasonal effect on performance, such as the effects of the holiday shopping season. For example, the average Internet retailer had 7.6 percent of orders returned during the first quarter of 2000, but by the second quarter this figure dropped to 5.7 percent. It is important for an e-business to understand the seasonality of its industry in order to come to reasonable conclusions about performance.

Comparing actual financial results to the financial plan portion of its business plan is one way an e-business can measure its performance. Another aspect of measuring an e-business's performance is measuring the effectiveness of its Web site. Part of that measurement includes setting **benchmarks**, which are performance-based objectives for the e-business's Web site, and then measuring the Web site's actual performance against those benchmarks. Typical benchmarks for a Web site's performance might include the number of unique visitors to the Web site or the number of user actions taken at the Web site, such as orders or registrations.

Setting performance-based objectives and then measuring actual performance against those objectives enables an e-business to make effective decisions for both its current operations and its long-term strategies. The process for setting benchmarks includes:

- *Determine the goal(s)*: The first step in benchmarking a Web site is to determine the Web site's operating goals. For example, if a Web site goal is to increase sales, then the e-business will look for an increase in sales attributable to online ordering. If a Web site objective is to reduce operating costs, then the e-business will look at the change in operating expenses over time, both in total dollars and as a percentage of sales.
- *Set benchmarks*: Once the Web site's operating goals are determined, the e-business must then select the performance-based objectives, or benchmarks, to be measured against. These benchmarks may be developed from the actual performance of similar companies or from industry averages. For a brick-and-click e-business with historical brick-and-mortar performance, the benchmarks may also be based on expected annual increases over that historical performance. When determining its Web site benchmarks, it is also a good idea for a new e-business to look at similar brick-and-mortar businesses as well as similar e-businesses.
- *Compare actual results to benchmarks*: Once the set of benchmarks has been determined, the e-business must then compare a Web site's actual results to the benchmarks, to determine if the Web site performed as well as, better than, or worse than the benchmarks.

- *Draw reasonable conclusions*: Whether the Web site meets its benchmarks or not, it is more important that the e-business's management arrive at meaningful conclusions on why the benchmarks were or were not met or exceeded. Because it is possible for an e-business to establish goals that are unobtainable, or to set goals that are too easily achieved, it may be necessary to reevaluate the original goals and benchmarks to determine if they were reasonable. Only through reevaluation can an e-business do a better job of setting future goals and establishing benchmarks upon which to measure those goals.

One measurement of Web site performance is return on investment.

Measuring Web Site Return on Investment

In general, any business has a goal of generating a return on the resources that they invest, and an e-business requires a large investment of resources (both money and time). Traditional return on investment calculations, called simply **ROI**, measure a business's profitability. Those traditional measurements should also be applied to an e-business's financial data.

However, an e-business must also be concerned with the ROI of its Web site operations, and this ROI may not be initially measured as increased profit. There are many other ways to achieve Web site ROI that do not necessarily increase profit in the short run, such as:

- *Customer satisfaction*: One of the primary returns on investment that an e-business generates is an increase in customer satisfaction. Although this may not be immediately measurable in dollars, it should lead to increased revenues and profits in the long term.
- *Increased sales*: The ROI that most e-businesses are looking for is an increase in sales. They want to know that the capital they are spending is generating additional sales, which they expect will lead to increased profits.

TIP

Although ROI is very important, many online marketing activities that may be crucial to an e-business's success may not be immediately measurable. For example, by having staff members participate in industry-specific chat rooms and online newsgroups, an e-business may be able to establish a reputation as a trusted source of information on industry issues. While it may be impossible to trace leads to sales from these types of activities, online participation in industry-specific groups could lead to the generation of new revenues now and in the future.

When trying to determine its Web site ROI, an e-business should realize that no single measure can define its Web site success. Various performance measures—such as pages viewed, the number of Web site visitors, the ratio of actual orders to Web site visitors, the optimization of advertising placement used to drive sales, and so forth—should be analyzed to determine an e-business's return on its Web site investment.

There are many tools available to measure Web site performance.

Using Online Measurement Tools

In order to determine if Web site benchmarks are being met and how the Web site is performing, it is necessary for an e-business to use online measurements, called Web metrics. There are two types of Web metrics. Basic metrics use a Web site's server log files to gather such information as date and time and files accessed at the Web site. Advanced metrics combine several basic measurements to gather more useful information such as the number of unique visitors. An individual e-business's needs determine the level of online measurement it requires; however, all e-businesses should, at minimum, take advantage of a log file analysis.

Log file analysis

Web servers record all of the events that they process to files called **log files**. Every time a browser makes a call to a Web server to download a Web page, view an image, or submit a form, this event is recorded in the Web server log file. In addition to the event, information such as date, time, IP address of the computer making the request, browser type, and even the referring URL are also recorded. From these log files, a treasure trove of information can be obtained. Two of the most significant pieces of Web server information used to gauge performance and to analyze the growth of Web site activity are unique visitors and page views.

The following are examples of pieces of information that can be obtained from Web server log files.

♦ *Visitors*: **Visitors** are the actual number of viewers to visit a Web site. A count of actual visitors is more meaningful than the number of hits. A hit is a recorded event in a Web site's server log for each element of a Web page downloaded to a viewer's browser. For example, if a viewer loaded a Web page with four graphics in his or her browser, the Web server would record five hits (one for the page and one for each of the four graphics). While early industry measurements were largely devoted to the number of hits at a Web site, a hit actually bears no relationship to the number of pages viewed or visits to a site.

♦ *Page views*: **Page views**, sometimes called **impressions**, indicate the number of times a given page has been viewed. Page views can be examined to determine which people are—and are not— looking at a Web site.

♦ *Page views per visitor*: By dividing the number of page views by the number of visitors, the number of **page views per visitor** can be determined. This is a useful measurement of how deep visitors are going into the Web site. A larger number means that visitors are looking at more pages and are likely spending more time at the Web site.

♦ *IP addresses*: IP addresses can be used to determine a viewer's origin by looking up the domains that correspond to the IP addresses. Analyzing IP addresses enables an e-business to identify approximately from what countries its Web site visitors are coming, as well as from what networks, such as AOL, MSN, or a university campus network.

♦ *Referring URLs*: A **referring URL** indicates how visitors reached the Web site— whether they keyed in the URL directly or whether they clicked a link from another Web site to view an e-business's Web site. If the referring URL is a search engine, the search terms that the visitor used can usually be determined from the URL.

♦ *Browser type*: The **browser type** indicates what type of Internet browser a visitor is using. An e-business can check this information to make certain that its Web site can be viewed properly by the browser that is most common among its visitors. An e-business should watch out for nonstandard browsers such as WebTV, because these browsers do not support all of the components that are supported by Microsoft Internet Explorer and Netscape Navigator. Viewers using WebTV to access the Web site may not have the same experience as viewers using the two most common Web browsers.

◆ *Conversion rate*: The **conversion rate** is the rate at which viewers are taking some action at a Web site. For example, by dividing the number of orders by the number of visitors, an e-business can determine a conversion rate related to sales. This conversion rate is an important measure of how well an e-business converts shoppers into buyers.

◆ *Errors*: Information about the number of errors recorded in a Web server log (log server errors) is very important. An example of an error a visitor might encounter is the "404 - File Not Found" error. An e-business can analyze these and other error messages to discover broken links or other problems at its Web site.

Figure 10-15 illustrates a log entry from the foodlocker.com server logs. This example shows a referring URL that indicates the visitor used the search engine at GoTo.com and searched for the term "bbq sauces."

http://www.goto.com/d/search/?Keywords=**bbq**%20**sauces**&type=topbar&x=16&y=9

Figure 10-15
foodlocker.
com server
log

In order to do a more sophisticated and advanced analysis of Web site performance, an e-business can use software or other services provided by e-business vendors.

Third-party analysis tools

Ad management networks, Web traffic analysis providers, media research firms, and software providers all offer products designed to combine information from Web server log files with customer databases and other sources, to provide information on ROI and other Web site success measures.

To help analyze the data contained in the log files, many companies such as WebTrends, netGenesis, Engage, Inc., and Accrue Software provide log analysis software that pulls the data from the server log files and reports it in an easy-to-understand format. E-businesses can also outsource their log analysis to companies such as WebSideStory and Primary Knowledge, Inc. (Figure 10-16). Some Web marketers provide log analysis as a service for clients using their ad network.

TIP

Some observers think that the conversion rate is the most powerful measurement of a Web site's performance. Not only does a high conversion rate increase sales and decrease the percentage of marketing costs to sales, the conversion rate may also reflect a Web site's nonquantitative aspects, such as usability and convenience.

Figure 10-16
Primary
Knowledge,
Inc.

Viewer tracking

It is possible not only to identify what parts of the Web site a viewer is accessing, but also to figure out how he or she traveled through a Web site. The path that a viewer takes through a Web site is often referred to as a **click trail**. A click trail can be identified by organizing information on the pages accessed from a specific IP address in sequential order.

However, a problem with analyzing a click trail is that an IP address may be from a firewall or proxy server that corresponds to many viewers on a network behind the firewall or server. To counter this, many Web sites use cookies to track an individual viewer as he or she travels through a Web site. A **cookie** is a message passed to a Web browser by a Web server. The Web browser stores the message in a text file on the viewer's hard drive. The message is then sent back to the Web server each time the Web browser requests a Web page from the Web server. Even if multiple viewers access the same site through the same proxy server, each viewer has a unique cookie, and therefore a unique session. This also helps an e-business to better determine the number of unique users. It is even possible to use cookies to track users across multiple sessions, when they return to the site later on.

Viewer-tracking software is available that determines if a visitor is returning or is visiting the site for the first time. This software can also track visitors as they click through to a Web site, and can identify the source of the click through clicking a banner ad, following a link from another site, responding to a special offer in an e-mail, or keying in a unique URL. The software then identifies a visitor when he or she returns to visit the Web site. In addition to tracking a visitor through an individual session, it is also important to track the frequency with which the visitor returns to the site. This frequency is a measure of a Web site's "stickiness." **Sticky Web sites** are Web sites whose content keeps visitors coming back again and again without relying on advertising to bring them back.

In addition to identifying the Web page where a viewer enters a Web site, it is also important to identify the last page he or she looks at before exiting the site. This information can tell an e-business whether a visitor is finding the information he or she expects to find at the Web site, or if a visitor gets frustrated and leaves before finding useful information. For example, if a large number of visitors exit a Web site after looking at a particular page, an e-business can infer that the visitors either found all of the information that they needed on that page or that there is a problem with the page that discourages visitors and prompts them to go elsewhere.

Another important piece of visitor data an e-business should examine is the keywords viewers use to search within the Web site. If the e-business has a search form on its Web site, a visitor can use the search form to find information or products by entering keywords that correspond to the desired information or products. Analyzing these keywords provides valuable information about a visitor's interest at the Web site. Sometimes a visitor is looking for something that is not available at the site; however, if enough visitors are searching for the same information or product, an e-business may want to add that product or information to its Web site.

E-CASE Logging Web Site Results

Boise Cascade Office Products Corporation, in Itasca, Illinois, is one of the largest B2B distributors of office products. It sells everything from pens and paper clips to office furniture and computer supplies, offering over 11,000 products in its online catalog.

Boise uses a log file analysis tool, WebTrends, to generate many detailed reports on Web site metrics. The IT staff uses these reports to evaluate network bandwidth. The marketing department uses the reports to review activity from promotions, to determine how visitors are entering the Web site, and to identify which customers are visiting the site most often.

One example of how Boise uses its analysis of Web site metrics is to determine where to place links to specific pages at the Web site. Boise offers a service called The WaterCooler that sends e-mail notices to customers about new products, promotions, and events. By analyzing where visitors first enter the Web site, Boise realized that 30 percent of their Web site visitors were entering the site through the Internet ordering login page. They then added a WaterCooler link to that page to help increase the number of visitors who signed up for WaterCooler service.

Boise also uses Web metrics to determine customer preferences. Boise noticed that during a particular marketing campaign, more visitors clicked on a "find the hidden office product" link than on the "view the sample catalog" link. This told them that their visitors wanted to be entertained. Because of this, Boise is incorporating more interactive content on their site.

Finally, Boise used Web metrics to better configure their Web site for their customers. Boise created a screen resolution test that displayed different images, depending on the customers' screen width. By examining the log file reports, they were able to determine exactly what percentage of their customers used the different resolutions, and they could then design the site according to their customers' preferences. Before analyzing the data, Boise (Figure 10-17) had estimated viewers' screen resolution preferences on the basis of averages reported for all Web users, but as it turned out, that was not an accurate picture of their customer base.

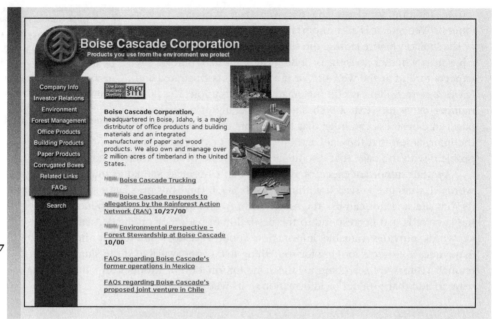

Figure 10-17

Boise
Cascade
Office
Products

Another way to measure Web site performance is with visitor traffic reports.

Web site traffic audit providers

Even though the e-business has the capability to generate its own traffic reports, it may be necessary at times to have a third party validate the traffic reports. If an e-business sells advertising, its advertisers will likely want an audit of traffic information by an independent third party to verify that the number of expected viewers are seeing the ads. Companies such as Media Metrix and Nielsen//NetRatings survey panels of viewers to determine how many times a viewer visits a specific Web site. Although this is fairly accurate for Web sites that have a large number of visitors, the samples are not large enough to accurately estimate traffic to smaller Web sites. Other companies such as ABC Interactive (Figure 10-18) independently analyze an e-business's Web server log files to determine the number of unique visitors.

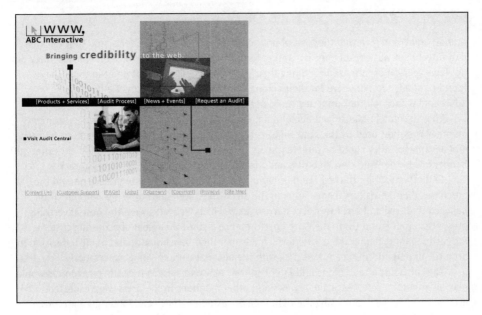

Figure 10-18
ABC
Interactive

Investors, as well as advertisers, may want to see an audit of an e-business's Web site traffic data. Whether the e-business is privately funded or is a public company, investors will want to know that it is reporting accurate visitor data.

Although objective measures of Web site performance are important, it is also important to remember that there is no one metric that satisfies every e-business. Also, Web metrics continue to evolve—from hits to page views to click-throughs to unique visitors and ROI. The challenge for an e-business is to sort through the mountains of available data and decide on meaningful Web metrics for its particular situation.

The final challenge for an e-business, of course, is to use those meaningful Web site metrics to evaluate its Web site performance and then to implement appropriate changes.

E-CASE IN PROGRESS **foodlocker.com**

foodlocker.com analyzes data from its log files on a daily basis to make sure they spot any potential problems before they become too severe. They examine the number of visitors, orders, average order size, and conversion rate at least every week to make sure they are growing their business at the projected rate. Also, since the business is a new business, they are trying to determine if there are any patterns in the traffic or sales, such as an increase at the end of the month, that could be used to help target marketing and promotion efforts.

Granoff and Olson reviewed their list of options and made some hard decisions. They decided that direct mail was not an effective option for a new e-business like Wine.com because they did not have years of experience in refining their mailing lists. Also, at that point in time, they found print ads too expensive for their company and too difficult to target effectively. Additionally, they felt that print ads could not be tracked effectively.

They decided to take advantage of online advertising, particularly banner advertisements on other Web sites used by the same viewers that Wine.com was targeting. Banner ads had several advantages: they could be targeted to an appropriate audience, viewers could click through banner ads to the Wine.com Web site, and the ads' performance was very easy to track.

Granoff and Olson tracked the number of visitors who clicked through each ad and the purchases that the visitors made. Some of the ads worked well and some did not, but Olson believed that trial and error was necessary to learn which Web sites were the best advertising prospects. Granoff and Olson measured performance against their goal of acquiring customers for each separate banner ad, decided which ads met their benchmarks, and made a decision to drop the underperforming ads and stick with the ads that were meeting expectations.

They also suspected, and would later confirm, that referrals from existing customers would play an important role. Magazine reviews and other mentions in the press were expected to be even more effective in attracting the right type of customer. They would later go on to place print advertisements. Visitors and sales proceeded to nearly double every year.

It was not until 1999, after it raised $30 million in venture capital funding, that the company was able to launch its first major marketing campaign. They launched a national radio and television campaign that September, in anticipation of the upcoming holiday season. They were also finally in a position where they felt ready to deliver on the promise that they make in their advertising, after spending a lot of time over the previous year getting their distribution system into place. They also had enough experience in targeting their customers that they were able to move into direct-mail advertising. For the 1999 holidays, Wine.com sent out a 16-page holiday catalog to subscribers of upscale lifestyle magazines. Their 1999 holiday marketing campaign was too successful, in fact, and Wine.com had trouble keeping up with holiday orders.

Granoff and Olson realized that some marketing and advertising tools that were not right for them at Wine.com's launch became effective tools when the e-business had grown and its systems were in place to handle the increased sales that marketing and advertising could deliver.

Summary

- Before making a new e-business Web site available to the public, it may be a good idea to launch it in beta mode and have it tested by viewers who can provide critical feedback on the Web site's content and usability.
- During beta launch testing, an e-business should begin marketing its Web site.
- Web site marketing techniques include submitting the Web site's URL and other information to major search engines and directories.
- Using meta tags and other Web site design elements can enhance the possibility that an e-business's Web site will appear high in a search hit list.
- An effective public relations program can help establish and maintain an e-business's public image.
- An e-business should consider participating in industry newsgroups or Internet special interest groups as an indirect way to market its products and services.

- Affiliate programs are a low-cost way to drive potential customers to an e-business's Web site.
- Other low- or no-cost promotion techniques available to an e-business include exchanging links with other e-businesses, becoming part of a Web ring, and creating good word of mouth about its products and services.

- An e-business may choose to purchase advertising in various traditional media and at other Web sites, using banner ads.
- Opt-in e-mail advertising can also be an effective tool for an e-business.
- An e-business must ultimately measure its overall financial performance and its Web site's operating performance.

CHECKLIST — Marketing and Advertising Your E-Business

- ❏ Have you submitted your Web site URL to all the major and relevant minor search engines and directories?
- ❏ Have you added relevant meta tag keywords to the HTML coding in your Web pages?
- ❏ Are your Web page titles brief and descriptive?
- ❏ Have you arranged a link exchange with complementary Web sites?
- ❏ Do you have a PR professional on staff, or have you hired a PR professional to prepare your press releases and other important public communications?
- ❏ Are you answering your customer e-mails in a timely manner?
- ❏ Do you participate in relevant industry newsgroups or other special interest groups?
- ❏ Have you checked out the possibility of participating in an industry Web ring?
- ❏ Are you using banner ads, and can you measure their effectiveness?
- ❏ Are you using relevant targeted e-mail advertising?
- ❏ Do you regularly review your Web server's log file or have an outside firm prepare a log file analysis on a regular basis?
- ❏ Have you set benchmarks for your Web site, and are they being met?
- ❏ Do you regularly review search results hit lists using keywords relevant to your e-business?

Key Terms

affiliate program	directory	referring URL
benchmark	hit	ROI
beta launch	impressions	search engine
beta test	log files	spiders
browser type	meta search engine	sticky Web site
click-through rate	meta tag	visitors
click trail	page views	Web ring
conversion rate	page views per visitor	
cookie	portal	

Review Questions

1. Which of the following depends on user-submitted information?

   a. Search engine
   b. Directory
   c. Beta launch
   d. Affiliate program

2. A benchmark is:

   a. A circle of related Web sites.
   b. ROI.
   c. A performance-based objective.
   d. Small segments of HTML code.

3. The click-through rate is the percentage of viewers who:

   a. Click on a banner ad.
   b. Ignore a banner ad.
   c. Search for a banner ad.
   d. Save a banner ad.

4. Traditional advertising media do not include:

   a. Print media.
   b. Outdoor signs.
   c. Radio and television.
   d. Banner ads.

5. Which of the following actions will likely not improve search hit list positioning?

   a. Clear and descriptive Web page text
   b. Using frames
   c. Using static HTML pages
   d. Using inbound links from other Web sites

6. An Internet newsgroup is a perfect place for an e-business to directly advertise its products and services. **True or False?**

7. An e-business's Web site operations will always hit its benchmarks. **True or False?**

8. Web site ROI should only be determined by marketing efforts that are measurable. **True or False?**

9. Sticky Web sites keep viewers coming back again and again. **True or False?**

10. It is a good idea for an e-business to periodically review search engine results, using its predefined keywords and meta tag keywords, to see how high its URL is returned in a hit list. **True or False?**

Exercises

1. Using the SearchEngineWatch.com Web site, the Spider's Apprentice Web site, or other relevant Web sites, research the similarities and differences in how these search tools index Web pages or catalog Web sites: HotBot, Alta Vista, LookSmart, Northern Light, Yahoo!, and Magellan Internet Guide. Then write a one-page paper briefly describing each site, identifying whether the site is a search engine, a directory, or a hybrid.

2. Using Internet search tools or other relevant sources, identify two e-businesses that provide Web site tracking services. Then write a one-page paper describing the e-businesses and comparing the services they offer.

3. Using Internet search tools or other relevant sources, identify three e-business Web rings. Then write a one- or two-page paper describing each Web ring and its participants.

4. Using Internet search tools and your Web browser features, review the meta tags on the home page of three B2C Web sites that sell similar products. Then create a list of at least 10–15 of the meta tag keywords being used.

5. Using Internet search tools or other relevant sources, locate three e-business-related newsgroups or Internet special interest groups. List the groups and any instructions provided on how to participate in the groups.

CASE PROJECTS

◆ 1 ◆

You are getting ready to submit your new C2C Web site to several directories. Create a name and brief description of your e-business. Then prepare a 25-word description and a list of 15 keywords you plan to submit to the directories.

◆ 2 ◆

You and your partners have just reviewed and approved a prototype of your new e-business Web site. Your partners are eager to launch the Web site for the general public. However, you think that first doing a beta launch test would be a better idea. You are meeting with your partners and board of advisors tomorrow to discuss the Web site launch. Using Internet search tools or other relevant sources, research the advantages of doing a beta launch test. Then write a one-page outline to use in tomorrow's meeting to support your argument.

◆ 3 ◆

Your new e-business B2C Web site is not attracting the number of visitors you would like, and you decide that exchanging links with complementary Web sites will help. First, create a name and brief description of your e-business. Then, using Internet search tools or other relevant sources, identify five Web sites that are complementary to your e-business. Write a one- or two-page paper describing each complementary Web site and the reasons for your choices.

Useful Links

adsGuide
http://www.ad-guide.com/

Advertising Secrets – Advertising Associations
http://www.advertisingsecrets.com/ad_associations.html

Advertising World – University of Texas at Austin
http://advertising.utexas.edu/world/

American Marketing Association
http://www.ama.org/

Analog
http://www.statslab.cam.ac.uk/~sret1/analog/

Audit Bureau of Circulations
http://www.accessabc.com/

Barkley's Comprehensive Technology Dictionary
http://www.oasismanagement.com/frames/
    TECHNOLOGY/GLOSSARY/
    index.html?TECHNOLOGY/GLOSSARY/index.html

BBB*OnLine*
http://www.bbbonline.org/

BtoB
http://netb2b.com/

Business 2.0
http://www.business2.com/

CNET News
http://www.news.com/

ClickTrade
http://www.clicktrade.com/

ClickZ
http://www.clickz.com/

Dan Janal – Internet Marketing
http://www.danjanal.com/

E-Commerce – ZDNet Best Practices
http://www.zdnet.com/enterprise/
    e-business/bphome/

eFuse
http://www.efuse.com/

eLab – Vanderbilt University
http://www.elabweb.com/

FAST Web site
http://www.fastinfo.org/measurement/pages/
    index.cgi/audiencemeasurement

Hunting With Spiders, Indexes, and Search Engines
http://www.cs.unca.edu/~davidson/hunt.html

IAB – Tools of the Trade – List of Links
http://www.iab.net/tools/content/toolcontent.html

ICONOCAST Inc.
http://www.iconocast.com/

Marketing and PR Sources On-line
http://www.online-pr.com/markpr.htm

Promo Magazine* – Promotional Marketing
http://www.promomagazine.com/

PRWeek* – U.S. Edition
http://www.prweekus.com/us/index.htm

Search Engine Forums
http://searchengineforums.com/bin/Ultimate.cgi

Searchengines.com
http://www.searchengines.com/

SmartAd – Internet Advertising and Marketing Resources
http://am.net/ad/

Spider Food
http://spider-food.net/

StatMarket
http://www.statmarket.com/

Submit It! – Tips For Announcing Web Sites To Search Engines
http://www.submit-it.com/subopt.htm

USC – Digital Commerce Center
http://www.ec2.edu/dccenter/index.html

Web Advertising Resources – Builder.com
http://builder.cnet.com/Business/Advertising/ss07.html

Web Marketing Info Center
http://www.wilsonweb.com/webmarket/

Websurfer Marketing/Advertising Directory
http://www.gowebsurfer.com/directory/business/
  Marketing/

Yahoo! GeoCities Internet Banner Services
http://www.geocities.yahoo.com/addons/interact/
  mbe.html

Zenith Media – Marketer's Portal
http://www.zenithmedia.com/map00.htm

Links to Web Sites Noted in This Chapter

ABC Interactive
http://www.abcinteractiveaudits.com/

Accrue Software
http://www.accrue.com/

AltaVista
http://www.altavista.com/

Amazon.com
http://www.amazon.com/

America Online
http://www.aol.com/

Boise Cascade Office Products Corporation
http://www.bcop.com/

Boston Consulting Group
http://www.bcg.com/

ClickTrade
http://www.clicktrade.com/

Dogpile
http://www.dogpile.com/

Engage, Inc.
http://www.engagetech.com/

eToys
http://www.etoys.com/

Forrester Research
http://www.forrester.com/

Google
http://www.google.com/

GoTo.com
http://www.goto.com/

HotBot
http://hotbot.lycos.com/

LinkExchange
http://adnetwork.bcentral.com/

LookSmart
http://www.looksmart.com/

Magellan Internet Guide
http://magellan.excite.com/

Media Metrix
http://www.mediametrix.com/

MetaCrawler
http://www.metacrawler.com/

Napster
http://www.napster.com

NetCreations
http://www.netcreations.com/

NetGenesis
http://www.netgenesis.com/

Newsgroup Netiquette
http://www.vonl.com/vtab24/news102.htm

Nielsen//NetRatings
http://www.nielsen-netratings.com/

Northern Light
http://www.northernlight.com/

Open Directory Project
http://directory.mozilla.org/

PC Data Online
http://www.pcdataonline.com/

Primary Knowledge, Inc.
http://www.primaryknowledge.com/

Search Engine Watch
http://www.searchenginewatch.com/

Shop.org
http://www.shop.org/

Spider's Apprentice
http://www.monash.com/spidap.html

SubmitIt!
http://www.submitit.com/

WebSideStory
http://www.websidestory.com/

WebTrends
http://www.webtrends.com/

WebTV
http://www.webtv.com/

Yahoo!
http://www.yahoo.com/

Yahoo! WebRing
http://dir.webring.yahoo.com/rw

yesmail.com
http://www.yesmail.com/

For Additional Review

Ali, Jaffer. 2000. "Market Meltdown: Ad Dollars Hang in the Balance," *ChannelSeven.com*, April 24. Available online at: http://www.channelseven.com/adinsight/commentary/2000comm/comm20000424.shtml.

Allen, Cliff. 2000. "Leaving Tracks," *ClickZ*, July 4. Available online at: http://www.clickz.com/cgi-bin/gt/article.html?article=1989.

Allen, Cliff. 2000. "Tracking Traffic," *ClickZ*, June 27. Available online at: http://www.clickz.com/cgi-bin/gt/article.html?article=1958.

Berry, John. 1999. "The World According to E-Biz Metrics," *InternetWeek*, October 4, 38.

Berry, Lyn. 2000. "Online Marketing Evolves Away from Banner Ads," *Denver Business Journal*, 51(43), June 9, 8A.

Black, Jason. 2000. "Old Ads, New Metric," *Internet World*, 16(5), August, 28.

Blankenhorn, Dana. 1999. "Software, Services Determine Online ROI," *BtoB*, August 27. Available online at: http://www.btobonline.com/cgi-bin/article.pl?id=1787.

Brown, Kalisha. 2000. "Simplicity Makes Web Sites 'Sticker,'" *Denver Business Journal*, 51(45), 29A.

Bruemmer, Paul J. 2000. "Looking for the Best ROI," *ClickZ.com*, June 28. Available online at: http://www.clickz.com/cgi-bin/gt/print.html?article=1956.

Bulik, Beth Snyder. 2000. "Most Visible Players: Celebrity Endorsements of Dot-Coms Haven't Exactly Ensured Their Success. But Three Sports Legends Could Work That Magic for MVP.com," *Business 2.0*, August 22. Available online at: http://www.business2.com/content/magazine/marketing/2000/08/22/17645.

Business 2.0. 2000. "Counting Heads," September 12. Available online at: http://www.business2.com/content/magazine/filter/2000/08/22/17250.

Business 2.0. 2000. "Web Ad Spending at $21 Billion in 2004," June 22. Available online at: http://www.business2.com/content/research/numbers/2000/06/22/13068.

Carpenter, Phil. 2000. *eBrands: Building an Internet Business at Breakneck Speed*. Boston, MA: Harvard Business School Press.

Cross, Kim. 1999. "Whither the Banner: Two Online Marketeers Wrangle Over One of the Web's Vexing Questions," *Business 2.0*, December 1. Available online at: http://www.business2.com/content/magazine/marketing/1999/12/01/20494.

Deise, Martin V. et al. 2000. *Executive's Guide to E-Business: From Tactics to Strategy*. New York, NY: John Wiley & Sons, Inc.

Desmond, John. 2000. "Passing the SECOND TEST," *Software Magazine*, 20(1), February, 34.

Dorf, Bob. 1999. "Survey Says: Most E-Business ROI Not Measured," *Inside1to1*, Pepper & Rogers Group, July 22. Available online at: http://www.1to1.com/articles/i1-072299/index.html.

Edwards, Len. 2000. "I've Gone Through a Learning Curve That's Staggering: Len Edwards of Avon Products Discusses Marketing Via Web Site," *Business Week*, 3699, 148.

Emert, Carol. 2000. " Catalogs Thrive in an Online Age: Mail-order Merchants Prosper Despite Competition from E-tailers," *San Francisco Chronicle*, January 3. Available online at http://www.sfgate.com/cgi-bin/article.cgi?file=/chronicle/archive/2000/01/03/BU101061.DTL.

Fischler, Michael. 2000. "Referrer Madness," *ClickZ*, July 10. Available online at: http://www.clickz.com/cgi-bin/gt/article.html?article=2015.

Fleishman, Glenn. 2000. "Love Those Links," *Business 2.0*, October 16. Available online at: http://www.business2.com/content/magazine/marketing/2000/10/16/20798.

Flory, Joyce. 2000. "Debunking the Myths of Web Site Promotion," *Marketing Health Services*, 20(2), Summer, 31.

Forbes.com. 2000. "Lyin' Eyeballs," August 7. Available online at: http://www.forbes.com/forbes/2000/0807/6604118a_print.html.

Gantz, John. 2000. "Get the Most Out of Web Site Evaluations," *Computerworld*, September 4, 32.

Gerace, Thomas and Klein, Lisa. 1996. "Virtual Vinyards," *Harvard Business School*, April 8.

Gurley, J. William. 2000. "The Most Powerful Internet Metric of All," *CNET:News*, February 21. Available online at: http://www.news.com/Perspectives/Column/0%2C176%2C403%2C00.html.

Haar, Steven Vonder. 1999. "Web Metrics: Go Figure.Too Many Yardsticks, Too Little Time. Start Making Sense," *Business 2.0*, June 1. Available online at: http://www.business2.com/content/magazine/marketing/1999/06/01/19674?template=article_pf.wm.

Hallford, Joshua. 2000. "E-Retailers Pull Back on TV Ads," *The Standard*, August 30. Available online at: http://www.thestandard.com/article/display/0,1151,18107,00.html.

Hartman, Amir et al. 2000. *Net Ready: Strategies for Success in the E-conomy*. New York, NY: McGraw-Hill.

Hawley, Noah. 2000. "Brand Defined," *Business 2.0*, June 13. Available online at: http://www.business2.com/content/channels/marketing/2000/06/13/12401.

Haylock, Christina Ford and Muscarella, Len. 1999. *NET SUCCESS: 24 Leaders in Web Commerce Show You How to Put the Internet to Work for Your Business*. Holbrook, MA: Adams Media Corporation.

Heidt, Jeremy. 2000. "Web-Site Tracking Standards Are Garnering Acceptance," *Sacramento Business Journal*, 17(14), June 16, 29.

Hodges, Jane. 1999. "In Search of the Perfect Buzz: Looking to Invest in Quality Public Relations? Stand in Line. And You May Have to Pay in Stock," *Business 2.0*, December 1. Available online at: http://www.business2.com/content/magazine/marketing/1999/12/01/20495.

Hodges, Jane. 2000. "Revenge of the Original Spammers: Dot-coms Used to Think Direct Marketing Was Beneath Them. Now It's All the Rage," *Business 2.0*, August 8. Available online at: http://www.business2.com/content/channels/marketing/2000/08/08/17941.

Hoy, Richard. 2000. " A Traffic Analysis Solutions for Small Business: Part 1," *ClickZ*, June 9. Available online at: http://www.clickz.com/cgi-bin/gt/article.html?article=1858.

Hoy, Richard. 2000. " A Traffic Analysis Solution for Small Business: Part 2," *ClickZ*, June 16. Available online at: http://www.clickz.com/cgi-bin/gt/article.html?article=1899.

Janal, Dan. "Improve Your Web Site's Marketing Effectiveness with Log Files," *eFuse*. Available online at: http:// www.efuse.com/Grow/marketing-logs.html.

Janal, Dan. 1998. *Online Marketing Handbook*. New York, New York: Van Nostrand Reinhold.

Junnarkar, Sandeep. 1999. "Virtual Vineyards Harvests Venture Cash," *CNET News.com*. June 18. Available online at http://news.cnet.com/news/0-1007-200-343827.html?st.ne.180.gif.1.

Kalakota, Ravi. 1999. *e-Business: Roadmap for Success*. Reading, MA: Addison Wesley Longman, Inc.

Keynote Systems. 2000. "Customer Success Stories," Available online at: http://www.keynote.com/company/customers/customers.html.

Kotler, Philip. 1997. *Marketing Management: Analysis, Planning, Implementation, and Control*. Upper Saddle River, New Jersey: Prentice Hall.

Kuchinskas, Susan. 2000. "Battered Banner, Breathe On," *Business 2.0*, July 11. Available online at: http://www.business2.com/content/channels/marketing/2000/07/11/14112.

Ladley, Eric. 2000. "Waving Banner Ads Goodbye," *ISP Business News*, 6(25), June 19.

Lee, Hane C. 2000. "Napster Won't Remain the Same," *The Standard*, November 6. Available online at: http://www.thestandard.com/article/display/0,1151,19934,00.html.

Mann, Charles C. 2000. "Promises, Promises — Can Napster Really Remain the Same?" *The Standard*, November 2. Available online at: http://www.thestandard.com/article/display/0,1151,19915,00.html.

Materna, Jessica. 2000. "Glitzy Banner Ads Droop on Internet Sideline," *San Francisco Business Times*, 15(8), October 6, 25.

Moro, Wendy. 2000. "5 Steps to Building Buzz," *Business 2.0*, March, 182.

Mottl, Judith N. 2000. "E-Retail Customer Service: It's More Than Just E-Mail — Online Stores Look for New Ways to Get and Stay Closer to Their Customers," *InformationWeek*, September 25, 180.

Nickell, Joe Ashbrook. 1999. "Marketing Makeover: As Banner ROI Slips into Oblivion, Rich-Media Ads Show New Signs of Life," *Business 2.0*, October 1. Available online at: http://www.business2.com/content/magazine/marketing/1999/10/01/16522.

Page, Larry and Brin, Sergey. 2000. "Merits of a Beta Launch," *Business 2.0*, March, 180.

Regan, Keith. 2000. "Lies, Damned Lies, and Unique Visitors," *E-Commerce Times*, June 21. Available online at: http://www.ecommercetimes.com/news/viewpoint2000/view-000621-1.shtml.

Rifkin, Glen and Lambert, Ken. 2000. "Marketing Your Startup," *Business 2.0*, March, 181.

Rosen, Anita. 2000. *The E-Commerce Question and Answer Book: A Survival Guide for Business Managers*. New York, NY: American Management Association (AMACOM).

Ross, Sid. 2000. "Declining Clickthrough Rates on Banner Ads Tell Only Part of Story," *ADWEEK Eastern Edition*, 152(6), September 11, 88.

Sales & Marketing Management. 2000. "The Incredible, Indelible Banner," June, 38.

Sandoval, Greg. 2000. "E-commerce Firms Step Back from Early Strategies," *CNET News.com*, March 21. Available online at http://home.cnet.com/category/0-1007-200-1580087.html.

Schibsted, Evantheia and Jeffers, Michelle. 2000. "What's New and Now in Ecommerce and Ebusiness Marketing and Advertising," *Business 2.0*, August 8. Available online at: http://www.business2.com/content/magazine/marketing/2000/08/08/15442.

Seckler, Valerie. 2000. "First Impressions Prompt a Swift ELuxury Makeover," *Women's Wear Daily*, October 2, 1.

Senyak, Zhenya Gene. 1999. "All About Promotion," *Home Office Computing*, 17(10), October, 106.

Stein, Lincoln D. 1999. "How Shall I Measure Thee? Let Me Count the Ways," *Web Techniques*, 4(7), July, 16.

Strother, Neil. 2000. "Are Surfers Stuck on Your Site?" *ZDNet AnchorDesk*, March 21. Available online at: http://www.zdnet.com/anchordesk/story/story_4589.html.

Sweeney, Terry. 2000. "Advertisers Seek More Bang for Their Web Bucks — Agencies Develop Sophisticated Ways to Gauge an Ad's Success at Brand Building, Driving Sales," *InformationWeek*, October 2, 130.

Tchong, Michael. 2000. "Dotcom Marketing Works," *Iconocast*, July 13.

Thompson, Maryann Jones. 1999. "The Measures of Web Success," *The Standard*, February 15. Available online at: http://www.thestandard.com/article/display/0,1151,3501,00.html.

Thylmann, Oliver. 2000. "Understanding Web Site Statistics," *ClickZ*, August 10. Available online at: http://www.clickz.com/cgi-bin/gt/print.html?article=2204.

Thylmann, Oliver. 2000. "Understanding Web Site Statistics: Part 2," *ClickZ*, August 16. Available online at: http: //www.clickz.com/cgi-bin/gt/article.html?article-2233.

Tillett, L. Scott. 2000. "An Alternative to Banner Ads," *InternetWeek*, March 27, 21.

Timmers, Paul. 1999. *Electronic Commerce: Strategies and Models for Business-to-Business Trading*. New York, NY: John Wiley & Sons, Inc.

Treese, G. Winfield and Stewart, Lawrence C. 1998. *Designing Systems for Internet Commerce*. Reading, Massachusetts: Addison Wesley Longman, Inc.

Tweney, Dylan. 1999. "Net Prophet," *InfoWorld*, 21(2), January 11, 58(1).

Ward, Eric. 1999. "NETSENSE: On Web, ROI Can't Always Be About Money," *BtoB*, August 27. Available online at: http://www.btobonline.com/cgi-bin/article.pl?id=1777.

Wieder, Tamara. 2000. "E-Commerce Benchmarking," *Computerworld*, August 7, 58.

Wilson, Tim. 1999. "Service Metrics Pinpoints Web Site Performance Problems," *InternetWeek*, July 1. Available online at: http://www.internetwk.com/story/INW19990701S0003.

Wonnacott, Laura. 2000. "SITE SAVVY: Despite the Doom and Gloom of Naysayers, Banner Ads Are Thriving," *InfroWorld*, 22(33), August 14, 68.

Yerxa, Gregory. 2000. "TESTORS MAKE THE GRADE – We Let Five Web-site Performance-Monitoring Services Test Some of Our Favorite Sites," *InformationWeek*, May 1, 121.

Creating a Web Site with Microsoft FrontPage

In this appendix, you will learn to:

Start the FrontPage application

Describe the FrontPage window components

Describe FrontPage views

Use a FrontPage wizard or template to create a Web site

FrontPage Overview

A Web site consists of related Web pages that are linked together with hyperlinks. FrontPage organizes a Web site's hyperlinked Web pages into units called FrontPage Webs. Each FrontPage Web is stored in its own folder on your hard drive or on a server. Usually, hyperlinked Web pages covering the same content are part of one FrontPage Web. Also, groups of pages administered by the same person may be included in the same FrontPage Web.

Starting FrontPage

Before you can begin to work with FrontPage, you need to open the application. When you open the application, a new, blank Web page opens as well. To open the FrontPage application and a new, blank Web page:

1. Click the **Start** button on the taskbar.
2. Point to **Programs**.
3. Click **Microsoft FrontPage**.

The FrontPage application opens with a blank Web page. Your screen should look similar to Figure A-1, which identifies the specific components of the FrontPage application window.

Figure A-1
FrontPage
application
window

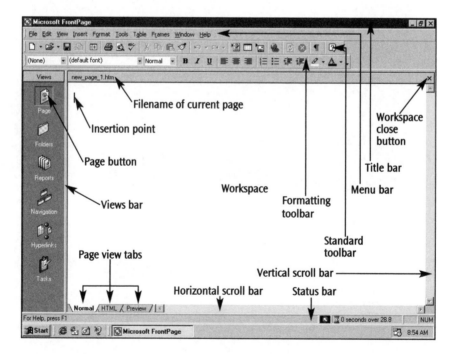

FrontPage 2000 features personalized menus and toolbars, which "learn" the commands you use most often. When you first install FrontPage, only the most frequently used commands appear immediately on a short version of the menus, and the remaining commands appear after a brief pause. Commands that you select move to the short menu, while those you don't use appear only on the full menu. The Standard and Formatting toolbars may appear on the same row in FrontPage. In this position, only the most commonly used buttons of each toolbar are visible. All the other default buttons appear on the More Buttons drop-down lists. As you use buttons from the More Buttons drop-down list, they move to the visible buttons on the toolbar, while the buttons you don't use move into the More Buttons drop-down list. If you arrange the Formatting toolbar below the Standard toolbar, all buttons are visible. Unless otherwise noted, the illustrations in this appendix show the full menus and the Formatting toolbar below the Standard toolbar.

Menu bar

The menu bar, located below the title bar, has ten drop-down menu commands that contain groups of additional, related commands. For example, the File menu contains commands for opening, closing, previewing, and printing Web pages. You can use the mouse or the keyboard to select a command from the menu bar. The activities in this appendix instruct you to select menu bar commands with the mouse.

Standard toolbar

The Standard toolbar is located under the menu bar and is made up of buttons that represent commonly used commands. For example, the Standard toolbar contains buttons for opening, saving, previewing, and printing a Web page. The Standard toolbar allows you to perform commands quickly by clicking the button that represents that command. You can customize the Standard toolbar (or any other toolbar) by adding or deleting buttons.

Formatting toolbar

The Formatting toolbar is located under the Standard toolbar in Figure A-1 and is made up of buttons that represent commonly used formats. With buttons on the Formatting toolbar, you can modify Web page text appearance, for example, by changing the font or text alignment.

Views bar

The Views bar, located under the Formatting toolbar on the left side of the window, provides shortcuts you can use to view FrontPage Webs. You use Page view to create or edit a Web page. You organize Web site files and folders in Folders view. Reports view provides a way to analyze and manage the content of a Web site. Navigation view provides tools you use to design your Web site structure. You can view and edit hyperlinks in Hyperlinks view. Finally, you can create and edit an electronic "To Do" list for the Web site in Tasks view.

Workspace

The large area to the right of the Views bar is called the workspace. You key heading text, bulleted and numbered lists, and pictures, and insert other Web page components in the workspace as you create or edit a Web page. You also view the Web structure, hyperlinks relationships, and tasks in the workspace.

Insertion point

The blinking vertical bar in the upper-left corner of the workspace is the insertion point. The insertion point marks the location where text is entered in a Web page.

Scroll bars

Scrolling changes the view of the current Web page in the workspace. The vertical scroll bar appears at the right side of the workspace and scrolls the current Web page up and down. The horizontal scroll bar appears below the workspace and scrolls the current Web page left and right.

TIP

When you install FrontPage, a folder called "My Webs" is installed on the C:\ drive in the My Documents folder. FrontPage stores any new Webs in this folder by default. Each Web you create is stored in its own subfolder inside the My Webs folder. For the activities in this appendix, save your Web in the location specified by your instructor.

The individual Web subfolders contain two additional subfolders created by FrontPage: the _private subfolder and the images subfolder. The _private subfolder contains files FrontPage uses to manage your Web and should be left alone. The images subfolder can be used to store picture, sound, or video files included at the Web site. Moving these files into the images folder reduces the clutter in the Web subfolder.

Page view tabs

FrontPage has three ways to look at the current Web page as you create or edit it in Page view: Normal, HTML, and Preview. The Normal, HTML, and Preview tabs are located to the left of the horizontal scroll bar in Page view. The Normal tab provides a WYSIWYG (What You See Is What You Get) view. You create and edit your Web pages in the Normal tab much as you create and edit a word-processing document. FrontPage inserts the appropriate HTML code as you create or edit a Web page in the Normal tab. HTML is the abbreviation for Hypertext Markup Language, the code or tags used to create Web pages. You use the HTML tab to view or edit the actual HTML tags. The Preview tab shows how the current Web page will look in a Web browser.

Status bar

The status bar appears at the bottom of the window above the taskbar and displays messages as you are working on a Web page. For example, when you open a Web page, the status bar displays a message telling you that the page is being opened and how long it takes for the page to download when someone accesses it with a Web browser.

Using Different FrontPage Views

FrontPage provides additional ways to view and manage a Web. For example, you can view different reports about the status of a Web, view a diagram of the hyperlinks in a Web, and view a list of folders and files in the Web. You do this by switching from Page view to the appropriate view with the shortcut buttons on the Views bar.

Using Reports view

Reports view provides information about the status of a Web site. For example, you might want to know how many files are in the current Web, or if there are broken hyperlinks in the current Web. A broken hyperlink is one that no longer connects to a valid location. You can get answers to these and other Web management questions in Reports view.

Using Folders view

To easily manage the folders and files in a FrontPage Web, you must be able to see a list of all folders and files. One way to view a list of folders and files is to display the Folder List, using the Folder List button on the Standard toolbar. Another way to view a list of folders and files for a Web is in Folders view.

Using Navigation view

A FrontPage Web consists of interrelated pages that are organized in a treelike structure. The home page appears at the top of the tree, and the related pages branch out at different levels below the home page. You use Navigation view to review the Web's tree structure. You can also modify that structure in Navigation view by adding or deleting pages, or moving pages to a new position in the tree.

Using Tasks view

Creating, editing, and maintaining a FrontPage Web consists of many related tasks. To help manage a Web effectively, you can create an electronic "To Do" list of outstanding items to be completed.

When completing some actions, FrontPage can automatically add a task to the task list for you. For example, when spell checking a FrontPage Web, you can have FrontPage create a task for each page that contains misspelled words. You can also manually add tasks to the list.

Getting Help in FrontPage 2000

It is easy to get online Help when working in FrontPage. From the Help menu, you can open the Help window, convert the mouse pointer to a Help pointer, and load the Microsoft Office Web page. You can also use the Microsoft FrontPage Help button on the Standard toolbar or press the F1 key to open the Help window. You can convert the mouse pointer to a Help pointer by pressing the SHIFT + F1 keys, and then you can click a menu command or toolbar button with the Help pointer to view its ScreenTip Help.

To close FrontPage:

1. Click **File** on the menu bar.
2. Click **Exit**.

Using a Wizard or Template to Create a Web Site

A wizard is a series of dialog boxes that assists you in performing certain tasks by asking you a series of questions in a step-by-step process. A template is a model document that contains page settings, formats, and other elements. Using a wizard or template to create a new Web saves time by automatically providing features you use to organize page content.

FrontPage contains a variety of wizards and templates you can use to create Web sites. Each wizard or template has a different goal or focus, such as building a corporate presence or providing customer support.

To view the Web wizards and templates:

1. Start FrontPage.
2. Click **File**.
3. Point to **New**.
4. Click **Web**.

The Web Sites tab in the New dialog box opens. Your screen should look similar to Figure A-2.

Figure A-2
New dialog box

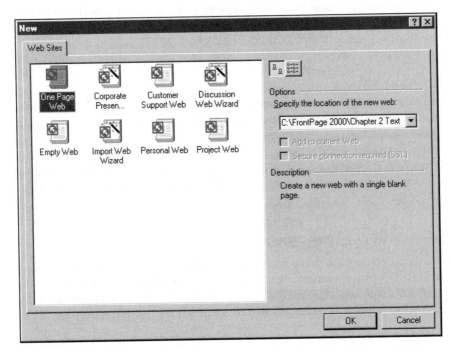

In this dialog box, you first select the desired wizard or template and then specify where the Web is stored. To use the Corporate Presence Wizard to create a sample Web:

1. Click the **Corporate Presence Web Wizard** icon to select it.
2. Key the path specified by your instructor and the Web folder name **Sample_Web** in the Specify the location of the new web: text box.
3. Click **OK**.

In a few seconds, the first Corporate Presence Web Wizard dialog box opens. Your dialog box should look similar to Figure A-3.

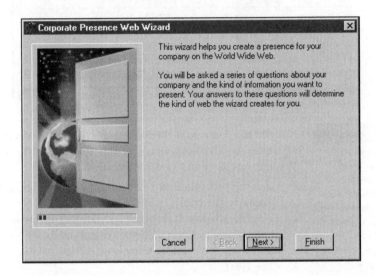

Figure A-3
First wizard step

You can cancel the wizard, go to the next step, return to the previous step, or finish the wizard process with buttons at the bottom of the dialog box. To go to the next step:

1. Click **Next**.

The second wizard dialog box opens.

You select the type of pages to include in the Web in this dialog box. A home page is required, but you can include or exclude the other listed pages by adding or removing the check mark from the check box to the left of the page name. To include the Products/Services page and the Feedback page:

1. Click the **What's New**, **Table of Contents**, and **Search Form** check boxes to remove the check marks, if necessary.
2. Click the **Products/Services** and **Feedback Form** check boxes to insert check marks, if necessary. Your dialog box should look similar to Figure A-4.
3. Click **Next**.

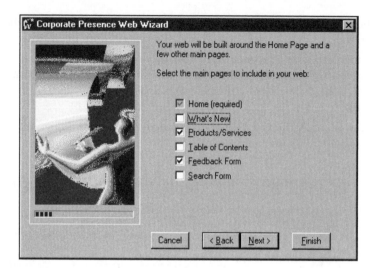

Figure A-4
Second wizard step

TIP

The wizard dialog boxes remember the selections made the last time the wizard was used. The options selected when the wizard dialog boxes open on your screen may vary from those shown in the book. Also, the number of wizard dialog boxes that open will vary, depending on the options you select as you go through the wizard process.

The third wizard dialog box opens. You select certain topics to appear on the home page in this dialog box. To select the Mission Statement and Contact Information topics:

1. Click the **Introduction** and **Company Profile** check boxes to remove the check marks, if necessary.

2. Click the **Mission Statement** and **Contact Information** check boxes to insert a check mark, if necessary. Your dialog box should look similar to Figure A-5.

3. Click **Next**.

Figure A-5
Third wizard step

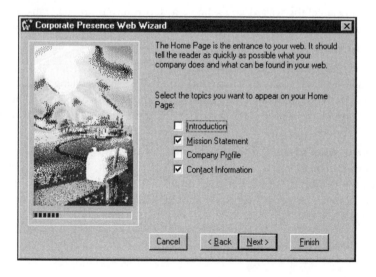

The fourth wizard dialog box opens. To create one additional Products/Services page:

1. Key **1** in the Products text box.
2. Press the **Tab** key.
3. Key **0** in the Services text box. Your dialog box should look similar to Figure A-6.
4. Click **Next**.

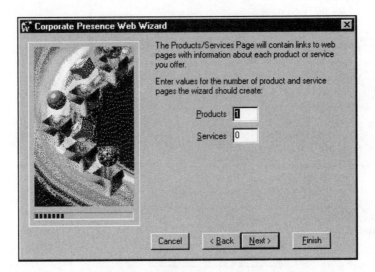

Figure A-6
Fourth wizard step

The fifth wizard dialog box opens. Your dialog box should look similar to Figure A-7.

1. Remove the check mark from each check box, if necessary.
2. Click **Next**.

Figure A-7
Fifth wizard step

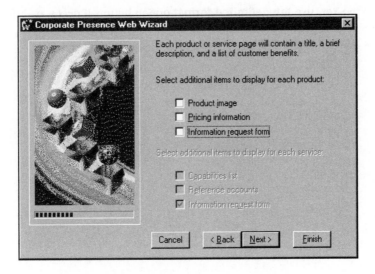

The sixth wizard dialog box opens. You now select the elements to appear on the Feedback Form page. You want to include all the elements, to capture as much information as possible about potential customers. To include all the elements:

1. Click each of the check boxes to insert a check mark, if necessary. Your screen should look similar to Figure A-8.
2. Click **Next**.

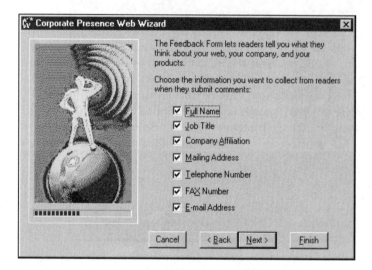

Figure A-8
Sixth wizard step

The seventh wizard dialog box opens. You want to select the format that separates individual information elements by tabs. This makes it easier to import the data into other applications such as Access and Excel. To select a data format:

1. Click the **Yes, use _tab-delimited format** option button, if necessary. Your dialog box should look similar to Figure A-9.
2. Click **Next**.

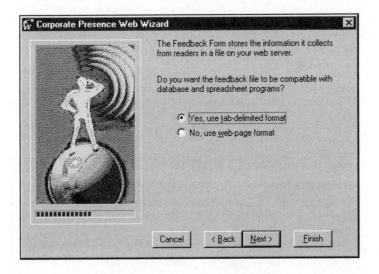

The eighth wizard dialog box opens. You can automatically add important elements to each page in this dialog box. One design objective could be to have the company logo on each page. Another design objective could be to have navigational hyperlinks to other pages at the site. Additionally, a well-designed Web page should include contact information for the Webmaster, the individual who manages a Web site; a copyright notice; and the page modification date. To add the desired elements:

1. Click the lower **Links to your main web pages** check box to remove the check mark, if necessary.
2. Click the remaining check boxes to insert a check mark, if necessary. Your dialog box should look similar to Figure A-10.
3. Click **Next**.

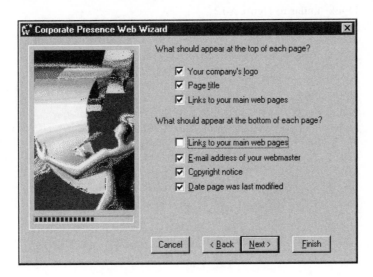

The ninth wizard dialog box opens. A published Web site is one that is uploaded and stored on a Web server. Some Web authors add incomplete Web pages to a published Web site and mark them with text and an icon indicating that the Web page is "under construction." Other Web authors think that only completed Web pages should be added to a Web site. To turn off the "under construction" icon:

1. Click the **No** option button. Your dialog box should look similar to Figure A-11.
2. Click **Next**.

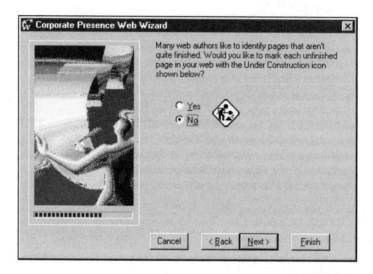

Figure A-11
Ninth wizard
step

The tenth wizard dialog box opens. You can add fictitious company name and address information in this dialog box. To add the company information:

1. Key name and address information in the appropriate text boxes. Your dialog box should look similar to Figure A-12.
2. Click **Next**.

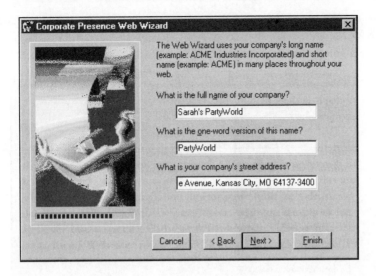

Figure A-12
Tenth wizard
step

The eleventh wizard dialog box opens. You add additional contact information in this dialog box. You can add fictitious telephone numbers and e-mail addresses. To add the information:

1. Key phone numbers and e-mail addresses in the appropriate text boxes. Your dialog box should look similar to Figure A-13.
2. Click **Next**.

Figure A-13
Eleventh
wizard step

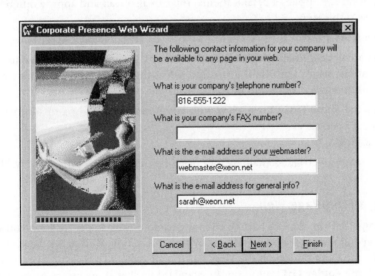

The twelfth wizard dialog box opens. Your dialog box should look similar to Figure A-14.

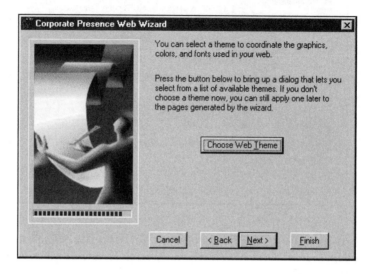

Figure A-14
Twelfth wizard
step

You can automatically apply a theme or design scheme with coordinated graphics, colors, and fonts to all the pages of the Web by clicking the Choose Web Theme button and then selecting the theme from a list of available themes. If you do not select a theme, FrontPage applies a default theme. You can also wait and apply a different theme later. To skip this step:

1. Click **Next**.

You are almost finished with the wizard steps. If you want FrontPage to automatically create items for the task list, you can select that option in the final wizard dialog box. To add the tasks items and finish the Web:

1. Click the **Show Tasks View after web is uploaded** check box to add the check mark, if necessary.
2. Click **Finish**.

The sample Web site containing four pages is complete. You can now view the pages.

Saving a FrontPage Web

The completed sample Web was saved in the Sample_Web folder location you specified in the second wizard dialog box. To verify that the sample Web pages were saved:

1. Click the **Page** button on the View bar, if necessary.
2. Click the **Folder List** button on the Standard toolbar, if necessary, to view the list of folders and files (Web pages) in the Web.
3. Observe that the Sample_Web folder contains the _private and images default folders, the home page, the feedback pages, and two products pages.

4. Double-click the **index.htm** home page file in the Folder List to open the page in the Normal tab in Page view.
5. Scroll the index.htm page and observe the elements added during the wizard process.
6. View the index.htm page in the HTML and Preview tabs.
7. Open and view the remaining pages in the Web.
8. Switch back to the index.htm page using the **Window** menu.

The pages in the Sample_Web contain partial navigation bars, sets of text or button hyperlinks to other pages in the Web, placeholders for text and graphics, and comments that indicate where you should key certain text, such as the mission statement. At this point the basic framework for a Web has been created; however, to complete this Web, the individual pages should be edited to include the appropriate text, graphics, and links. Other pages might be added and a hierarchical navigational structure determined. Using the FrontPage Wizard to create a Web site is simply a first step in the Web creation process.

After the FrontPage Web is created and edited, the next step is to publish it to a Web server that others can access to view the Web pages in their Web browser. Your instructor may provide additional information on how to publish your sample FrontPage Web to a Web server.

To close the sample Web and FrontPage:

1. Click **File**.
2. Click **Close Web**.
3. Click **File**.
4. Click **Exit**.

Glossary

3FLs Third-party fulfillment logistics (fulfillment houses that target e-businesses).

3PLs Third-party logistics (fulfillment houses).

Accredited investor An individual investor with a minimum net worth of $1 million or an individual income of at least $200,000 per year or household income of $300,000 per year.

Action plan The portion of a marketing plan that describes the actions to be taken to promote the business's products and services.

Affiliate program Marketing method where an e-business pays a referral fee or commission on sales made by customers referred from another Web site.

Angel investment club Individual angels who come together as a group to identify new investment opportunities and sometimes combine their individual investments with other club members to spread the investment risk.

Angel investor An individual with money and time who enjoys the excitement and risk of investing in the early stages of a new business.

Application service providers (ASPs) Companies that deliver and manage software applications and other computer services for multiple customers from remote data centers.

Audit An independent review of an e-business's records to verify that they are accurate and conform to accepted practice.

Authentication The process of identifying an individual or e-tailer usually based on a combination of username and password.

B2B An e-business model where the e-business sells its products and services to other businesses or brings multiple buyers and sellers together.

B2B exchange An e-business model where multiple business buyers and sellers are brought together in a central marketspace.

B2C An e-business model where the e-business sells its products and services directly to the consumer.

B2G An e-business model where the e-business sells its products and services to local, state, and federal government agencies.

Back-end systems Those processes such as accounting, inventory control, marketing, and order fulfillment not directly accessed by customers.

Backup procedure Making duplicate copies of critical system and data files to protect against hardware or software failure or a catastrophic event.

Bandwidth Data transmission rate.

Benchmarks Established reference points used to monitor a computer network's operations; performance-based objectives.

Beta launch Testing a Web site by making it available to a few visitors before marketing it to the public.

Beta test User testing of software applications before the applications are marketed to the public.

Board of advisors Individuals with valuable industry or business experience that act as advisors to a new business.

Brand A combination of name, logo, and design that identifies a business's products or services and differentiates those products and services from its competitors.

Brick-and-click Business that sells products and services from a physical location and on the Internet.

Brick-and-mortar Business conducted in a physical building.

Browser type The type of Web browser (Internet Explorer, Navigator, WebTV, and so on) a visitor uses to view pages at a Web site.

Browser-safe palette A Web browser's recognizable RGB color set.

Burn rate The rate at which a startup business is using its cash reserves.

Business description A brief explanation of a business's background and concept included in a business plan.

Business incubator Traditionally, a nonprofit organization such as a government agency or university that nurtures new businesses by providing access to investors, technology, and expertise.

C2B An e-business model where consumers name a price for a product or service and the targeted business accepts or declines that price.

C2C An e-business model where consumers sell items or services directly to other consumers.

Certificate authority A trusted third-party organization that guarantees the identity of the sender and issues digital certificates.

Charge card A rectangular piece of plastic used instead of cash or checks to pay for goods and services; card holders pay the card issuer in full each month upon receipt of a statement of charges.

Chargeback A consumer's refusal to pay a charge on her or his credit card account because of returned products, billing errors, non-delivery of product, or fraud.

Check A written order on a bank or other financial institution to pay money belonging to the signer of the check to the check's presenter.

Cipher text Data translated into a secret code before being transmitted.

Click trail The recorded path a viewer takes through a Web site.

Click-through rate The percentage of viewers who click on a banner ad to view the advertised Web site.

Coinage Stamping small round pieces of precious metals as a symbol of purity and weight.

Co-located server Space and Internet connectivity rented for servers owned by an e-business.

Commercial business incubator A for-profit business incubator.

Competitive analysis A portion of a business plan that identifies direct and indirect competitors including their name and a summary of each competitor's market strengths, weaknesses, objectives, and strategies.

Consumer characteristics Target market characteristics involving consumer purchasing and usage traits such as frequency of purchase or usage, method of usage, or frequency of purchasing similar products to those offered by a business.

Conversion rate The rate at which viewers are taking some action at a Web site such as placing an order.

Cookie A message passed to a Web browser by a Web server and stored on a computer's hard drive as a text file.

Copyright A form of legal protection for the author of published or unpublished original work.

Corporation A taxable entity owned by shareholders who are protected from liability.

Cover sheet The first page of an e-business plan that identifies the e-business and the preparer of the plan.

Cracker A malicious hacker.

Credit card A rectangular piece of plastic used instead of cash or checks to pay for goods and services; card holders pay the card issuer in full or in part upon receipt of a monthly statement.

Cryptography The art of protecting information by encrypting it.

Customer relationship management (CRM) A system that uses software and the Internet to help an enterprise manage its customer base.

Cyber cash *See* electronic cash.

Cybersquatters Individuals or businesses that illegally register domain names that are closely related to an existing business with the plan to resell the domain name to its rightful owner.

Data mining A process of analyzing data stores to discover previously unknown relationships among data.

Data spill Accidental publication at a Web site of sensitive information such as customers' names and addresses.

Data warehouse A database that contains huge amounts of data.

Debit card A rectangular piece of plastic used instead of cash or checks to pay for goods and services; cards are issued in connection to an existing bank account and charges are immediately deducted from that bank account.

Decryption (decrypted) Retuning data to its original unencrypted form called plain text.

Defacement Vandalism that physically changes Web pages.

Demographic characteristics Specific, observable, and objective identifiers shared by those in a target market.

Denial of service attacks (DoS) An attack on a computer network that is designed to disable the network by flooding it with useless traffic.

Digital cash *See* electronic cash.

Digital certificate An electronic message attachment that verifies the sender's identity.

Digital signature A unique code attached to an electronically transmitted message that identifies the sender.

Digital wallet *See* electronic wallet.

Directory A Web search tool whose index is created based on manual submissions of company information and Web site URLs.

Disaster recovery The process of recovering computer system and/or data files after a catastrophic event.

Discount rate A merchant account fee charged to a business for each credit, debit, or charge transaction processed.

Distributed denial of service attacks (DDoS) A DoS attack launched from multiple computers.

Domain name   Identifies a Web site on the Internet including the server that handles Web browser requests, the specific organization associated with the domain name, and the general category in which the organization operates.

Domain name brokers   Legitimate businesses that buy generic domain names that they can resell.

E-business   A broad spectrum of business activities on the Internet including buying and selling products and services, customer service and support, business partner collaboration, and enhancing internal productivity.

E-business plan   A document used to guide an entrepreneur starting a new e-business and sometimes used to seek financing for that business; an e-business plan contains information about the startup e-business including a description of the e-business idea; information on products and services to be sold; industry and marketplace analyses; marketing, operating, and financial goals and objectives; an identification of critical risks; and an exit strategy.

E-commerce   *See* electronic commerce.

Electronic cash (e-cash)   A method of transmitting a unique electronic number or other identifier that carries a specific value to pay for goods or services purchased online.

Electronic checks (e-checks)   An electronic version of a paper check.

Electronic commerce   Buying and selling products and services across a telecommunications network; *see* e-business.

Electronic wallet (e-wallet)   Encryption software that stores payment information much like a physical wallet and may reside on the user's PC, the card issuer's server, or an e-business's server.

Encryption   The translation of data into a secret code called cipher text.

Enterprise   An organization that uses computers and a network to interact with employees, suppliers, and customers.

Enterprise resource planning (ERP)   A system that integrates all aspects of a business including planning, manufacturing, human resources, accounting, finance, sales, and marketing.

Entrepreneur   Someone who assumes the risks associated with starting and running his or her own business.

Entrepreneurial process   A series of steps to (1) decide if someone is an entrepreneur; (2) if yes, decide to purchase or start a business; (3) if starting a business then define the business idea, create a business plan, and secure financing; (4) if purchasing or starting a business operating that business; and (5) harvesting the business.

Entry page   *See* splash page.

E-retail, e-tail, e-tailer   B2C retail e-businesses.

Executive summary   The third or fourth page of a business plan that provides a quick overview of the entire plan.

Exit strategy   A section of a business plan that describes the long-term plans an entrepreneur has for a new business and how potential investors will recover their investment.

Extranet   Two or more intranets connected via the Internet to allow participating business partners to view each other's data and to complete business transactions.

Fail-over system   A spare computer system that can immediately begin processing operations in the event of a failure with the primary computer system.

Fair use   The limited use of copyrighted material under certain circumstances.

Filter   A process or device that screens incoming information and allows only the information meeting certain criteria to pass through to the next area.

Firewall   Software or hardware used to isolate a private computer or computer network from the public network.

First-mover advantage   An advantage inherent in being the first business of its kind in the marketplace.

Font   A collection of characters that have the same appearance.

Font styles   Font attributes such as bold, italic, and underline used to add emphasis to text.

Forms   An electronic version of a paper form with text boxes, check boxes, option buttons, and drop-down lists used on a Web page to gather information.

Frames   Sections of a Web browsers display area that can contain different Web pages.

Friends and family investors   The network of family members, friends, and their family members and friends who invest in a new business.

Front-end systems   Those processes such as viewing information and ordering products that can be interfaced with and controlled by customers.

Fulfillment houses   Independent companies that warehouse, pick, package, and ship products for other businesses.

General partnership   Multiple owners of a for-profit business.

Geographic characteristics   Characteristics of a target market based on physical location such as country/region, state, city/town, size of population, climate, and population density.

GIF images   A compressed image file format used for simpler Web images and animated images.

Hacker   Originally a slang term for a gifted software programmer, now a slang term for someone who deliberately gains unauthorized access to individual computers or computer networks.

Hit   The Web page URL and page title (and other information) returned by a search tool in response to a keyword search.

Host   A computer on the Internet.

Hot-swappable components   Components such as hard drives that can be installed while a computer system is running.

Hyperlink   Text or a picture that is associated with the location (path and filename) of another page.

Identify theft   Theft of sensitive customer information used by cybercriminals to obtain new credit or make large credit purchases.

Impressions   *See* page views.

Independent software vendors (ISVs)   Companies that develop and sell software applications for a variety of operating system platforms.

Industry   Businesses that make or sell similar, complementary, or supplementary products or services.

Industry analysis   A portion of a business plan that describes an industry's size, characteristics, trends, growth factors, barriers to entry, government regulations, distribution systems, competitors, technological influences, and other related topics.

Insourcing   Work done by a business's employees.

Internal hyperlink   A connection between two Web pages at the same Web site.

Internet   A public worldwide network of networks that connects many smaller private networks.

Internet accelerator   A commercial business incubator that gets an e-business up and running quickly.

Internet access providers (IAPs)   *See* Internet Service Providers.

Internet Service Provider (ISP)   Provides access to a host computer on the Internet.

Intranet   An internal network using Internet technology to allow employees to view and use internal Web sites not accessible by the outside world.

Intrusion detection   Analyzing real-time data to detect, log, and stop unauthorized computer network access.

Issues analysis   A portion of a business plan that identifies the threats or opportunities a new business faces.

JPEG images   A compressed image file format used for complex images such as photographs.

Keiretsu model   A commercial business incubator that offers entry into a network of existing businesses that profit from doing business with each other.

Legacy systems   Existing systems.

Limited partnership   A partnership in which the general partners assume management and unlimited liability for the partnership and other partners have no management responsibility and are legally liable for their capital contribution.

Log files   A record of all events on a Web server.

Logic bomb   A virus whose attack is triggered by some event such as the date on a computer system's clock.

Macro   A short program written in an application such as Microsoft Word or Excel to accomplish a series of keystrokes.

Macro virus   A virus that infects Microsoft Word or Excel macros.

Marketing budget   The portion of a business plan that estimates the costs for all marketing strategies.

Marketing objectives   The portion of a marketing plan that describes measurable, time-bound goals that should lead to sales.

Marketing plan   A business plan component that establishes, directs, and coordinates marketing efforts for a specific period of time.

Marketing strategies   The portion of a marketing plan that describes how the products and services will be priced, promoted, and distributed.

Marketspace   Electronic marketplace.

Merchant account   An account at a financial institution set up by a business to process credit, debit, and charge card payments from their customers.

Meta search engine   A Web search tool that allows users to search the indexes of multiple individual search tools at one time.

Meta tag   Small segments of HTML code that are not visible on a Web page but are read by search engines.

Middleware   Special data translation programs used to assist in the exchange of data between systems.

Mirrored server   A backup server that duplicates the processes and data of a primary server.

Mission statement   A statement of challenging but achievable actions a business takes to realize its vision.

Mobile commerce (m-commerce)   Using Web-enabled cell phones, PDAs, and other devices to purchase products or services.

Money A unit of account, a common measure of value, a medium of exchange, and a means of payment.

Money order An order for the payment of a specified amount of money, usually issued and payable at a bank or post office and often used by individuals who do not have bank accounts or in circumstances where checks are not accepted in payment.

Navigation bar A series of icon or text internal hyperlinks to major pages at a Web site.

Navigational outline A hierarchical outline showing all the levels of links between the home page or another major page and the page currently being viewed.

Network A group of two or more computers linked by cable or telephone lines.

Network effect The increasing value of a network to each participant as the number of total participants increases.

Operations plan The section of a business plan that describes the business location, necessary equipment, required labor, and other processes related to the sale of products or services including Web site operation.

Order fulfillment Getting the ordered product into the hands of the customer.

Order picking Selecting products from warehouse shelves and bins.

Outsourcing Work done for a business by people or organizations other than the business's employees.

Page views The number of times a given page has been viewed.

Page views per visitor The number of page views divided by the number of visitors.

Partnership A legal business form that consists of two or more owners and is governed by a signed agreement that details the fundamentals of the partnership.

Password A code, often a common word, used to gain access to a computer network.

Penetration testing Using real world hacking tools to text computer security.

Perquisite (perk) An incidental benefit extended to employees.

Personal firewall Firewall software used to protect individual computers from unauthorized access when the computers have an always-on connection to the public network.

Person-to-person (P2P) payment systems Internet payment systems based on sending a stated value from a bank account or credit card to someone via e-mail who then deposits that value to his or her bank account or credit card.

Pitch document A short marketing document based on a business plan's executive summary used to market a startup business to potential investors.

Plain text Decrypted text.

Portal A Web site offering a wide array of services such as search engines, directories, and shopping services.

Primary records Records that provide supporting documentation for key e-business activities such as sales, order fulfillment, and payment.

Private placement memorandum Discloses the benefits and risks of an investment to potential private investors.

Products and services section The portion of a business plan that describes the products and services the business will offer including what they are, what they do, and their customer benefits.

Protocol A standard or agree-upon format for electronically transmitting data.

Psychographic characteristics Target market characteristics that relate to attitude, beliefs, hopes, fears, prejudices, needs, and desires.

Referring URL The URL of the Web page a viewer clicked to get to another Web site.

Risk assessment *See* issues analysis.

ROI Return on Investment.

Router A device that connects two or more networks and forwards information between networks.

Sans serif Without a serif; fonts that do not have a serif.

Scalability The ability of a business to continue to function well regardless of how large the business becomes; also the ability of a server to handle increased traffic loads without crashing.

Search engine A Web search tool whose index is created automatically by software programs, called spiders, that browse the Web looking for new pages.

Secondary records Records containing important information not used in daily operations but generated from an e-business's activities such as customer data and Web site logs.

Secure Electronic Transactions (SET) A security protocol developed by Visa and MasterCard for presenting credit card transactions on the Internet.

Secure Sockets Layer (SSL) A security protocol that provides server-side encrypted transactions for electronic payments or other secure Internet communications.

Self-incubators Startup businesses that share office space, ideas, and a network of advisors during the early stages of each business.

Serif The small tail at the end of characters in certain fonts that helps a reader's eye follow a line of text.

Server Special computer that provides user access to shared resources such as files, programs, and printers.

Service mark A distinctive symbol, word, or phrase used to identify a business's services and distinguish them from other businesses' services.

Site map A Web page that shows each page at a Web site and how all the pages are linked together.

Smart card A small electronic device approximately the size of a credit card that contains electronic memory and is used for a variety of purposes including storing medical information, network identification, and electronic cash.

Sole proprietorship An individual who starts and operates a business without the formalities associated with other legal organizational forms.

Spiders Software programs that browse Web pages to automatically update search engine indexes with new URLs.

Splash page A Web page that usually contains big, flashy graphics, and perhaps sounds used to create a showy entrance to a Web site.

Staging server A temporary server used to test Web sites before they are published to their final destination server.

Sticky Web site Web sites whose content keeps viewers coming back again and again.

Storyboard A blueprint for design of a Web site.

Sweat equity An entrepreneur's time, effort, and own money used to get a new business started.

System integrators Companies that oversee the implementation of software applications and configure special application functions.

Table of contents The third or fourth page of an e-business plan that provides page number references for the major sections and subsections of the plan.

Target market The group of potential customers that share a common set of traits that set the group apart from other groups.

Term sheet A list of the major points of proposed financing being offered by an investor.

Thumbnail A small version of a larger image used on a Web page to speed up download times.

Tiled image A small image that repeats over and over on a page.

Title page The second page of an e-business plan that repeats information from the cover sheet and adds the preparers' contact numbers and the name of the person receiving the plan copy.

Top-of-page hyperlink Text or graphic image that is linked to a position at the top of the current page.

Trademark A distinctive symbol, word, or phrase used to identify a business's products and distinguish them from other businesses' products.

Trojan horse A special type of virus that emulates a benign application by appearing to do something useful or entertaining.

Trust protections A method of securing data exchanged with outside business partners that verifies the data is sent from an e-business server.

Turnkey package A ready-to-publish package that includes a Web site and store front that can easily be modified with custom content.

Value chain All the primary and support activities performed to create and distribute a company's goods and services.

Value-added reseller (VAR) A company that modifies an existing software product by adding its own "value" such as a special computer application and then resells the product.

Venture capital (VC) firms Investors that raise hundreds of millions of dollars from other organizations such as endowments, insurance companies, and pension funds to fund new businesses.

Vertical market A specific industry in which similar products or services are developed an sold using similar methods.

Virtual inventory Inventory of products belonging not to the e-business but to its partners who own, warehouse, and ship the products.

Virtual malls B2C e-businesses that host many different online e-tailers.

Virus A small program that inserts itself into other program files that then become "infected" in the same way a virus in nature embeds itself in normal cells.

Vision statement A business's statement of long-term dreams and goals.

Visitors The actual number of viewers to visit a Web site.

Web browser Software application used to access and view Web pages.

Web hosting company Hosts commercial Web sites from Internet data centers.

Web integrators Companies that consult on Web strategies, Web application development, and the integration of Web-based applications with older systems.

Web mining Data mining techniques applied to data from Web sources such as Web server logs and Web transactions.

Web pages Documents that can contain text, graphics, video, audio, and hyperlinks stored on Web servers and viewed in a Web browser.

Web ring A circular chain of related Web site links.

Web servers A computer that stores Web pages.

Web site A collection of related Web pages.

Wireless Application Protocol (WAP) A standardized protocol that allows for the delivery of Internet content via cellular networks to small-screen devices such as cell phones.

World Wide Web (Web) A subset of the Internet consisting of computers called Web servers that store documents linked together by hypertext links.

Worm A special type of virus that doesn't alter program files directly but replaces a document or program with its own code and then uses that code to replicate itself.

Index

C2B model. *See* consumer-to-business model
C2C. *See* consumer-to-consumer model
Cambridge Incubator (CI), 168
 discussed, 154, 174
CarSmart.com, 11
Carsten, Jack, 158
cash flow statement, in financial plan, 133
catalog merchants, 16
Catchword, 201
CATEX, 21
CERN, 5
certificate authority, 77
charge card, 73
chargeback, 74
Charles River Ventures, 162
charts, in business plan, 127
Chaum, David, 82
checks. *See also* electronic payment
 in general, 71-72
Chemdex, 201
CI. *See* Cambridge Incubator
ciphertext, 76
Cisco Systems, 57, 83
 intrusion detection, 287
 mission statement, 123
 Web site, 246-247
click trail, Web site, 366
ClickTrade, 353
Clinton Administration, 78
CMG. *See* College Marketing Group
Cobaltcard, discussed, 98
coinage, 71. *See also* money
ColdFusion, 256
College Marketing Group (CMG), 168, 169
Collins, Brian, 254
color. *See also* Web site design
 browser-safe palette, 245
 for Web site, 245-247
.commerx PlasticsNet, 19
communication ports, 288. *See also* security
competition
 differentiating from, 172
 in e-business, 39
 first-mover advantage, 47
 hiding information from, 281
 identifying in business plan, 127
 name identification, 47
Competitive Advantage (Porter), 12
Computer Security Institute (CSI),
 279, 284, 285

Confinity, 103
consultants, 137
Consumer Reports, 15
consumer-to-business (C2B) model, in general,
 26-27
consumer-to-consumer (C2C) model
 in general, 25-26
 expert information exchange, 26
content liability, 191
contract, for Web site design, 258
cookie, 366
Coolboard, 171
copyright. *See also* legal issues
 fair use doctrine, 189, 240
 in general, 189-190
corporation
 "C" corporation, 141
 "S" corporation, 141
cost reduction, electronic, 57
Costco, 50
Counterpane Internet Security, 295
Cox Enterprises, 168
crackers, 274
credit card. *See also* e-business;
 electronic payment
 in general, 72-73
 prepaid, 96-102
credit card fraud. *See also* security
 disposable credit card numbers, 79-80
 security issues, 281-282
CRM. *See* customer relationship management
cryptography, 76
CSI. *See* Computer Security Institute
CSK Auto, 142
customer
 communicating with, 352
 feedback from, 344, 351
customer relationship. *See also* audience; relationships
 improving responsiveness, 53-54
 protecting, 272, 352
 returns, 325
customer relationship
 management (CRM), 207
CyberCash, 75, 81
Cybergold, 83
Cybermoola, discussed, 99
Cyberplex, 259
cybersquatters, 203

escrow services. *See also* electronic payment
   i-Escrow, 80
ESPN, 254
eTour, 168
eToys.com, 16
   discussed, 331, 353
exit strategy. *See also* e-business plan
   describing in business plan, 140
Exodus Communications, 212, 218
expert information exchange, 26
extranet, 17

facilities plan, suggestions for, 196
fail-over system, 318. *See also* security
Failure Magazine, 39
fair use doctrine, copyright, 189
FBI. *See* Federal Bureau of Investigation
Federal Bureau of Investigation (FBI), 284
Federal Trade Commission (FTC), 193, 324
FedEx, 324
fees, for merchant account, 74, 79
Fernandes, Gary, 60
filter, security, 275
financial plan. *See also* e-business plan
   in business plan, 133-136
      break-even analysis, 135
      financial ratios analysis, 135
      statement of financial assumptions, 135
financial ratios analysis.
      See also e-business plan
   in financial plan, 135
Fingerhut. *See also* fulfillment
   discussed, 51-52, 327
firewall, 287. *See also* security
   personal, 292, 293
first-mover advantage, 47.
      See also competition
Fisher, Jon, 217
Flanders, Vincent, 242
Flatiron Partners, 161
Flooz, 83
font, 247
font style, 248
foodlocker.com, 27, 120, 124, 126, 127, 129,
      130, 132, 138, 139, 192, 202
   discussed, 58-59, 80, 116, 131, 162, 215,
         234, 261-262, 281, 311, 324, 332, 349,
         358, 361, 365, 369
forms, for Web site, 253-254
Forrester Research, 360

forums, in general, 352-353
frames. *See also* Web site
   in Web site, 252-253
friends and family, financing from, 154-156
front-end system
   business records maintenance, 314-315
   in general, 310-311
   integration with back-end, 311-314
      legacy systems, 313
      middleware, 313
      trust protections, 313
FTC. *See* Federal Trade Commission
fulfillment, 18. *See also* marketing
   Fingerhut, 51-52, 327
   in general, 319-320
   international issues, 325-326
   inventory management, 320
   order picking, 320-321
   outsourcing, 326-329
   processing returns, 324-325
   shipping and delivery, 322
   3PLs, 327
   virtual inventory, 332
fulfillment house, 326
   choosing, 330-332
funding. *See* e-business plan; startup financing

Gallery Furniture, 252
GAO. *See* General Accounting Office
Gartner Group, 78
General Accounting Office (GAO), 194
general partnership, 140.
      See also e-business organization form
geographic characteristics, 128
GIF images, 250
GIF Wizard, 250
Glaser, Teresa, 70
Global Crossing Ltd., 212
Global Food Exchange, 162, 164
goals, determining, 362
Goerke, Eileen, 70
Golub, Ben, 88
Golub, Bill, 88
Golub Corporation, 88
Goodstein, Marcia, 167
Goodyear, 278
Google, 345, 346
Gossamer Threads, 243
GoTo.com, 345, 351
Gov.com, 24

Granoff, Peter, 344, 370
graphs, in business plan, 127
Greenlight.com, 143
grocery sales, 8
GroceryWorks.com, 320, 321
    discussed, 59–60, 131
Gross, Bill, 167
Grossman, Michael, 254

hackers, 274, 279, 281, 282, 293.
        See also security
Hagar, Kelby D., 38, 59
Half.com, discussed, 52–53
Hammurabi, 72
hard drive. *See also* security; technology
    crash, 315
    hot-swappable, 318
    RAID, 318
Harmonized Tariff System, 326
Heidrick & Struggles International, 197
Hester, Staci, 155
High Tech Start Up (Nesheim), 50
hits, 350
host, 4
Hotel Resource, 18
Hotmail, 46
hyperlink, 5, 242
    internal, 240
    text and icon, 244
    top-of-page, 242

i-Escrow, 80
IAP. *See* Internet access provider
ICANN. *See* Internet Corporation for Assigned
        Names and Numbers
ICVERIFY, 75
idealab!, 167, 168, 170
identity theft, 282
images
    GIF, 250
    JPEG, 250
    thumbnails, 250
income statement. *See also* e-business plan
    in financial plan, 133
incubation. *See* business incubators
independent software vendors (ISVs), 210
industrial espionage. *See also* security
    in general, 279–281

industry analysis. *See also* e-business plan
    in business plan, 125–127
Infineum, 241
infomediary, 10, 11, 44
information privacy, 191–193
information technology, 197. *See also*
        technology
initial public offering (IPO), 45, 164
Inktomi, 7
insourcing, 137
insurance, 296
Interliant, 214
intermediary, 9–10. *See also* infomediary
International Data Corporation, 328
International Risk Management Institute
        (IRMI), 316
International Trademark Association, 190
Internet, 43. *See also* World Wide Web
    demographics, 5–7
    in general, 3–4
        network, 3
        server, 3
    history, 4–5
        ARPANET, 4
        MILNET, 4
        National Science Foundation, 4
    host, 4
    protocols, Transmission Control
            Protocol/Internet Protocol, 4
    tax issues, 194–195
Internet accelerator. *See also* business incuba-
        tors
    discussed, 168–170
Internet access provider (IAP), 211
Internet Corporation for Assigned Names and
        Numbers (ICANN), 203
Internet service provider (ISP), 4.
        See also Web hosting company
    bandwidth, 213
    in general, 211
Internet Tax Freedom Act, 194
Internet Truckstop, 10
intranet, 16
intrusion detection. *See also* security
    security, 287
inventory management, 206. *See also*
        fulfillment
    fulfillment, 320
IPO. *See* initial public offering
IRMI. *See* International Risk Management
        Institute

ISP. *See* Internet service provider
issues analysis. *See also* e-business plan
   in business plan, 139–140
iStart Ventures, 168
ISVs. *See* independent software vendors

Jarrard, Jim, 292
Jones, Leo, 291
JPEG images, 250
Jupiter Media Metrix, 327

Kanarick, Craig, 188
Katalyst, 168
keiretsu providers.
    See also business incubators
   in general, 168–170
Kelly, Patrick, 312
keywords, 348, 351, 367
Klein, John, 291
Kleiner Perkins Caufield & Byers, 161
Knight, Stephen, 154, 174
Korn/Ferry International, 197
Kucirek, Scott, 172

leadership, 38
Lefcourt, Jenny, 114, 144
legacy systems, 313
legal issues. *See also* security
   content liability, 191
   copyright, 189–190, 240
   employee hire, 196–199
   in general, 188
   information privacy, 191–193
   leasing commercial office space, 195–196
   service mark, 190
   taxation, 194–195
   trademark, 189–190
LevelEdge.com, 165
Levy, Jeff, 168
Lightfoot, Franklin & White, LLC, 235
limited liability company (LLC), 141
limited partnership, 141
link, internal hyperlink, 240
link exchange. *See also* marketing
   in general, 354
LinkExchange, 354
linkLINE Communications, 275

LLC. *See* limited liability company
log file analysis. *See also* performance measurement
   in general, 364–365
     browser type, 364–365
     conversion rate, 365
     IP address, 364
     page views, 364
     page views per visitor, 364
     referring URL, 364
     visitors, 364
LookSmart, 346
Lorenzo, Alain, 246
Losefsky, Ron, 163

m-commerce. *See also* electronic payment
   discussed, 102
     Wireless Application Protocol, 102
McAfee, VirusScan, 287
McConnell, Lisette, 70
McPhee, Lynn, 155, 171
macro virus, 277. *See also* security
Mafiaboy, 276
Magellan Internet Guide, 346
magnetic ink character recognition (MICR), 89
management, professional recruiters, 197
management plan. *See also* e-business plan
   board of advisors, 137
   in business plan, 137–139, 172
     insourcing, 137
     outsourcing, 137
Mancuso, Charles, 166
marketing. *See also* advertising; branding; customer
   affiliate programs, 353
   consumer communication, 352
   directories, 345–352
   forums, 352–353
   in general, 345
   innovative, 43, 46–47
   link exchange, 354
   newsgroups, 352–353
   public relations and, 352
   search engines, 345–352
     meta search engines, 347
   "viral marketing," 46, 356–357
   Web rings, 355
   word-of-mouth, 356
marketing budget, 130
marketing objectives, 129

risk management issues, in general, 295-297
Robertson, Michael, 190
RocketCash, discussed, 99
ROI. *See* return on investment
Rosen, Ben, 167
RosePlace.com, 165
router. *See also* security
 security, 275
RSW software, 259
RTF. *See* rich text format

Safeway, 60
St. Joseph's Hospital, 317
SalariesReview.com, 199
sans serif font, 247
SAP, 210
Sapient, 218
SBA. *See* Small Business Administration
SBANC. *See* Small Business Advancement
 National Center
scalability, 43
 in general, 44-45
Scheurer, William, 97
SCORE. *See* Service Corps of Retired
 Executives
search engines
 in general, 345-352
 hits, 350
 keywords, 348, 351, 367
 meta search engines, 347
 meta tags, 349
 "spamming," 351
 submission service, 349
Secure Electronic Transactions (SET), 77
Secure Sockets Layer (SSL), 289.
  See also security
 in general, 76
security. *See also* legal issues
 audits, 289-291, 298
 backup system, 315-316
 benchmarks, 286
 communication ports, 288
 credit card fraud, 78-80, 281-282
 data protection, 289
 data spill, 283
 denial of service attack, 274-276
 disaster recovery, 316-318
  fail-over system, 318
  options, 318-319
 e-business vulnerability, 284-285

electronic payment, 76-78
 authentication, 77
 certificate authority, 77
 ciphertext, 76
 cryptography, 76
 decryption, 76
 digital certificate, 77
 digital signature, 77
 digital wallet, 78
 e-wallet, 78
 encryption, 76, 289
 plaintext, 76
 protocol, 76
 Secure Sockets Layer, 76, 289
firewall, 287
 personal, 292, 293
 in general, 272-274, 285-286
  filter, 275
  router, 275
hackers, 274, 279, 281, 282, 293
identity theft, 282
industrial espionage, 279-281
intrusion detection, 287
network and Web site, 286-288
password, 286
patches, 288
PC risks, 292-293
penetration testing, 289-291
providers
 in general, 294
 products, 294
 services, 294-295
risk management issues, 295-297
transaction security, 289
viruses, 276-277
Web site defacement, 277-278
Segue Software, 259
Sendmail, Inc., 158
serif font, 247
server, 3
 co-locating, 214
 mirrored server, 319
 staging server, 259
 Web server, 5
Service Corps of Retired Executives
 (SCORE), 114
service mark, 190
services, new services creation, 56-57
SET. *See* Secure Electronic Transactions
Seventeen, 254
Severiens, Hans, 158

shipping. *See also* fulfillment
    fulfillment, 322
    international, 325-326
SHOP2gether.com, 19
Shop4.com, 10
SI. *See* system integrator
Sieble Systems, 216
Sigma Partners, 161, 162
SIGs. *See* special interest groups
Slatalla, Michelle, 70
Small Business Administration (SBA), 114
Small Business Advancement National Center
    (SBANC), 115
smart card. *See also* credit card;
    electronic payment
    discussed, 86-88
SmartShip.com, 323
Smith, Jack, 46
software. *See also* technology
    independent software vendors, 210
    payment-processing, 75-76
    Web development, 255-256
Sohl, Jeffrey, 158
sole proprietorship, 140
sound, for Web site, 252
Southwest Airlines, discussed, 54
Southwestern Bell, 250
"spamming," 351
special interest groups (SIGs), 353
Spencer Stuart, 197
spiders, 345
Spielberg, Steven, 167
Spinwave, 250
splash pages, 243
SSL. *See* Secure Sockets Layer
Stamps.com, 323
Staples, 8, 18
startup financing. *See also* business
    incubators; e-business plan
    burn rate, 197
    early revenue stream, 160
    in general, 154
        sweat equity, 154
    pitching your ideas, 171-173
        pitch document, 171
        private placement memorandum, 173
        term sheet, 173
    sources
        angel investors, 157-160
        friends and family, 154-156
        venture capital investors, 161-164

storyboard, 237-239
StoryServer, 256
SubmitIt!, 349
Sucky Pages, 242
Sun Microsystems, 57, 83
Swann, Christopher, 162, 164
sweat equity, 154
Symantec AntiVirus Research Center
    Encyclopedia, 277
system integrator (SI), in general, 218

target market, 128. *See also* marketing
Tattered Cover, 7-8
tax issues. *See also* legal issues
    audit, 315
    in general, 194-195
    international, 326
    records, 314
TCP/IP. *See* Transmission Control
    Protocol/Internet Protocol
TechEx, 21
technology selection. *See also* software
    application service provider, 216-218
    customer relationship management
        system, 207
    data mining, 207-210
    defining enterprise, 205
    enterprise resource planning
        system, 205-206
    in general, 204-205
    independent software vendor, 210
    Internet service provider, 211
    system integrator, 218
    technology service provider, 211
    value-added reseller, 210
    Web hosting company, 212-216
    Web integrator, 218
TeleCheck, 89
term sheet, 173
text
    rich text format, 254
    for Web site, 247-249
Thiel, Peter, 103
Third Voice, 43
Thornton, Christopher, 72
3PLs, 327. *See also* fulfillment
thumbnails, images, 250
Ticketmaster, discussed, 50-51
tiled image, Web site, 249